Disaster Resilience and Green Growth

Series Editors

Anil Kumar Gupta, National Institute of Disaster Management, New Delhi, India

SVRK Prabhakar, Climate Change Adaptation, Institute of Global Environment Strategies, Kanagawa, Japan

Akhilesh Surjan, College of Indigenous Futures, Arts and Society, Charles Darwin University, Darwin, Australia

In recent years, it has become clear that environmental changes and disaster risks are closely linked. Climate change, changes in land use, and the depletion of natural resources make communities and infrastructure more vulnerable. This has created a need for comprehensive and collaborative solutions that bring together different fields. Effective management of disaster risks and adaptation to climate change requires coordinated efforts across infrastructure, social systems, and ecological resilience. Factors related to the environment and human activities greatly influence risk exposure. Therefore, risk management must consider these elements. Bridging fields such as environmental science, engineering, economics, and policy can lead to effective disaster resilience.

Traditional economic growth models often overlook environmental sustainability, resulting in more resource depletion and vulnerability. In response, new green growth strategies focus on sustainability, including nature-based solutions and climate actions to promote circular economies and low-carbon development, enhancing the approach of disaster resilience. Current policy discussions emphasize the importance of building resilient systems in infrastructure, industry, and public health as part of the sustainability agenda post-2030. Furthermore, new energy technologies present both opportunities and risks, requiring strategies to manage their impacts effectively. Chemical and industrial safety risks in addition to natural hazard induced technological disasters as complex emergencies are also a growing concerns in disaster management and sustainability in the emerging world.

The scope of book series relates all the SDG in implicit ways but is focused particularly on SDG 2, 3, 8, 9, 11, 12, 13 and 15. The book series "Disaster Resilience and Green Growth" aims to share research and insights on these topics surrounding safety and resilience of human life and society in the changing world, fostering innovation and knowledge-sharing for a sustainable future.

Deepak Kumar • Nick P. Bassill
Editors

Artificial Intelligence and Machine Learning for Climate Disaster Management

Springer

Editors
Deepak Kumar
Atmospheric Sciences Group
Department of Geosciences, College of Arts & Sciences, Texas Tech University
Lubbock, Texas, USA

Center of Excellence in Weather & Climate Analytics, Atmospheric Sciences Research Center (ASRC)
University at Albany
State University of New York (SUNY)
Albany, New York, USA

Nick P. Bassill
Center of Excellence in Weather & Climate Analytics, Atmospheric Sciences Research Center (ASRC)
University at Albany
State University of New York (SUNY)
Albany, New York, USA

State Weather Risk Communication Center (SWRCC), Atmospheric Sciences Research Center (ASRC)
University at Albany
State University of New York (SUNY)
Albany, New York, USA

ISSN 2662-4885 ISSN 2662-4893 (electronic)
Disaster Resilience and Green Growth
ISBN 978-981-95-1587-5 ISBN 978-981-95-1588-2 (eBook)
https://doi.org/10.1007/978-981-95-1588-2

This Springer imprint is published by the registered company Springer Nature Singapore Pte Ltd.
The registered company address is: 152 Beach Road, #21-01/04 Gateway East, Singapore 189721, Singapore

Dedicated to Beloved Parents

Preface

The growing frequency and intensity of climate-related disasters pose unprecedented challenges to communities, ecosystems, and economies worldwide. Addressing these challenges requires innovative approaches that go beyond traditional methods of disaster management. Artificial intelligence (AI) and machine learning (ML) have emerged as powerful tools capable of transforming how we predict, mitigate, and respond to climate disasters.

This book, entitled *Artificial Intelligence and Machine Learning for Climate Disaster Management*, explores the intersection of cutting-edge technology and climate resilience. It delves into the ways AI and ML are being harnessed to analyse vast datasets, enhance early warning systems, optimize resource allocation, and improve decision-making during crises. The integration of these technologies helps us to achieve solutions for smarter, faster, and more effective disaster management. A comprehensive examination of real-world applications, challenges, and prospects inspires researchers, policymakers, and practitioners to utilize the power of AI and ML as key enablers for a sustainable and disaster-resilient future. There are several ways in which AI and ML are being utilized and they comprise Early Warning Systems, Disaster Response Planning, Damage Assessment, Risk Assessment and Vulnerability Mapping, Climate Modeling and Simulation, Resilience and Adaptation Strategies, Energy Management, Supply Chain Management, Community Engagement and Education, Satellite Imagery for Disaster Response, Vulnerability Mapping and Resilience Planning, Data Integration in Emergency Response, Climate Adaptation Policies, Climate-Induced Migration, and Ethical and Privacy Considerations in Disaster Management.

This book focuses on the importance of AI and ML being utilized in climate disaster management to enhance preparedness, response, recovery, and mitigation efforts. The integration of these technologies with climate disaster management greatly improves decision-making, resource allocation, and overall effectiveness in addressing the impacts of climate-related disasters. It offers a roadmap for leveraging geospatial information to navigate the stormy seas of our changing climate and emerge more resilient and prepared on the other side. It provides a comprehensive and enlightening guide to harnessing the power of geospatial information and services for climate-resilient disaster management.

This book navigates through the theoretical foundations, practical applications, and ethical considerations that accompany the integration of data analytics, AI, ML, and climate change for disaster management with a collection of expert-contributed chapters. This book is authored by leading experts in the field, and it is a timely exploration of how cutting-edge technology can shape our response to climate-related disasters. The book embarks on a journey through the intricate web of geospatial data, uncovering its indispensable role in disaster preparedness, response, and recovery. Readers will gain a deep understanding of the intersection between climate change and disasters and how geospatial solutions provide a lifeline in this evolving landscape.

This empowerment is rooted in the understanding that collective action is fundamental to mitigating the far-reaching consequences of AI, ML, climate change, and data analytics for disaster management. This discussion proceeds beyond the surface level into a more comprehensive perspective on climate-induced disaster management. A more complete picture and more in-depth knowledge of the problem explore the political, economic, and social aspects of climate change. The comprehensive understanding can give us a better understanding of environmental issues impacting climate change and sustainability including the health and happiness of the common man, which in turn allows us to take action based on accurate information and find long-term solutions.

The book provides an in-depth exploration of the concepts and explanations related to artificial intelligence, machine learning, and climate change. It explores the foundational principles of AI and ML, explaining how these technologies work and their potential applications. It discusses their intersection with climate change, examining how AI and ML can be leveraged to monitor environmental changes, predict climate patterns, and develop innovative solutions to address global environmental challenges. Each topic is explained comprehensively, with real-world examples and case studies to enhance understanding and application.

Chapter 1: Artificial Intelligence and Machine Learning for Climate Disaster Management

This chapter explores the impact of artificial intelligence (AI) and machine learning (ML) in changing climate disaster management with advanced tools and techniques for predicting, mitigating, and responding to natural disasters. These technologies provide huge amounts of data from various sources like remote sensing, weather stations, and social media, to enhance the accuracy and efficiency of disaster management processes. AI and ML algorithms analyse complex datasets to identify patterns and trends that human analysts might overlook. The forecasts enable authorities to implement proactive measures, such as reinforcing infrastructure, pre-positioning emergency supplies, and issuing early warnings to vulnerable communities. Models like predictive models can forecast the likelihood of extreme weather events such as hurricanes, floods, and droughts based on historical data and current conditions. AI and ML play a crucial role in real-time disaster response in addition to prediction. These technologies process real-time data from various sensors and sources to provide situational awareness during any disaster. AI-enabled drones and satellite imagery assess damages to identify affected areas and prioritize rescue and relief operations. Machine learning algorithms analyse social media posts to gauge public sentiment, detect emergencies, and direct resources to areas in need. AI and ML contribute to long-term resilience and recovery efforts with the analysis of post-disaster data to evaluate the effectiveness of response strategies and identify areas for improvement. They model different scenarios to help policymakers develop robust disaster management plans and policies. AI and ML support community engagement by providing accessible and actionable information to the public through chatbots and mobile applications. This chapter discusses how the adoption of AI and ML in climate disaster management faces challenges such as data privacy concerns, the need for high-quality data, and the requirement for skilled personnel to develop and maintain these systems. It explores the challenges that are essential to fully realize the benefits of AI and ML for building climate resilience for ensuring effective disaster management. It discusses how AI and ML will enhance the prediction, response, and recovery phases of climate disaster management to help authorities protect communities from the adverse impacts of climate-related disasters.

Chapter 2: The Climate Science, Designing, and Implementation of Climate Policy

This chapter explores the complex relationships between climate science, policymaking efforts, and the development of effective adaptation to change as well as resilience strategies based on scientific literature and documents related to policymaking. It breaks down the intricate issues caused by climate change consequences such as increased temperatures, extreme weather events, sea level surges, and changes in rainfall patterns. The study highlights the importance of incorporating scientific knowledge into policy structures to strengthen adaptive capabilities and enable resilience on local, regional, and global scales. It presents a thorough review of critical aspects of climate science, such as modelling on climates and evaluations regarding risks along with vulnerabilities. This part ventures to discover ways by which such scientific understanding can be used in driving the development and implementation of policies. It also emphasizes cases and examples of successful development of climate adaptation resilience to provide insight into the vital role of interdisciplinary cooperation and stakeholder involvement in long-term strategies. The interconnection of the discipline of climate science and policy strives to close gaps in present knowledge to increase societal and ecological strength against these threats.

Chapter 3: Strategic Integration of Green Innovation and Information Systems for Sustainable Development

This chapter focuses on the impact of Strategic Integration of Green Innovation and Information Systems for Sustainable Development on the effective management of resources critical for sustainable development and environmental stewardship. Green innovation focuses on developing eco-friendly technologies and practices that minimize environmental impact while fostering economic growth. This explains the strategic integration of green innovation and information systems and articulates a new approach to sustainable development. These innovations can be seamlessly implemented, monitored, and optimized with advanced information systems. It also drives transparency and accountability by providing stakeholders with accessible, data-driven insights. Information systems enable the collection and analysis of real-time data, enhancing decision-making and resource management to support sustainability objectives. This highlights the effect of green innovation on business performance. This presents the primary quantitative methods with the survey and analysis of the data collected from the participants. The entire analysis has been conducted with the SPSS tool. This depicts the role of green innovation in explaining the impact of business performance within the business. This discusses the synergy between green innovation and information systems as a powerful framework for addressing global environmental challenges to advance economic development and pave the way for a sustainable future.

Chapter 4: Artificial Intelligence and Machine Learning to Assess the Effects of Covid-19 on Land Surface Temperature (LST) and Air Quality Index

This chapter discusses an empirical study to bring out the impact of Covid-19 on Land surface temperature and its relationship with environmental quality. The study emphasizes the land surface temperature (LST) and air quality as important remote sensing parameters to identify the positive effect of the lockdown on the environment due to Covid-19 pandemic in Dehradun City. The trend analysis of land surface temperature collected for the last 12 years reflects dramatic variations. The satellite imagery of Landsat 8 OLI and Landsat 5 TM of April in the years 2009, 2019, 2020, and 2021 was explored to focus on the effect of variation in LST on environmental quality to estimate the relationship between land surface temperatures and air quality. The findings of this study show that there is a significant effect on LST during lockdown. The result shows a sudden decrease in air quality with an unexpected decrease in surface temperature in April 2020 due to the lockdown in Dehradun city and again an increase in March 2021 after Covid-19 lockdown. It shows the improvement of environmental quality in Dehradun city during lockdown, but after lockdown, it reaches the same. The study will help the decision-makers of various organizations in urban planning, climate change analysis, and forest fire risk management study for the growth of sustainable development.

Chapter 5: Trend Analysis of Long-Term Temperature Data for Prediction of Heat Waves Through Statistical Analysis Using Extreme Value Theory for Climate Disaster Management

This chapter examines trend analysis of long-term temperature data for the prediction of heat waves through statistical analysis using extreme value theory for climate disaster management. This section explains the term "heat wave" in the context of hot weather. It explains the rising global warming, which is expected to become more intense, frequent, and prolonged. The future heatwaves are expected to occur more frequently, endure longer, and be stronger in almost all inhabited places. The proposed statistical framework applies the extreme value theory (EVT) for the estimation of the probability of uncommon occurrences that have not yet been seen and have not been recorded in the past. This chapter discusses the possibility of determining the locations and chances of future heat waves of varying sizes. The results will help in predictive maintenance to mitigate and adapt to extremely hot conditions and also provide support in the near real-time prediction of heat events in resource optimization and provide better decision-making to achieve sustainable developmental goals.

Chapter 6: Risk Assessment and Vulnerability Mapping of Soil Erosion for Climate Resilient Disaster Management with Geospatial Techniques

This chapter discusses the risk assessment and vulnerability mapping of soil erosion for climate-resilient disaster management with geospatial techniques. It explains soil erosion as a significant issue that poses a threat to the environment. The chapter discusses the most popular analytical model, i.e. Revised Universal Soil Loss Equation, for calculating the average annual soil loss. It presents a case study to predict possible yearly soil loss in Chikodi Taluk. The map of estimated soil loss will be a useful tool for developing sustainable development strategies.

Chapter 7: A Geospatial Approach for DEM-Based Surface Hydrological Modelling of Sub-Watersheds Towards Climate Disaster Management

This chapter deals with the geospatial approach for DEM-based surface hydrological modelling of sub-watersheds towards climate disaster management. It discusses that the management, development, and prioritization of watersheds are essential for preserving the water resources with the surrounding soil and vegetation. The focus of the study for the watershed management of Bandal River sub-watersheds is based on morphometric analysis method with a Digital Elevation Model (DEM) based surface hydrological modelling. Several other significant hydrological and topographical components of sub-watersheds, namely, slope, aspect, elevation, and drainage density, have been analysed. This study concludes that the Bandal River sub-watersheds have dendritic and parallel drainage patterns with moderate to high relief. The watershed's dendritic and parallel drainage patterns facilitate understanding of a variety of topographical aspects, such as runoff and infiltration rate. The study will benefit various managers and decision-makers of sustainable watershed management.

Chapter 8: Urban Heat Island Mitigation Through Green Infrastructure for Sustainable Development

This chapter discusses urban heat island mitigation with the use of green infrastructure for sustainable development. It reiterates that urban heat islands (UHI) pose a significant concern to the well-being of urban residents and the sustainability of urban environments due to urbanization and industrialization worldwide, and it has drawn considerable attention recently. This warming effect, where urban areas experience higher temperatures than rural areas, is primarily caused by changes in the energy balance within cities. The UHI effect lowers air quality, increases water demand, increases energy consumption owing to cooling requirements, and increases morbidity and death from heat stress. Mitigation strategies are a top priority right now because of the possible detrimental effects of UHI on human life. Implementing "green infrastructure", or adding natural and semi-natural features to urban areas, is one viable option. This chapter addresses UHI sources and effects while evaluating how green infrastructure can help lessen its negative effects. This chapter covers the benefits of green infrastructure regarding social, economic, and environmental aspects, such as increased community well-being, better air and water quality, and increased urban biodiversity. The integration of cutting-edge technology, such as Geographic Information Systems (GIS) and remote sensing, is being investigated to maximize the layout and arrangement of green infrastructure components. It summarizes that cities can create more resilient, sustainable, and pleasant urban environments in response to the escalating challenges posed by climate change.

Chapter 9: Identification of Soil Erosion Susceptible Areas with Analytical Hierarchy Process (AHP) Modelling for Climate Disaster Management

This chapter explains the identification of soil erosion susceptible areas with analytical hierarchy process (AHP) modelling for climate disaster management. It explores the previous literature to provide an idea on easterly flowing rivers Panchaganga and Warna. The work delineates the land degradation vulnerability sites using eleven thematic layers, viz. drainage density, slope, land cover, normalized difference vegetation index (NDVI), soil texture and depth, rainfall, land surface temperature (LST), multi-resolution index of valley bottom flatness (MRVBF), terrain ruggedness index (TRI), and soil organic carbon (SOC). The land degradation risk zones have been analysed using the Analytic Hierarchy Process (AHP), and Geographical Information Techniques (GIS), and the obtained results have been validated with the help of fieldwork observations and Google Earth image evidence. The study area shows most of the sites where creep movement took place in the last few years.

Chapter 10: Role of Artificial Intelligence and Machine Learning for Climate Disaster Management and Earth Resource Management

The current chapter explains the role of artificial intelligence and machine learning for climate disaster management and earth resource management. It focuses on the integration of artificial intelligence (AI) in earth resource management (ERM) and climate disaster management (CDM) to enhance precision, efficiency, and sustainability. It discusses advancements in emerging AI technologies such as Explainable AI (XAI) and Transfer Learning to address the challenges of model interpretability and cross-domain application to ensure that AI-driven solutions are transparent, ethical, and adaptable across various environmental contexts. This chapter emphasizes on the potential of AI to enhance the sustainability and resilience of resource management and disaster response. The integration of AI into our analytical processes can improve, manage, and mitigate the impacts of climate-related events for a more sustainable and resilient future.

Chapter 11: From Data to Planning: Innovations Through LiDAR Technology and Data Science in Sustainable Urban Resource Management

This chapter discusses how data planning through LiDAR technology and data science can help in sustainable urban resource management. It focuses on promoting sustainable urban development and improving community well-being. Integration of digital technologies has become essential for modern cities. Advanced remote sensing technologies like airborne laser scanning (ALS) can generate detailed three-dimensional representations of landscapes through point cloud datasets. However, current urban representation often overlooks vegetation, focusing on buildings, limiting overall comprehensiveness. This chapter establishes a framework and methodology for a comprehensive near-realistic representation of urban areas to enhance governance and support sustainable development. The study emphasizes developing methods for efficiently representing complex urban objects, extracting geometric properties from diverse elements, and implementing a flexible schema to ensure consistency and compatibility for application across multiple domains. The proposed framework is implemented on ALS data over Thiruvananthapuram, Kerala, India, as a case study. The chapter also demonstrates how the proposed framework can contribute to achieving the three pillars of Sustainable Development Goal 11: mapping and monitoring, community involvement, and data-driven decision-making.

Chapter 12: Beyond the Office Walls: Exploring Nature Connectedness for Climate Stress and Disaster Management

The purpose of this chapter is to explore the link between job-related stress management and nature connection. It reveals how nature mitigates stress and becomes increasingly important for businesses to seek novel approaches in improving worker productivity and general health. This study investigates the potential gains of nature connection regarding diminishing workplace stress. The study recognizes the relationship between nature relatedness and stress levels among employees in the workplace and outlines effective ways to integrate nature into the work environment. It explores the concept of nature connectedness and the promise it exhibits for employees at work. The chapter explores the relationships between stress and the workplace; a discussion of different stressors elucidates how the organization of the workplace is a vital factor for employee satisfaction and productivity. It has been discovered that mindfulness practices are helpful stress-reduction strategies that may also be applied to handling challenging work-related situations. The chapter elucidates the critical need to create workplaces that incorporate natural features, placing special emphasis on the role of outdoor settings. The research investigates the effects of workplace green space on workers' well-being and illuminates the subsequent gains for companies seeking to enhance the quality of professional life. The chapter concludes by anchoring in multiple viewpoints and offering a crisp distillation of methods professional organizations can employ to foster work environments that encourage enhanced productivity and collaboration. It distinguishes the productive effects of mindfulness and the resource of greenery, ultimately offering a blueprint for upgrading workplace stress and promoting improved work-life balance in modern professional settings.

Chapter 13: Challenges and Future Trends in Climate Disaster Management

This chapter illustrates the challenges and future trends in climate disaster management. This explores the challenges and future trends in applying AI and ML for climate disaster management. It highlights the key challenges like data quality and availability which can impede model accuracy. Model interpretability and transparency are critical issues, as many AI-driven systems function as "black boxes", making it difficult to validate predictions and build trust. Ethical and social considerations, such as ensuring equity and fairness, play a significant role in the responsible deployment of these technologies. Despite these challenges, several future trends highlight the potential of AI and ML in climate disaster management. The development of higher-accuracy climate models and real-time monitoring systems will improve predictive capabilities, allowing for more effective early warning systems and impact assessments. The establishment of regulatory and ethical frameworks will guide the responsible development and application of AI and ML in this field. We can develop more effective strategies to manage and mitigate the impacts of climate disasters by resolving these challenges of future trends.

This book is being formulated with the intent to uncover the upcoming research directions in which AI and ML offer significant benefits in climate disaster management. Readers will gain a deep consideration of the intersection of AI and ML for climate change and disasters. It will facilitate as an invaluable resource for policymakers, disaster managers, researchers, and anyone concerned with safeguarding communities against the ravages of climate-related disasters. Real-world case studies and practical insights illustrate how geospatial technology empowers early warning systems, enhances risk assessment, and streamlines resource allocation. It delves into the nuances of climate adaptation planning, shedding light on innovative strategies for building resilient communities and infrastructure. This book is intended for a diverse audience, ranging from scholars and researchers to practitioners and policymakers involved in environmental science, geography, and data science. The interdisciplinary nature of the content ensures that readers with varied backgrounds can glean valuable insights into the role of artificial intelligence, machine learning, climate change, data analytics and disaster management in shaping the future of sustainable earth resource management.

As editors, we are immensely grateful to the contributors who have shared their expertise and experiences, fostering a collaborative effort to illuminate the potential of these technologies in addressing the ecological imperatives of our time. We hope this compilation serves as a guiding compass for those navigating the exciting and dynamic resource management. May this book inspire and inform, contributing to a future where technology becomes an ally in our shared commitment to preserving and nurturing the resources.

Lubbock, Texas, USA | Deepak Kumar
Albany, New York, USA | Nick P. Bassill
April 2025

Acknowledgements

When you want something, the entire Universe conspires in helping you to achieve it. Paulo Coelho

The editors would like to acknowledge the help of all the people involved in this project and, more specifically, the authors and reviewers who took part in the review process. Without their support, this book would not have become a reality. The editors wish to acknowledge the valuable contributions of the reviewers regarding the improvement of quality, coherence, and content presentation of chapters. Editing this book, *Artificial Intelligence and Machine Learning for Climate Disaster Management*, has been a journey filled with learning, collaboration, and countless moments of inspiration. It is with immense gratitude that we acknowledge the individuals and organizations who have supported and contributed to this work.

We extend our heartfelt thanks to the researchers, scientists, and practitioners in the fields of artificial intelligence, machine learning, and climate science. We had several stimulating scientific discussions with many insights into theories, particularly the concepts of artificial intelligence (AI), machine learning (ML), and climate change (CC). We would look forward to continuing our work in the future since it was not only scientifically exciting but also immensely satisfying on a personal level. Our discussions were extremely inspiring and stimulating and may result in joint research papers.

We are grateful for all the facilities at the "Beautiful campus" for providing first-class infrastructure and offering research facilities, creative thinking, and scientific writing. This work could not have been completed without the great support and encouragement of several people. We are deeply grateful to our mentors and colleagues, whose guidance, feedback, and encouragement have been invaluable throughout this endeavour. Since at this moment it is difficult to memorize all of them, we would like to apologize for any person we forgot to mention. Regrettably, we cannot acknowledge by name all my fantastic friends, because the list would not fit in here, but we gratefully thank each one of them including wonderful friends at our workplace for the great and joyous moments we shared.

We are thankful to the institutions and organizations that have funded and supported research in AI, machine learning, and climate resilience. Their dedication to innovation and sustainability is truly commendable. Thank you for creating a foundation that allows ideas to flourish and impact to grow. We would like to thank NOAA OAR for fellowship support under the Cooperative Agreement Grant # NA21OAR4590360. We would like to thank their colleagues working at the Center of Excellence in Weather and Climate Analytics, Atmospheric Science Research Center (ASRC), State University of New York at Albany and anonymous reviewers for their insightful comments and suggestions. Earnest appreciations are extended to the reviewers for their constructive comments improving the chapters.

Lastly, we sincerely express our special and heartfelt respect, gratitude, and gratefulness to my family members and parents for their endless support and blessings. We admire the almighty for being so kind to us during this period of trials and tribulations, whose grace has sustained us to reach this level of life. It has been a long journey when many obstacles stood in the way; yet the almighty sorted me through, turning every failure into successful and enjoyable moments.

Deepak Kumar
Atmospheric Sciences Group
Department of Geosciences, College
of Arts & Sciences, Texas Tech University
Lubbock, Texas, USA

Center of Excellence in Weather &
Climate Analytics, Atmospheric Sciences
Research Center (ASRC)
University at Albany, State University
of New York (SUNY)
Albany, New York, USA

Nick P. Bassill
Center of Excellence in Weather &
Climate Analytics, Atmospheric Sciences
Research Center (ASRC)
University at Albany, State University
of New York (SUNY)
Albany, New York, USA

State Weather Risk Communication
Center (SWRCC), Atmospheric
Sciences Research Center (ASRC)
University at Albany, State University
of New York (SUNY)
Albany, New York, USA

Contents

Editors and Contributors

About the Editors

Deepak Kumar is an academic researcher with a multidisciplinary background and is currently working as Research Scientist with the Atmospheric Sciences Group at Texas Tech University. Previously, he worked at SUNY Albany's Atmospheric Sciences Research Center, focusing on weather and climate analytics and the Urban-Climate-Energy Nexus. He combines humanities, social sciences, and technology to support climate resilience and policymaking. He is recognized among the World's Top 2% Scientists for 2025 by Stanford University and Elsevier, for his work in Artificial Intelligence, Image Processing, Geomatics, and Engineering Sciences. His expertise spans Geospatial Sciences, Computational Modeling, Climate Change, and Sustainability. Before moving to the U.S., he was an Assistant Professor at Amity University, India. His research covers remote sensing, geoinformatics, urban climate modeling, and climate impacts on socio-environmental systems. He has led two government-funded projects, holds seven patents, and has mentored numerous Ph.D. and postgraduate students. He currently co-supervises two doctoral scholars externally. He actively presents at conferences with over 75 peer-reviewed publications and nine academic books. He has reviewed 230+ manuscripts and serves as an editor for high-impact journals. His work supports key UN Sustainable Development Goals, including clean energy, reduced inequalities, sustainable cities, and climate action. Deepak is committed to transdisciplinary research and innovative solutions for global challenges.

Nick P. Bassill over several years of experience in climate computing research and its impact assessment. He is currently working as Director of the State Weather Risk Communications Center along with Director of Research & Development at the Center of Excellence in Weather & Climate Analytics, Atmospheric Sciences Research Center (ASRC), University at Albany, State University of New York (SUNY), Albany, New York, USA and Adjunct Professor at the Atmospheric Sciences Research Center (ASRC), Department of Atmospheric & Environmental

Sciences, University at Albany, State University of New York (SUNY), Albany, New York, USA. Dr. Bassill received his PhD in Atmospheric and Oceanic Sciences from the University of Wisconsin-Madison, with a research focus on ensemble and parameterization studies. Afterwards, he worked as a post-doctoral researcher at the University of Utah, where he studied aspects of Hurricane Sandy's unusual track (including what led some models to make an incorrect forecast). Afterwards, he was hired as a Post-Doctoral researcher with the New York State Mesonet to begin building an operational analysis system that would utilize Mesonet observations. Concurrently, he is actively involved in building meteorological products for and doing basic research with the New York State Mesonet. Currently, Nick predominantly spends his time working with the Center of Excellence on a variety of projects. He has published various research papers in international journals of high repute. He has also conducted sessions on Numerical weather prediction, with a specialization in WRF modelling, Data assimilation, Weather forecasting, Tropical cyclones, Data analysis of both gridded model data and weather observations and the New York State Mesonet. He also has experience in providing scientific leadership to high-profile strategic computational sustainability and conservation initiatives with several years of research experience. Managed cross-sectoral and interdisciplinary teams of professionals to deliver on complex research projects. His core area of Interest covers Financial Markets, Energy and utilities, Emergency Management, Risk Management, Climate Computing, AI and Machine Learning, and Industry 4.0. He has presented several research papers at international and national conferences on these topics. Also, he has experience in providing scientific leadership to high-profile strategic sustainability and conservation initiatives with several years of research experience.

Contributors

Dhende Atish Department of Geology and Geoinformatics, School of Earth Sciences, Punyashlok Ahilyadevi Holkar Solapur University, Solapur, Maharashtra, India

Ruchi Badola Ganga Aqualife Conservation Monitoring Centre, Wildlife Institute of India, Dehradun, Uttarakhand, India

Nick P. Bassill Center of Excellence in Weather & Climate Analytics, Atmospheric Sciences Research Center (ASRC), University at Albany, State University of New York (SUNY), Albany, New York, USA

State Weather Risk Communication Center (SWRCC), Atmospheric Sciences Research Center (ASRC), University at Albany, State University of New York (SUNY), Albany, New York, USA

Deepali Bidwai Independent Researcher, Arcueil, France

Gurav Chandrakant Department of Geology, Rajaram College, Kolhapur, Maharashtra, India

Santosh Chougale Department of Marine Geology, Mangalore University, Mangalagangothri, Karnataka, India

Kulkarni Dhaval Department of Geology and Geoinformatics, School of Earth Sciences, Punyashlok Ahilyadevi Holkar Solapur University, Solapur, Maharashtra, India

Syed Ainul Hussain Ganga Aqualife Conservation Monitoring Centre, Wildlife Institute of India, Dehradun, Uttarakhand, India

Sonal Jain KIIT School of Economics & Commerce (KSEC), Kalinga Institute of Industrial Technology (KIIT) Deemed to be University, Bhubaneswar, Odisha, India

Rakesh Kadaverugu CSIR- National Environmental Engineering Research Institute, Nagpur, Maharashtra, India

Academy of Scientific and Innovative Research (AcSIR), Ghaziabad, Uttar Pradesh, India

Samanpreet Kaur Department of Soil and Water Engineering, Punjab Agricultural University, Ludhiana, Punjab, India

Ambrina Sardar Khan Department of Environmental Sciences, Integral University, Lucknow, Uttar Pradesh, India

Farhina Sardar Khan Integral Business School, Integral University, Lucknow, Uttar Pradesh, India

Department of Commerce, Integral University, Lucknow, Uttar Pradesh, India

Deepak Kumar Atmospheric Sciences Group, Department of Geosciences, College of Arts & Sciences, Texas Tech University, Lubbock, Texas, USA

Center of Excellence in Weather & Climate Analytics, Atmospheric Sciences Research Center (ASRC), University at Albany, State University of New York (SUNY), Albany, New York, USA

Maya Kumari Amity School of Natural Resources & Sustainable Development (ASNRSD), Amity University Uttar Pradesh (AUUP), Noida, Uttar Pradesh, India

Anugrah Rohini Lall Department of Humanities and Social Sciences, Graphic Era Deemed to be University, Dehradun, Uttarakhand, India

Ashish Mani Amity School of Natural Resources & Sustainable Development (ASNRSD), Amity University Uttar Pradesh (AUUP), Noida, Uttar Pradesh, India

Ganga Aqualife Conservation Monitoring Centre, Wildlife Institute of India, Dehradun, Uttarakhand, India

Richa Manocha Amity School of Business, Amity University Uttar Pradesh (AUUP), Noida, Uttar Pradesh, India

Syed Shahid Mazhar Integral Business School, Integral University, Lucknow, Uttar Pradesh, India

Pramod Kumar Mishra Acharya Narendra Deva University of Agriculture and Technology, Kumarganj, Ayodhya, Uttar Pradesh, India

Simran Mittal Amity School of Business, Amity University Uttar Pradesh (AUUP), Noida, Uttar Pradesh, India

Abhishek Pandey Division of Agricultural Engineering, ICAR-Indian Agricultural Research Institute, New Delhi, India

Anandakumar M. Ramiya Department of Earth and Space Sciences, Indian Institute of Space Science and Technology, Thiruvananthapuram, Kerala, India

Prashant Shukla Division of Agricultural Engineering, ICAR-Indian Agricultural Research Institute, New Delhi, India

Mukesh Siag Department of Soil and Water Engineering, Punjab Agricultural University, Ludhiana, Punjab, India

Prateek Srivastava Department of Botany, University of Allahabad, Prayagraj, Uttar Pradesh, India

Sagar Tomar CSIR- National Environmental Engineering Research Institute, Nagpur, Maharashtra, India

Academy of Scientific and Innovative Research (AcSIR), Ghaziabad, Uttar Pradesh, India

Gijs van den Dool Independent Researcher, Arcueil, France

Jayati Vijaywargiya Department of Earth and Space Sciences, Indian Institute of Space Science and Technology, Thiruvananthapuram, Kerala, India

Chapter 1
Artificial Intelligence and Machine Learning for Climate Disaster Management

Deepak Kumar and Nick P. Bassill

Abstract Artificial Intelligence (AI) and Machine Learning (ML) are revolutionizing climate disaster management by providing advanced tools and techniques for predicting, mitigating, and responding to natural disasters. These technologies provide huge amounts of data from various sources, including remote sensing, weather stations, and social media, to enhance the accuracy and efficiency of disaster management processes. AI and ML algorithms can analyse complex datasets to identify patterns and trends that human analysts might overlook. The forecasts enable authorities to implement proactive measures, such as reinforcing infrastructure, pre-positioning emergency supplies, and issuing early warnings to vulnerable communities. Models like predictive models can forecast the likelihood of extreme weather events such as hurricanes, floods, and droughts based on historical data and current conditions. AI and ML play a crucial role in real-time disaster response in addition to prediction. These technologies process real-time data from various sensors and sources to provide situational awareness during any disaster. AI-enabled drones and satellite imagery assess damages to identify affected areas and prioritize rescue and relief operations. Machine learning algorithms analyse social media posts to gauge public sentiment, detect emergencies, and direct resources to areas in need. AI and ML also contribute to long-term resilience and

D. Kumar (✉)
Atmospheric Sciences Group, Department of Geosciences, College of Arts & Sciences, Texas Tech University, Lubbock, Texas, USA

Center of Excellence in Weather & Climate Analytics, Atmospheric Sciences Research Center (ASRC), University at Albany, State University of New York (SUNY), Albany, New York, USA

N. P. Bassill
Center of Excellence in Weather & Climate Analytics, Atmospheric Sciences Research Center (ASRC), University at Albany, State University of New York (SUNY), Albany, New York, USA

State Weather Risk Communication Center (SWRCC), Atmospheric Sciences Research Center (ASRC), University at Albany, State University of New York (SUNY), Albany, New York, USA

D. Kumar, N. P. Bassill (eds.), *Artificial Intelligence and Machine Learning for Climate Disaster Management*, Disaster Resilience and Green Growth,
https://doi.org/10.1007/978-981-95-1588-2_1

recovery efforts. The analysis of post-disaster data evaluates the effectiveness of response strategies and identifies areas for improvement. They model different scenarios to help policymakers develop robust disaster management plans and policies. AI and ML support community engagement by providing accessible and actionable information to the public through chatbots and mobile applications. The adoption of AI and ML in climate disaster management faces challenges, such as data privacy concerns, the need for high-quality data, and the requirement for skilled personnel to develop and maintain these systems. These challenges are essential to fully realize the benefits of AI and ML in building climate resilience and ensuring effective disaster management. AI and ML are transformative technologies that enhance the prediction, response, and recovery phases of climate disaster management. These advanced tools will help authorities to protect communities and ecosystems from the adverse impacts of climate-related disasters.

Keywords Artificial intelligence · Machine learning · Disaster · Climate · Infrastructure · Planning

1.1 Introduction

Artificial Intelligence (AI) and Machine Learning (ML) have emerged as contemporary technologies having immense potential to revolutionize climate disaster management (Basheer Ahammed and Pandey 2019; Yadav 2023). The frequency and severity of climate-related disasters, such as hurricanes, floods, wildfires, and droughts have increased due to climate change, and it has become imperative to adopt innovative approaches for better preparedness, response, and recovery efforts (Kumari and Rana 2023; Shahvar et al. 2025). AI and ML offer good insights to improve the accuracy, efficiency, and effectiveness of disaster management strategies. AI and ML help to process and analyse vast amounts of data from diverse sources for climate disaster management (Aravinda et al. 2024; Zhou and Zhou 2021). The sources of data include remote sensing technologies, weather stations, social media, and historical climate data. The application of AI and ML algorithms with these datasets can identify patterns to detect anomalies for better predictions, which may be overlooked by human analysts (Chai et al. 2021; Imhoff et al. 2000). This capability is valuable for predicting extreme weather events and assessing their potential impacts on vulnerable communities and infrastructure.

One of the primary advantages of AI and ML in climate disaster management is their role in enhancing early warning systems. Predictive models developed with ML algorithms can forecast the likelihood and intensity of natural disasters, and these provide authorities and communities with timely alerts and warnings (Ceccato et al. 2018; Hou et al. 2022). In real scenarios, AI-enabled systems can analyse real-time data from satellites and weather stations to predict the path and intensity of hurricanes for better preparation and evacuation planning. AI- and ML-driven early warning systems can save lives, reduce property damage, and minimize the economic impact of disasters.

Machine learning algorithms can analyse social media posts to gauge public sentiment, detect emergencies, and direct resources to areas in need. This real-time information enables a more coordinated and efficient disaster response, ensuring that resources are allocated where they are needed most (Sharmila et al. 2019; Sonthi et al. 2022). During a disaster, these technologies can process real-time data from various sensors and sources to provide situational awareness. AI-driven drones and satellite imagery can assess damage, identify affected areas, and prioritize rescue and relief operations. These also contribute to long-term resilience and recovery efforts (Ouafiq et al. 2023; Shaikh et al. 2022). The analysis of the post-disaster data along with the technologies can evaluate the effectiveness of response strategies and identify areas for improvement. They model different scenarios to help policymakers develop robust disaster management plans and policies (Manunta et al. 2010; Navarro and Carbonell 2008). These can support community engagement by providing accessible and actionable information through chatbots and mobile applications. These tools can educate the public about disaster risks and preparedness measures, fostering a culture of resilience. The use of AI and ML as a powerful tool will enhance the prediction, response, and recovery phases of climate disaster management. These will support authorities to protect communities and ecosystems from the adverse impacts of climate-related disasters and it will ultimately contribute to a safer and more sustainable future.

1.2 Artificial Intelligence (AI) and Machine Learning (ML) for Climate Disaster Management

AI refers to the ability of machines to simulate human intelligence processes, whereas ML is a subset of AI that allows systems to learn and improve from experience without explicit programming (Fan et al. 2025; Kumari et al. 2025). In the context of disaster management, these technologies have been developed to analyse vast amounts of data from various sources, such as satellite imagery, weather models, and sensor networks. This allows for more accurate predictions and faster decision-making (Rad et al. 2011; Yadav et al. 2015). The advent of AI and ML has enabled automated pattern recognition in complex datasets, making it possible to predict the likelihood and intensity of disasters such as floods, wildfires, and storms with higher precision (Tshimanga et al. 2024; Wang et al. 2024). The improvement of Artificial Intelligence (AI) and Machine Learning (ML) for climate disaster management has progressed rapidly in recent years and these are driven by the need to improve disaster prediction, response, and recovery. The advancements in deep learning approach have significantly enhanced the ability to process geospatial and environmental data for smarter, data-driven solutions to climate-induced disasters (Wang et al. 2024; Zhou and Zhou 2021). AI and ML offer numerous advantages in climate disaster management, improving both the effectiveness and efficiency of disaster preparedness and response. One of the key benefits is the ability to predict natural disasters well in advance with more accuracy.

AI algorithms analyse historical weather data, satellite images, and real-time environmental data to provide early warning systems, which will provide communities more time to prepare and evacuate. ML models are trained to recognize patterns and anomalies in climate data, which assists in identifying potential risks such as flooding, hurricanes, and heat waves. Moreover, these technologies facilitate optimized resource allocation during a disaster. AI-driven algorithms analyse the real-time data from disaster zones for damage assessment to direct emergency services to high-priority areas with effective distribution of available resources (Kumar et al. 2020; Zhang et al. 2013). The trustworthiness of AI and ML in climate disaster management is essential for their widespread adoption and use in high-stakes environments. While these technologies are powerful, their reliability depends on the quality and accuracy of the data they are trained on. Therefore, if the data is incomplete, outdated, or biased, then the predictions and recommendations made by AI and ML models can be flawed, which can potentially lead to misguided decisions during any disaster (Ma et al. 2024; Othman et al. 2020). Thus, ensuring the trustworthiness of AI and ML requires efforts to maintain high standards of data quality and to validate the models used. As these technologies are continually evolving, regular updates and retraining of models are necessary to adapt to changing climate patterns and emerging disaster scenarios (Kumari et al. 2025). So, when these technologies are used for public safety then the transparency in AI decision-making processes is also crucial. The need for human oversight is also important, as AI and ML should complement, rather than replace, human expertise.

AI and ML can be used for wider segments in climate disaster management, and to support each phase of disaster management viz. *prevention, preparedness, response, and recovery*. In the prevention phase, AI can analyse climate models and predict areas at high risk for future natural disasters, allowing for better urban planning and infrastructure development (Abbas et al. 2024a; Quesada-Román et al. 2024; Simarmata et al. 2021). In preparedness, AI-powered early warning systems use real-time data to forecast extreme weather events and issue timely alerts to vulnerable populations, enabling evacuation and resource mobilization. During the response phase, AI and ML can assist in real-time decision-making by processing data from sensors, satellites, and social media to identify the most affected areas and prioritize relief efforts (Yeşilköy et al. 2024; Zhou et al. 2024). ML algorithms can help assess the severity of damage and predict the most efficient routes for rescue teams to follow. In recovery, AI can be used to monitor reconstruction progress, assess long-term environmental changes, and optimize recovery strategies to ensure resilience against future disasters. The integration of AI and ML into these areas has the potential to transform climate disaster management to make it more data-driven, responsive, and resilient (Aravinda et al. 2024; Samantray and Gouda 2024).

1.3 Significance of Artificial Intelligence and Machine Learning in Climate Disaster Management

Artificial Intelligence (AI) and Machine Learning (ML) have become integral to modern climate disaster management, significantly enhancing the effectiveness and efficiency of disaster preparedness, response, recovery, and mitigation strategies (Al-Hamdan et al. 2016; Barbierato et al. 2019; Igun and Williams 2018; Tran et al. 2024). The growing frequency and intensity of climate-related disasters, such as floods, hurricanes, wildfires, and droughts, necessitate advanced technological tools to predict, manage, and mitigate these events (Jian et al. 2020; Song et al. 2024). AI and ML offer powerful capabilities that can transform the way disaster management agencies plan, respond, and recover from climate disasters.

(a) *Improved Prediction and Early Warning Systems*

AI and ML in climate disaster management can improve prediction accuracy and early warning systems. AI and ML algorithms can process vast amounts of data from satellites, weather stations, and sensors to predict extreme weather events and environmental hazards with high precision (Kuo et al. 2024; Pranto et al. 2024). These technologies can identify patterns and trends in climate data that may not be immediately apparent to human analysts, allowing for earlier and more accurate forecasting of events such as storms, floods, and heatwaves (Bernetti et al. 2020; Venter et al. 2020). The early warning capabilities with the help of AI and ML can enable authorities to issue timely alerts, giving communities ample time to prepare and evacuate if necessary, reducing loss of life and property damage.

(b) *Real-Time Data Analysis and Decision-Making*

AI and ML provide real-time analysis of large, complex datasets to support rapid decision-making during any disaster. AI systems can analyse satellite images and data from drones to detect fire outbreaks and track their progression in real-time for monitoring wildfires. AI and ML automate data analysis for data-driven decisions to improve the overall efficiency of the response (Conde and De Mata Muñoz 2019; Singh and Malhotra 2019). This allows disaster response teams to respond quickly and allocate resources to the most affected areas. Similarly, ML algorithms analyse data from weather sensors to track flooding events, identify vulnerable areas, and prioritize rescue operations.

(c) *Enhancing Resource Allocation and Logistics*

Managing resources like food, medical supplies, and rescue teams can be a daunting task during a disaster. However, AI-driven systems can optimize the distribution of these resources by analysing real-time data and predicting where they will be needed most. AI and ML enhance resource allocation and logistics during and after a climate disaster (Amani et al. 2020; Song et al. 2024; Talha et al. 2025). ML algorithms can assess the severity of the situation, consider transportation routes, and identify supply shortages, ensuring that resources are allocated efficiently and effectively. It helps to minimize delays by optimization the logistics of relief operations for the effectiveness of disaster response to ensure that aid reaches the areas in need.

(d) *Post-disaster Damage Assessment and Recovery*

Conventionally damage assessment has been time-consuming and a labour-intensive process with teams of surveyors for affected areas. AI and ML is a crucial role in damage assessment and recovery efforts after a disaster. This allows for faster and more accurate assessments, enabling authorities to prioritize recovery efforts and allocate resources where they are needed most. Image recognition algorithms and drones automate to quickly analyse aerial imagery and identify areas of significant damage with AI-powered tools (Mondol et al. 2025; Selvam and Al-Humairi 2025). Besides, it is used to model and simulate different recovery scenarios to help communities and governments plan to implement more effective rebuilding strategies that enhance resilience to future disasters.

(e) *Climate Change Adaptation and Mitigation*

AI and ML also contribute to climate change adaptation and mitigation efforts apart from disaster management. AI systems assist in the identification of areas at risk for targeted climate adaptation strategies (Buscail et al. 2012; He 2022; Wang et al. 2016). The analysis of historical data and modelling of future scenarios to predict the long-term impacts of climate change on specific regions. AI and ML contribute to reducing greenhouse gas emissions with optimization of energy usage, improving waste management, and advancing renewable energy technologies (Brown 2018; Gupta et al. 2024; Zhuang et al. 2020). It helps to design more resilient infrastructure, optimize water management, and recommend agricultural practices to reduce vulnerability to climate change (Oyounalsoud et al. 2024; Singh and Manoharan 2024).

(f) *Capacity Building and Education*

AI and ML support capacity-building and education for climate disaster management. These technologies can train disaster management professionals, local governments, and communities to respond effectively to climate disasters (Charlesworth and Warwick 2020; Nkwunonwo 2013). AI-driven educational platforms deliver personalized learning experiences, teaching individuals about disaster risks, preparedness measures, and recovery strategies (Ayub et al. 2024; Beernink et al. 2022). These help to build more knowledgeable and resilient inhabitants, empowering communities to take proactive steps to reduce their exposure to climate hazards.

(g) *Data Integration and Interoperability*

AI and ML enhance data integration, interoperability, and analysis to ensure data from different platforms in a unified manner. Climate disaster management integrates data from various sources like weather stations, satellites, sensors, and social media. This improves situational awareness to help responders for more informed decisions to coordinate more effectively with other agencies and organizations (Maheswari et al. 2024). AI systems process and synthesize information from these diverse sources to provide disaster management agencies with a comprehensive understanding of the situation.

Artificial intelligence and machine learning in climate disaster management enhance prediction accuracy, improve real-time data analysis, optimize resource allocation, and accelerate recovery to transform disaster management practices (Dtissibe et al. 2024; Pandey et al. 2024). These technologies enable more proactive, data-driven approaches that save lives, reduce economic losses, and enhance resilience to future climate disasters. Climate change poses new challenges, so AI and ML in disaster management will offer resolutions to help societies adapt, mitigate risks, and recover more effectively in an ambiguous future climate.

1.4 Characteristics of AL and ML for Climate Disaster Management

Artificial Intelligence (AI) and Machine Learning (ML) bring unique characteristics to enhance climate disaster management for more accurate predictions, efficient resource allocation, and better decision-making (David et al. 2024; Isaac et al. 2024). These technologies are especially valuable in the context of climate change, where the increasing frequency and severity of natural disasters require adaptive, data-driven solutions.

AI and ML have the most prominent characteristics of data-driven decision-making. These decisions rely on vast datasets to provide insights and solutions. It processes a wide variety of data sources including satellite images, climate models, weather forecasts, sensor data, and social media updates (Alotaibi and Nassif 2024; Ravanelli et al. 2018). It analyses the historical and real-time data to predict potential hazards, identify at-risk areas, and optimize response strategies. These approaches identify the patterns within large volumes of data to forecast natural hazards like hurricanes, wildfires, floods, and droughts. These have real-time monitoring and analysis capabilities to monitor changing conditions from incoming data instantaneously during a disaster (Manunta et al. 2010; Zeng et al. 2024). These systems continuously process data from multiple sources covering the weather stations, satellites, and ground sensors to provide up-to-date reports. This real-time analysis is critical for disaster response teams to generate instant decisions for evacuations, resource allocation, and emergency services. Machine learning models can learn from new data and improve over time. These can handle large amounts of data from diverse sources and scale up or down depending on the size and scope of the disaster (Motwani et al. 2021; Story 2012). This scalability of AI and ML systems can be used effectively in a wide range of disaster scenarios, from small-scale emergencies to large, complex events that require coordinated international efforts. Characteristics like data-driven decision-making, predictive capabilities, real-time monitoring, automation, adaptability, and scalability make these approaches indispensable in climate-induced disaster management.

1.4.1 Social Aspects

The assimilation of Artificial Intelligence (AI) and Machine Learning (ML) into climate disaster management brings technological advancements with significant social implications. These technologies influence how societies prepare for, respond to, and recover from climate-related disasters (Edeh et al. 2024). While AI and ML can significantly enhance disaster management, they also raise important social considerations, such as equity, accessibility, trust, and community involvement.

One critical social aspect of AI and ML is equity and inclusivity. AI and ML have the potential to improve disaster response systems, and it is essential to ensure their benefits for all segments of society, particularly vulnerable and marginalized communities (Abe et al. 2023; Bâra and Oprea 2024). In many disaster-prone regions, low-income, rural, or underserved people are the most affected by climate disasters. AI-driven systems consider these community-specific needs to avoid intensifying inequalities. AI models incorporate local knowledge with socio-economic data to accurately predict how different groups will be impacted by climate events (Kesavan et al. 2024; Talaviya et al. 2020). Failure to do so could lead to biased predictions and insufficient support for these populations during crises. Disaster management systems through AI and ML must be accessible to all regardless of digital literacy or access to technology.

Another social consideration is trust and transparency. As AI and ML systems become more embedded in disaster management, public trust in these technologies becomes crucial. Communities must have confidence that AI-driven predictions and responses are accurate, reliable, and unbiased (Jitkajornwanich et al. 2024; Rijal et al. 2024). This trust can be built through transparency in how AI models are developed and how decisions are made. When AI systems are used to allocate resources or make critical decisions during a disaster, it is essential to have clear, understandable explanations for the processes and outcomes. Involving local communities in the development and evaluation of AI models can help build trust and ensure that these technologies reflect the needs and concerns of the people they aim to protect.

Community participation is another vital social aspect of AI and ML in disaster management. While these technologies provide powerful tools for predicting and responding to disasters, they should not replace human judgement and community engagement. AI and ML enhance decision-making, but local knowledge and community participation are essential for crafting effective disaster management strategies (Mohajane et al. 2021). Involving citizens in data collection (through crowdsourcing) and in the development of AI models can lead to more accurate predictions and better-targeted responses (Isaev et al. 2024). Disaster recovery plans should consider the social dynamics and needs of communities, ensuring that AI tools complement, rather than replace, human and social capital.

Strict data protection regulations and clear guidelines for the ethical use of AI are necessary to address these concerns and maintain public trust. The use of AI and ML raises questions about privacy and data security (Liu et al. 2021a; Venkatesan et al.

2019). Disaster management systems rely on collecting and analysing large amounts of data, including personal information, to make accurate predictions and decisions. Ensuring ethics and security in data usage becomes vital to prevent misuse or breaches of privacy. Communities must be assured that their data is handled responsibly and their participation in disaster management processes will not result in harm or exploitation by any AI systems or other means.

AI and ML offer transformative solutions for climate disaster management, but their social aspects must be carefully considered. Ensuring equity, transparency, privacy, community involvement, and capacity building will be essential for these technologies to fulfil their potential in creating more resilient and inclusive disaster management systems (Liao et al. 2024). The implementation of AI and ML in climate disaster management presents opportunities for capacity building and education. As these technologies become more prominent, there is an increasing need to train local communities and disaster management professionals in their use (Abbas et al. 2024b). Enhancing digital literacy, particularly in underserved areas, will help ensure that the benefits of AI and ML are harnessed by those who need them. Educational programs emphasize the social and ethical aspects of these technologies as crucial in preparing the next generation of disaster management experts who can use AI and ML responsibly.

1.4.2 Technological Aspects

AI and ML are revolutionizing the field of climate disaster management through their technological capabilities to process vast amounts of data, analyse patterns, and make real-time decisions. These technologies provide advanced tools for predicting, monitoring, and responding to climate-related disasters (Sun et al. 2024). The technological aspects of AI and ML, such as data processing, modelling, automation, and integration with other systems enhance disaster management efforts (Fadlullah and Kato 2021; Kumar et al. 2024; Paswan et al. 2024). One significant technological aspect is the data processing and analysis capabilities of AI and ML. Climate disaster management involves dealing with large volumes of data, including satellite imagery, weather forecasts, sensor data, and geographical information (Sahani et al. 2019). AI and ML can process and analyse this data much faster and more efficiently than traditional methods.

Another key technological feature is the predictive modelling capabilities of AI and ML. Machine learning algorithms can be trained on large datasets of historical climate data to forecast natural disasters with increasing accuracy (Maji et al. 2024). AI models can predict the intensity and path of hurricanes, or the likelihood of flooding based on real-time weather and environmental data. These predictive models enable early warning systems to provide authorities and affected communities with vital information before a disaster strikes.

Automation is another important technological aspect of AI and ML in disaster management. Both technologies can automate processes that traditionally require manual intervention to speed up decision-making and resource deployment. During post-disaster scenarios, the automated damage assessment helps responders identify the most affected areas and prioritize recovery efforts efficiently (Kow et al. 2024; Senanayake et al. 2024). Automation enhances the efficiency of disaster response and recovery efforts to ensure the effective use of limited resources.

The integration of AI and ML with other disaster management technologies often works in tandem with other tools like geographic information systems (GIS), remote sensing technologies, and the Internet of Things (IoT) (Quan et al. 2017; Saha et al. 2021; Xu et al. 2024). GIS provides spatial data to analyse the AI models, while IoT devices like sensors for monitoring water levels or soil conditions, feed real-time data into AI systems to provide updates on the growing disaster conditions (Elhag and Abdurahman 2020; Kumar et al. 2023; Paswan et al. 2024). The seamless integration of these technologies enhances the ability of disaster managers to assess risks, monitor live conditions, and make data-driven decisions throughout the disaster lifecycle. In the case of wildfires, IoT sensors might provide real-time temperature and humidity data to predict the spread of fire to provide recommendations for evacuation routes (Barnawi et al. 2021; Mahalakshmi and Latha 2019).

The technological aspects of AI and ML in climate disaster management provide powerful tools for improving preparedness, response, and recovery. The capabilities of data processing, predictive modelling, automation, and integration with other systems, and technologies are transforming the approach of disaster management.

1.4.3 Economic Aspects

The combination of AI and ML for climate disaster management offers a range of economic benefits to significantly enhance the efficiency and effectiveness of disaster preparedness, response, and recovery efforts. The use of these technologies by governments, organizations, and communities can reduce the economic impact of climate-related disasters, optimize resource allocation, and lower overall disaster management costs (Gao et al. 2019; Sreekeshava et al. 2024). However, these benefits come with upfront costs for technology development and implementation, which are required to be managed for a positive economic return in the long term.

One of the primary economic benefits of AI and ML in climate disaster management is their ability to reduce disaster-related costs (Maity et al. 2024). The improvement in disaster prediction and early warning systems mitigates the damage caused by extreme weather events. Accurate predictions enable timely evacuations, better preparation, and more efficient allocation of resources, to reduce property damage and the loss of life (Murugadoss et al. 2024). AI-driven models can predict the paths of a hurricane, and it gives governments and communities to evacuate at-risk areas and minimizing casualties and property destruction. The avoidance of such costs is a clear economic advantage, as early intervention can significantly lower the financial burden of post-disaster recovery (Costanzo et al. 2021). AI and ML enhance the efficiency of disaster response efforts, which leads to significant economic savings. These technologies automate labour-intensive processes like damage assessments and resource allocation. Satellite images analysed with AI algorithms can quickly assess the extent of damage in disaster areas to help responders prioritize their efforts (Chang et al. 2023; Maity et al. 2024). AI-driven logistics systems optimize the distribution of relief supplies, ensuring that resources are delivered most efficiently to reduce the cost of transportation and response time.

AI and ML help minimize economic losses with better planning and forecasting in the long-term recovery and reconstruction phase. The analysis of data from past disasters assists in designing more resilient infrastructure to withstand future climate events. AI models simulate the potential impacts of climate change on infrastructure to plan for more robust and climate-resilient buildings, roads, and utilities (Ahmad and Ali 2023). This proactive approach to resilience building helps avoid costly repairs and rebuilding efforts in the aftermath of disasters, ultimately reducing long-term recovery costs and fostering more sustainable economic growth in disaster-prone areas (Almalbis et al. 2023; Tanniru and Ramsankaran 2023).

AI and ML offer economic opportunities for innovation and job creation. There will be an increased need for skilled workers in fields like data science, software development, and remote sensing. The development and deployment of AI and ML solutions also create business opportunities for technology companies, particularly in the areas of climate modelling, risk analysis, and disaster recovery. It can contribute to the growth of the tech sector, generating new economic activities and employment opportunities in both developed and developing economies (Ian et al. 2023a). The economic adoption comes with initial costs for developing and implementing AI systems for disaster management, data collection, and system integration. Governments and organizations are required to allocate funding for the research and development of these technologies, as well as for training personnel to use and maintain them (Bakirman et al. 2023; Gupta et al. 2023). While the return on investment is expected to be high in the long term, the upfront costs could be a barrier for lower-income regions or developing countries with limited budgets. International partnerships, public-private collaborations, and funding mechanisms could be explored to help bridge the gap in resources to address this challenge.

There are costs associated with data privacy and security. The use of AI and ML in disaster management often involves the collection of vast amounts of personal and environmental data. Ensuring that this data is securely stored and protected from misuse requires investments in cybersecurity measures and compliance with data protection regulations (Abe et al. 2023; Zabihi et al. 2023). The potential economic costs of data breaches or misuse could undermine the effectiveness of AI-driven disaster management systems and damage public trust. The economic aspects of AI and ML for climate disaster management reveal both significant benefits and challenges (Oliveira et al. 2022; Rolnick et al. 2023). These technologies offer substantial economic returns in the long term with improvement in prediction accuracy, optimization in resource allocation, reducing disaster-related costs, and fostering innovation.

1.4.4 Environmental Aspects

The integration of AI and ML in climate disaster management has significant environmental implications. These technologies help to optimize prediction, monitoring, and response systems in minimizing the environmental impact of climate-related disasters, improving resource management, and supporting sustainability efforts (Oliveira et al. 2022). However, it requires careful consideration of their environmental footprint, as well as the energy consumption and environmental costs associated with AI and ML systems.

One of the primary environmental benefits of AI and ML in disaster management is their ability to improve the efficiency of resource use during and after disasters (Oliveira et al. 2022). AI-driven systems can optimize the allocation of resources, ensuring that aid and supplies reach the areas that need them most with minimal waste. This efficiency extends to energy use, water management, and the distribution of emergency supplies, reducing the carbon footprint of disaster relief efforts (Thangavel et al. 2023). ML algorithms can help optimize the logistics of transporting food, water, and medical supplies, lowering fuel consumption and transportation-related emissions (Chen et al. 2023). AI and ML support environmental protection in the early detection and response to climate-related environmental hazards, such as wildfires, floods, and hurricanes.

Another critical environmental benefit is AI and ML's contribution to climate change mitigation and adaptation. These technologies forecast future environmental risks, including the impact of climate change on ecosystems and vulnerable areas through predictive modelling (Garbulowski et al. 2021; Jian et al. 2020). The analysis of data on changing weather patterns, sea levels, and land use supports sustainable land management to prioritize areas for conservation and restoration. AI systems assist in identifying regions at high risk of deforestation or desertification for timely interventions to preserve biodiversity and natural resources (Mujetahid et al. 2023; Supriya and Gadekallu 2023). One of the main concerns is the energy consumption required to train AI models and run machine learning algorithms. The computational power needed for processing large datasets, running simulations, and analysing satellite images could consume significant amounts of electricity, contributing to environmental degradation if the energy used is derived from non-renewable sources (Algarni 2023; Grammalidis et al. 2023). The high demand for servers, data centres, and computing devices to support AI applications results in the generation of electronic waste, which can contribute to environmental pollution if not properly managed. The e-waste generated by the hardware used to run AI and ML systems poses another environmental challenge (Benis et al. 2017; Liang et al. 2020). The environmental aspects of AI and ML in climate disaster management highlight both significant opportunities and challenges. These technologies offer powerful tools for enhancing resource efficiency, supporting environmental protection, and contributing to climate change mitigation efforts (Ian et al. 2023b; Shi et al. 2018). However, their energy consumption and environmental footprint require careful consideration, especially in the context of increasing demand for AI applications. Sustainable energy solutions, efficient data management, and minimizing e-waste, AI and ML will have an essential role in promoting a more environmentally responsible approach to climate disaster management.

1.5 Climate Resilient Disaster Management

The social aspects of climate-resilient disaster management focus on how communities and societies are affected by and respond to climate-induced disasters. A core element is ensuring equity and inclusivity in disaster preparedness and response. Vulnerable and marginalized populations, such as low-income groups, women, children, and people with disabilities, are often disproportionately affected by climate disasters (Cilli et al. 2022). Climate-resilient disaster management aims to ensure that these groups are not left behind in disaster preparedness plans, relief efforts, and recovery initiatives.

Another key social aspect is community involvement and empowerment. Local knowledge and community-based approaches play a critical role in building resilience to disasters. Empowering communities to take part in risk assessments, decision-making processes, and disaster response activities increases local ownership and strengthens social cohesion (Sreelakshmi et al. 2022; Theodoraki et al. 2018). Likewise, promoting a culture of preparedness, where individuals and communities understand the risks and take proactive measures to reduce vulnerability, can greatly enhance resilience (Murray et al. 2023; Schaefer et al. 2021). This can be achieved through education and awareness campaigns that engage local populations and teach them how to act before, during, and after a disaster.

In addition to protecting natural ecosystems, there is a growing emphasis on sustainable resource management in disaster management practices. Sustainable agricultural practices can reduce the risk of crop failure due to droughts or floods, while reforestation efforts can prevent soil erosion after heavy rainfall. Climate-resilient disaster management also involves minimizing the environmental impact of recovery efforts by utilizing sustainable materials in rebuilding projects and reducing the carbon footprint of reconstruction activities (Mamatha et al. 2022). Climate-resilient disaster management increasingly relies on digital tools, so climate-resilient disaster management must balance social, technological, economic, and environmental considerations to be effective and sustainable. It is essential to ensure that vulnerable communities are supported and empowered.

1.5.1 Key Elements of Climate Disaster Management

Climate disaster management involves a series of harmonized efforts to minimize the impacts of climate-related events such as floods, hurricanes, wildfires, droughts, and heat waves. These efforts are organized into key modules that focus on preparation, response, recovery, and mitigation (Kumar and Shekhar 2014; Rolnick et al. 2023). Each module plays a critical role in enhancing resilience, protecting lives, and ensuring the sustainability of communities in the face of climate risks. Some key elements of climate disaster management include the following:

1. *Risk Assessment and Early Warning Systems*

   One of the foundational modules in climate disaster management is risk assessment. This involves evaluating the potential hazards, vulnerabilities, and risks associated with climate disasters in specific areas. It includes identifying climate-related threats such as floods, storms, heatwaves, or droughts, and assessing the exposure and vulnerability of the population and infrastructure (Isaev et al. 2024; Rolnick et al. 2023). Risk assessment is often supported by advanced technologies like Geographic Information Systems (GIS), satellite imagery, and machine learning (ML) algorithms that analyse past and present environmental data to predict future disaster risks (Orimoloye et al. 2022). Complementing risk assessment, early warning systems (EWS) constitute a vital element in comprehensive disaster management frameworks by delivering timely and accurate alerts regarding potential hazards. These systems play a pivotal role in enhancing disaster preparedness by facilitating proactive response strategies, which are essential for minimising the detrimental effects of imminent disasters. EWS enables the rapid dissemination of critical information to key stakeholders, including emergency responders, government agencies, and vulnerable communities. This swift communication ensures that appropriate preventative measures and resource mobilisation can occur promptly. The integration and synergy between risk assessment and early warning systems therefore significantly strengthen overall disaster resilience, enabling more efficient mitigation efforts and adaptive strategies that reduce vulnerability and improve recovery outcomes. These systems collect data from various sources, including weather stations, satellites, and sensor networks, to predict and monitor the progression of climate hazards. The early warnings allow governments and communities to take proactive measures, such as evacuations, resource mobilization, or reinforcing infrastructure, to reduce the impact of disasters.
2. *Disaster Preparedness and Mitigation*

   The preparedness component is engaged in ensuring that communities are ready to face potential climate-related disasters. This involves the development of disaster response plans, the establishment of emergency communication systems, and conducting training and drills for both responders and the general population (Linardos et al. 2022; Rahman et al. 2022). Preparedness activities include creating disaster response kits, securing shelters, and ensuring access to critical resources like food, water, and medical supplies. Similarly, the mitigation module reduces the long-term risks associated with climate disasters. This includes measures like the promotion of sustainable land use, restoring ecosystems, building climate-resilient infrastructure, and reducing carbon emissions (Millar et al. 2021; Zhou et al. 2021). Implementing sustainable agricultural practices can reduce the vulnerability of communities to droughts and food shortages.

Mitigation efforts try to enhance the capacity of governments and organizations to address the impacts of climate change through policymaking, innovation, and resource management.

3. *Disaster Response and Relief*

   The disaster response takes precedence after any disaster event. This stage involves the immediate actions taken to save lives, minimize harm, and prevent further destruction. Key activities include search and rescue operations, providing emergency medical care, distributing food and water, and establishing temporary shelters for displaced populations (Mawgoud et al. 2021). Efficient coordination between governmental agencies, local authorities, NGOs, and international organizations is required for a timely and effective response. AI and technology play an essential role in the improvement of situational awareness through real-time data and facilitating the deployment of resources to the areas of greatest need.

4. *Recovery and Reconstruction*

   Once the immediate impacts of a disaster have been addressed, the focus shifts to recovery. This aims to restore normalcy and rebuild communities affected by the disaster. Recovery efforts include restoring infrastructure, such as roads, bridges, and hospitals, and addressing the psychological and social needs of the affected population (Majidi et al. 2021; Sit et al. 2020). Economic recovery efforts are essential for the resumption of local businesses and livelihoods. Reconstruction is a long-term process that involves rebuilding communities in a way that makes them more resilient to future climate hazards (Chang and Guo 2020; Inapakurthi et al. 2021). This may include building homes and infrastructure to withstand extreme weather events, improving the robustness of public health systems, and promoting sustainable urban planning.

5. *Monitoring, Evaluation, and Adaptation:*

   The monitoring and evaluation is essential for assessing the effectiveness of disaster management efforts. This involves collecting data on the outcomes of preparedness, response, recovery, and mitigation activities, and identifying areas for improvement. Governments and organizations can refine their strategies and policies to better prepare for future disasters by analysing the results (Mosavi et al. 2018; Ramírez 2019). Adaptation is an ongoing process that focuses on adjusting systems and strategies in response to changing climate conditions (Bhatia and Ganguly 2017; Fujihara et al. 2021). This emphasizes the need for continuous learning and flexibility in disaster management approaches. Adaptation efforts may involve adjusting early warning systems, redesigning infrastructure, and implementing new risk management practices to cope with emerging threats. The goal of adaptation is to ensure that communities can thrive in a changing climate while minimizing vulnerability to future disasters.

6. *Capacity Building and Education*

   The capacity-building focuses on enhancing the skills, knowledge, and capabilities of individuals and organizations involved in climate disaster management. This includes training disaster response teams, educating local communities about disaster risks and preparedness measures, and promoting the development of policies and technologies that can improve climate resilience (Giupponi and

Biscaro 2015; Kapetas and Fenner 2020; Tshimanga et al. 2024). Education programs also help to raise awareness about climate change, its impacts, and the importance of reducing carbon emissions and other environmental practices to support disaster resilience.

Climate disaster management is a versatile process that involves a range of interconnected components designed to reduce the impact of climate-related disasters. Risk assessment and early warning systems help prepare for potential disasters, while preparedness, mitigation, and response efforts focus on reducing harm during and after a disaster. Recovery and reconstruction help communities rebuild and adapt to future risks, and monitoring and evaluation ensure that disaster management strategies remain effective. The integration of technology, education, and capacity building further enhances the ability of societies to cope with the growing challenges posed by climate change.

1.5.2 Key Difficulties of Climate Disaster Management

Climate disaster management plays a crucial role in mitigating the devastating impacts of climate-related hazards, such as hurricanes, floods, droughts, wildfires, and extreme heatwaves. As climate change accelerates, these disasters are becoming more frequent and intense, posing significant challenges to governments, organizations, and communities (Ma et al. 2024; Wang et al. 2024). Despite significant progress in disaster management practices, several key issues continue to hinder effective response, preparedness, recovery, and mitigation efforts. These issues require urgent attention and innovative solutions to ensure more resilient, adaptive, and sustainable communities.

1. *Inadequate Early Warning Systems and Predictive Tools*

   One of the key issues in climate disaster management is the lack of adequate early warning systems. Early warnings are essential for saving lives and minimizing property damage by giving people enough time to evacuate or take precautionary measures. However, early warning systems often lack accuracy or timeliness, especially in low-resource settings or rural areas. Some regions lack the infrastructure or technology to collect and process weather data, leading to delays in issuing warnings (Quesada-Román et al. 2024; Yadav 2023). Moreover, warnings may not always reach the people who need them most due to gaps in communication systems. These shortcomings leave communities vulnerable to sudden, extreme events such as storms, floods, or heatwaves, increasing mortality rates and economic losses.
2. *Lack of Preparedness and Insufficient Resources*

   Another critical issue is the insufficient preparedness for climate disasters, especially in regions that are underdeveloped or lack strong governance systems. Many communities have limited disaster preparedness plans or are unaware of how to respond to extreme weather events (Abbas et al. 2024a; Yeşilköy et al.

2024). Inadequate preparedness often leads to confusion, panic, and inefficient responses when a disaster strikes. Moreover, many disaster response agencies struggle with insufficient resources such as funding, trained personnel, equipment, and supplies, which hampers their ability to effectively manage and mitigate the impacts of disasters (Samantray and Gouda 2024; Zhou et al. 2024). Communities in poverty-stricken or conflict-affected regions often lack access to the necessary infrastructure and resources to build resilience against climate disasters, making them more vulnerable.

3. *Coordination and Integration Challenges*

   Effective disaster management requires coordinated efforts from various actors, including local governments, national agencies, non-governmental organizations (NGOs), and international bodies (Aravinda et al. 2024; Samarinas et al. 2023). However, coordination and integration challenges often arise due to conflicting priorities, bureaucratic hurdles, and inadequate communication between stakeholders. Fragmented responses lead to inefficiencies, duplication of efforts, and gaps in the allocation of resources. Some organizations or government entities may be more focused on short-term relief efforts rather than long-term recovery and resilience building (Efremova et al. 2023; Marzhan et al. 2018). This lack of integrated planning and collaboration results in delayed responses, suboptimal resource distribution, and missed opportunities to improve future resilience.

4. *Vulnerability of Marginalized Populations*

   Climate disasters disproportionately affect marginalized populations, including low-income communities, women, children, elderly individuals, and people with disabilities. These groups are often more vulnerable due to inequality, lack of resources, and social exclusion (Song et al. 2024; Tran et al. 2024). Poor communities may live in high-risk areas such as floodplains or on unstable hillsides, increasing their exposure to climate-related hazards. Marginalized groups often have limited access to critical information, healthcare, and disaster relief services, which worsens their ability to recover (Pranto et al. 2024). The inadequate effects of climate disasters highlight the need for inclusive disaster management practices to prioritize the needs of vulnerable populations and ensure their full participation in decision-making processes.

5. *Data Deficiencies and Inconsistent Information*

   Effective climate disaster management relies heavily on accurate and timely data to make informed decisions about risk assessment, resource allocation, and response strategies. However, there are often significant data deficiencies in many regions (Kumari et al. 2025; Shahvar et al. 2025). Data gaps exist in terms of geographic coverage, accuracy, and timeliness of weather forecasts, risk maps, and vulnerability assessments. Moreover, data integration remains a challenge, as information may come from different sources, such as satellites, weather stations, and social media, and may not be compatible or easily accessible. Inconsistent or incomplete data makes it difficult to assess the scale of a disaster or to predict future risks accurately, which in turn hampers effective decision-making.

6. *Environmental Degradation and Ecosystem Vulnerability*

Environmental degradation intensifies the effects of climate disasters and creates feedback loops which intensify the risks. Deforestation, soil erosion, overexploitation of natural resources, and urban sprawl have made ecosystems more vulnerable to climate hazards (Fan et al. 2025; Kuo et al. 2024). The loss of wetlands, mangroves, and forests reduces natural buffers to protect communities from floods, storm surges, and landslides. The destruction of coral reefs exposes coastal areas to the impacts of hurricanes and rising sea levels. Addressing environmental degradation and restoring natural ecosystems is critical for reducing disaster risks and enhancing climate resilience in many regions.

7. *Insufficient Climate Change Adaptation and Resilience Strategies*

   Many regions still lack comprehensive and long-term climate change adaptation strategies. While disaster response efforts focus on immediate relief, adaptation involves preparing for future climate risks by implementing policies, technologies, and infrastructure that reduce vulnerability (Masoumi et al. 2019; Mohamed and Abdel Samee 2022). This includes climate-resilient infrastructure, water management systems, and sustainable agricultural practices. However, in many cases, adaptation strategies are either underfunded, poorly designed, or insufficiently integrated into national development plans. Governments often focus on short-term recovery after disasters, neglecting the need for long-term resilience building. This lack of forward-thinking exacerbates the vulnerability of communities to future climate events.

8. *Climate Finance and Inequitable Funding*

   The financing of climate disaster management continues a significant challenge, particularly in developing countries. Despite increased recognition of climate risks, international climate finance often falls short of what is needed to support developing countries in both disaster preparedness and recovery. Financial constraints prevent many governments and organizations from adequately preparing for or responding to climate disasters (Kim 2016; Sahani et al. 2019). Funding tends to be poorly distributed, with wealthier areas or countries receiving more support while poorer, more vulnerable regions struggle to access resources. Addressing the funding gap and ensuring a more equitable distribution of resources is essential for creating resilient communities worldwide.

9. *Public Awareness and Education Deficits*

   A lack of public awareness and education about climate disaster risks and appropriate response measures is a persistent issue. Disaster preparedness education is often limited, and people may not know how to respond to early warning alerts or evacuate safely in the event of a disaster. Many communities, particularly in rural or underserved areas lack knowledge about how to prepare for climate-related hazards (Hamed et al. 2015; Selvam and Al-Humairi 2025). Public awareness campaigns and community-based education programs are crucial to ensuring that individuals and communities understand the risks they face and are equipped with the knowledge to take preventive actions.

Climate disaster management faces numerous challenges that require to be addressed to build more resilient, adaptive communities in the face of increasing climate risks. From inadequate early warning systems and insufficient resources to the disproportionate impacts on vulnerable populations and environmental degradation, these issues hinder effective disaster response and recovery efforts (Ayub et al. 2024; Singh and Manoharan 2024). These challenges require a comprehensive approach that integrates better planning, data collection, improved communication, financial investments, and inclusive policies towards a more sustainable, climate-resilient future.

1.5.3 Factors Affecting Climate Disaster Management

Climate disaster management is influenced by a variety of factors that determine the effectiveness and efficiency of disaster preparedness, response, recovery, and mitigation strategies. The increasing frequency and intensity of climate-related hazards such as floods, wildfires, hurricanes, and droughts make it essential to consider a wide range of factors when planning for and managing climate disasters (Algarni 2023; Puno et al. 2020). These factors include environmental, social, economic, technological, and political elements that collectively shape how well communities, governments, and organizations can handle the impact of climate disasters.

(a) *Environmental Factors:* Environmental conditions play a significant role in determining the severity and frequency of climate-related disasters. Climate change itself is a primary driver, with rising global temperatures leading to more extreme weather events, such as intense storms, heatwaves, and floods. Ecosystem degradation, such as deforestation, soil erosion, and loss of wetlands intensify the effects of disasters (Sahani et al. 2019; Wang et al. 2024). The destruction of mangroves and forests can leave coastal areas vulnerable to storm surges and flooding, while soil erosion increases the risk of landslides during heavy rainfall. Natural environmental factors, including the topography of an area, proximity to bodies of water, and geological conditions, also influence disaster risks and the effectiveness of mitigation efforts.

(b) *Social and Demographic Factors:* Social vulnerability is a key factor in climate disaster management. Poverty, marginalization, and inequality contribute to the heightened vulnerability of certain populations to climate disasters. Low-income communities are often located in high-risk areas, such as floodplains, informal settlements, or coastal regions, which increases their exposure to climate-related hazards (Puno et al. 2020; Sahani et al. 2019). Gender and age have a crucial role in determining how individuals experience and respond to disasters. Women, children, the elderly, and people with disabilities often face disproportionate challenges in disaster scenarios due to limited mobility, access to resources, or social networks. Therefore, disaster management strategies must account for these social factors to ensure an inclusive, equitable response that reaches the most vulnerable populations.

(c) *Economic Factors:* The financial capacity of governments, organizations, and communities is crucial to effective disaster management. Limited financial resources can impede disaster preparedness, response, and recovery efforts, especially in developing countries. Investments are needed in infrastructure, early warning systems, emergency response training, and community resilience programs (Maheswari et al. 2024). Economic development and access to resources such as insurance and financial safety nets can also determine how well individuals and communities can recover after a disaster (Dtissibe et al. 2024). Economic factors can influence the speed and scale of recovery, as more affluent areas may have the means to rebuild faster, while poorer regions may struggle for years to restore basic services and infrastructure.
(d) *Technological Factors:* Technology has a key role in enhancing the effectiveness of climate disaster management. Geospatial technologies like satellite imagery and remote sensing, provide valuable data for monitoring and forecasting weather patterns, tracking environmental changes, and assessing disaster risks. Artificial Intelligence (AI) and Machine Learning (ML) are remodelling disaster management with better predictions, real-time data analysis, and optimized resource allocation (Rajkumar et al. 2024). Communication technologies are essential for disseminating early warnings, coordinating relief efforts, and providing information to affected populations. However, challenges related to access to technology, digital literacy, and the reliability of infrastructure can limit the potential of these tools, particularly in remote or underserved areas.
(e) *Political and Governance Factors:* Effective climate disaster management relies on strong governance, political will, and cooperation between various levels of government. Political stability, transparency, and the commitment of leaders to prioritize climate resilience are essential for developing and implementing disaster management strategies. The coordination between national, regional, and local authorities is critical in ensuring that resources are allocated efficiently and that the needs of affected communities are met (Gupta and Hall 2021; Liu et al. 2021b; O'Donnell et al. 2017). Policies and regulations related to climate change adaptation, risk reduction, and disaster management frameworks influence the ability of governments to prepare for and respond to climate hazards (Dong et al. 2020; Riadi et al. 2018). Poor governance, lack of accountability, corruption, and weak institutional capacity can significantly hinder disaster management efforts, leading to inefficiencies and delays in response.
(f) *Cultural and Behavioural Factors:* Cultural beliefs, values, and attitudes significantly influence how communities respond to climate disasters. In some regions, there may be a lack of understanding or awareness of climate risks and appropriate disaster preparedness actions. Risk perception and behavioural responses often shape how individuals and communities prepare for or react during a disaster. Communities that have experienced previous disasters may be better prepared and more resilient, but complacency or denial of future risks can also hinder preparedness efforts. In areas where disaster education and awareness campaigns are lacking, individuals may fail to take necessary actions, such

as evacuating or securing their property in advance of a storm (Velterop et al. 2019).

(g) *Legal and Regulatory Factors:* The effectiveness of climate disaster management is shaped by the presence of strong legal frameworks and disaster laws. Regulations that mandate climate risk assessments, enforce building codes, and integrate disaster preparedness into urban planning can help reduce vulnerability and improve resilience. Insurance policies, disaster relief laws, and emergency management regulations guide the allocation of resources and ensure a more coordinated response (Anderson et al. 2022; Chu and Guo 2013). However, weak enforcement of laws or gaps in legal frameworks can hinder disaster management efforts. Land use policies that encourage unsafe development in high-risk areas, such as floodplains or coastal zones, can worsen the impacts of climate disasters.

(h) *International and Global Factors:* Climate disasters do not recognize national borders, and international cooperation is often necessary for effective disaster management. The global nature of climate change calls for collaboration between countries, international organizations, and humanitarian agencies to share knowledge, resources, and technologies. International funding, especially for vulnerable countries, is crucial for building resilience and providing support during and after a disaster (Goodman 2016). Global trade and supply chains are often disrupted during major climate events, and recovery can require international assistance and cooperation. However, differences in political interests, resource availability, and levels of development can complicate global coordination and response efforts.

(i) *Public Awareness and Education:* Public awareness and education are crucial in empowering communities to take proactive measures to reduce their vulnerability to climate disasters. When communities are informed about climate risks, preparedness measures, and response actions, they are more likely to act accordingly. Lack of awareness or misinformation can result in risky behaviours, such as ignoring early warnings or failing to evacuate in time (Ahmed et al. 2021). Thus, disaster management strategies should prioritize public education campaigns and community engagement to foster greater resilience and ensure that individuals are prepared to respond effectively to climate-related hazards.

Climate disaster management is a complicated procedure affected by environmental, social, economic, technological, political, and cultural factors (Pranto et al. 2024; Zhou and Zhou, 2021). Understanding and addressing these factors is essential for creating effective disaster management strategies that minimize the impact of climate disasters. Communities can improve their resilience to the increasing threats of climate change by concentrating on improving preparedness, fostering better coordination, ensuring equitable resource distribution, and leveraging technological advancements.

1.5.4 Strategies for Improved Climate Disaster Management

Strategies for improved climate disaster management are essential to enhance the preparedness, response, recovery, and resilience of communities facing climate-related disasters and some key strategies include the following:

(a) *Risk Assessment and Mapping*

Conducting comprehensive risk assessments and mapping vulnerable areas are critical. Vulnerability mapping helps prioritize areas for intervention and resource allocation. So, with historical data analysis, climate models, and geospatial information, authorities can recognize the regions at high risk of disasters such as floods, hurricanes, and wildfires.

(b) *Early Warning Systems*

Implementing and maintaining robust early warning systems is required for timely alerts and warnings. These systems leverage real-time data from remote sensing, weather stations, and other sources to detect and predict natural disasters. Effective early warning systems can save lives and reduce property damage by providing communities with the information they need to take proactive measures.

(c) *Infrastructure Resilience*

Strengthening infrastructure to withstand extreme weather events is necessary for reducing vulnerabilities. This includes designing and retrofitting buildings, roads, bridges, and other critical infrastructure to be climate resilient. Infrastructure projects should incorporate climate projections and adopt best practices for disaster risk reduction.

(d) *Community Engagement and Education*

Engaging local communities in disaster preparedness and response efforts fosters a culture of resilience. Educational programs, public awareness campaigns, and community-based initiatives can raise awareness about disaster risks and promote proactive measures. Empowering communities with knowledge and resources enhances their ability to respond effectively to disasters.

(e) *Ecosystem-Based Approaches*

Protecting and restoring natural ecosystems can enhance resilience to climate-related disasters. Wetlands, mangroves, forests, and other natural buffers absorb excess rainfall, reduce storm surges, and stabilize coastlines. Ecosystem-based approaches, such as reforestation and wetland restoration, provide valuable ecosystem services and contribute to disaster resilience.

(f) *Data Integration and Technology*

Leveraging advanced technologies like artificial intelligence, machine learning, and geospatial information systems can improve disaster management processes. These technologies can analyse vast amounts of data to identify patterns, predict future events, and optimize response strategies. Integrating data from multiple sources enhances situational awareness and decision-making.

(g) *Policy and Governance*

Effective governance and policy frameworks are significant for implementing and sustaining climate disaster management strategies. Governments have to prioritize climate resilience in their policies, allocate resources for disaster risk reduction, and promote collaboration among various stakeholders. Regulatory frameworks must encourage sustainable practices and support long-term resilience efforts.

(h) *Continuous Monitoring and Evaluation*

Regular monitoring and evaluation of disaster management strategies are essential for identifying areas for improvement. Post-disaster assessments and feedback loops help refine response plans, improve preparedness, and enhance overall resilience. Continuous learning and adaptation ensure that strategies remain effective in the face of evolving climate risks.

These strategies can enhance their preparedness, response, recovery, and resilience to climate-related disasters, ultimately contributing to a safer and more sustainable future.

1.6 Issues in Climate Disaster Management with AI and ML

AI and ML have a substantial capability for improving climate disaster management, but these have several issues and challenges to hinder their effectiveness in real-world applications (Liu et al. 2021c). These issues should be addressed to increase the reach of these technologies to their full potential for disaster preparedness, response, and recovery efforts. Some key issues associated with the use of AI and ML in climate disaster management include the following:

(a) *Data Availability and Quality*

AI and ML models depend heavily on the availability of high-quality, comprehensive, and real-time data to generate accurate predictions and responses. In many parts of the world, especially in developing regions or remote areas, access to reliable and consistent data is a significant challenge. Inadequate or incomplete data may lead to errors in predictions, unreliable models, and potentially misguided disaster response actions (G. Li et al. 2022; Li et al. 2018). Any missing or inaccurate weather data, satellite imagery, or sensor readings will significantly undermine the effectiveness of AI/ML models in forecasting and managing climate-related disasters like floods, hurricanes, and wildfires.

(b) *Bias and Fairness Issues*

AI and ML systems are trained on historical data, and if that data contains biases, the resulting models may perpetuate those biases. This can be especially problematic in climate disaster management, where certain communities may be disproportionately affected by disasters. If the data used to train AI/ML models does not account for the unique challenges faced by marginalized populations, such as rural or economically disadvantaged groups, the models may not perform well in those areas, leaving them underserved during crises (Cordeiro 2021; Leiva-Murillo et al. 2013). This raises problems of fairness and equity, as AI-driven responses may prioritize wealthier or more data-rich regions, exacerbating inequality.

(c) *Transparency and Interpretability*

One of the major concerns with AI and ML in disaster management is the lack of transparency and interpretability of the models, especially for complex systems like deep learning. In emergencies, decision-makers need to understand how the AI or ML system reached specific predictions or recommendations (Gao et al. 2024; Wang 2023). If these models operate as "black boxes", without clear explanations for their decisions, it can be difficult for users to trust the system, and it may lead to poor decision-making.

(d) *High Computational and Infrastructure Demands*

AI and ML systems require significant computational resources to process large amounts of real-time data, such as satellite imagery, weather patterns, or sensor inputs. In disaster-prone regions or developing countries, the necessary infrastructure (e.g., high-performance computing, cloud services, and data storage) may not be readily available, and it will make it difficult to deploy AI/ML solutions effectively.

(e) *Lack of Integration with Existing Disaster Management Systems*

Many current disaster management systems still rely on traditional methods, including manual data collection and response strategies. Integration of AI and ML with existing systems can be a complex and costly process and it requires system upgrades, training, and technical expertise (Benz et al. 2004; Li et al. 2024). The challenge lies in ensuring that AI/ML tools work in tandem with traditional disaster management approaches and local knowledge, as there may be resistance to adopting AI-driven systems due to concerns about disruption or unfamiliarity with the technology.

(f) *Ethical and Privacy Concerns*

The use of AI and ML in climate disaster management requires the collection and analysis of sensitive data like personal location data, health information, and demographic details. This raises serious privacy and ethical concerns. AI systems rely on mobile phones, GPS tracking, or social media data to assess disaster impacts or predict human movement may violate individual privacy rights if the data is not managed properly (Arundel et al. 2023; Liu et al. 2017). The widespread surveillance required for some AI-driven solutions may lead to concerns about data misuse, abuse, or potential discrimination against certain groups.

(g) *Over-Reliance on Technology*

There are risks of over-relying on technology and neglecting human judgement and local knowledge. Climate disaster management often involves dynamic, real-time decisions that consider local conditions, human behaviour, and socio-political factors and elements that AI and ML models may not fully capture (Amani et al. 2020). Disasters require adaptability, and human decision-makers are better positioned to consider the nuances and complexities of real-world situations (Chavan and Mulay 2023). Over-dependence on AI systems could result in decisions that lack context or fail to address specific local needs.

(h) *Limited Generalization and Regional Differences*

AI and ML models are typically trained using data from specific regions, and their ability to generalize to other areas with different environmental, social, and economic conditions can be limited. A flood prediction model developed for one river basin may not apply to a different basin with different topography, climate, or flood behaviour (Gerlak et al. 2018; Shafiei Shiva et al. 2019). When AI models are deployed in regions where the local conditions differ from those in the model's training data, their predictions may become inaccurate, leading to ineffective disaster responses. Customizing and retraining models for specific regions can be resource-intensive and time-consuming.

(i) *Lack of Skilled Personnel*

Implementing AI and ML solutions in climate disaster management requires specialized data scientists, climate experts, and technicians to develop, monitor, and optimize these systems. However, there is often a shortage of qualified personnel, especially in resource-poor settings, which limits the ability to effectively use AI and ML technologies (Nayak et al. 2018; Sharma et al. 2021). This lack of expertise in some regions hinders the adoption and maintenance of AI and ML-driven disaster management systems, leading to inefficiencies and poor performance.

AI and ML hold great promise for revolutionizing climate disaster management, but the associated issues range from data quality and bias to ethical concerns and technical barriers. Tackling these challenges requires investments in data infrastructure, technical expertise, transparency, and policies that prioritize fairness, privacy, and local involvement. These must be addressed to ensure that these technologies provide reliable, effective, and equitable solutions.

1.7 Strategies for Usage of AI and ML in Climate Studies

Artificial intelligence (AI) and machine learning (ML) revolutionize climate studies by providing new insights, enhancing predictive models, and supporting decision-making processes. Some key approaches for using AI and ML in climate studies:

(i) *Data Collection and Integration*

Satellite Data Analysis: AI and ML process and analyse large volumes of satellite data to monitor climate variables such as temperature, precipitation, and cloud cover. These tools can identify patterns and trends that might be missed by traditional methods.

Sensor Networks: Deploying sensor networks in various environments can collect real-time data on climate conditions. AI algorithms can integrate and analyse this data to provide accurate and timely information.

(ii) *Climate Modelling and Prediction.*

Enhanced Climate Models: AI and ML improve the accuracy of climate models by incorporating more data and identifying complex relationships between variables. This can lead to better predictions of climate changes and extreme weather events.

Downscaling Techniques: ML algorithms are used to downscale global climate models to regional and local scales, providing more precise information for specific areas and helping local communities prepare for climate impacts.

(iii) *Extreme Weather Event Prediction*

Early Warning Systems: AI analyses data from various sources, such as weather stations, satellites, and social media, to provide early warnings for extreme weather events like hurricanes, floods, and heatwaves. These systems help to save lives and reduce economic losses.

Impact Assessment: ML models assess the potential impact of extreme weather events on infrastructure, agriculture, and human health. This information guides emergency response and mitigation efforts.

(iv) *Climate Change Mitigation and Adaptation*

Renewable Energy Optimization: AI optimizes the operation and integration of renewable energy sources like solar and wind power into the grid, improving their efficiency and reducing greenhouse gas emissions.

Smart Agriculture: AI-powered precision agriculture techniques help farmers optimize water usage, reduce the need for chemical inputs, and improve crop yields. This enhances food security and reduces the environmental impact of agriculture.

(v) *Environmental Monitoring and Conservation*

Biodiversity Monitoring: AI analyses data from camera traps, drones, and acoustic sensors to monitor wildlife populations and detect changes in biodiversity. This information advises conservation efforts and protects endangered species.

Forest Management: ML algorithms process satellite imagery to monitor deforestation, forest degradation, and wildfires. This data guides forest management practices and supports reforestation efforts.

(vi) *Public Awareness and Education*

Climate Communication: AI-driven tools help communicate complex climate information to the public through interactive visualizations, personalized recommendations, and educational content. This increases public awareness and encourages sustainable behaviours.

Citizen Science: AI supports citizen science initiatives and enables the collection and analysis of climate-related data from volunteers. This enhances community engagement and contributes to scientific research.

The integration of AI and ML with climate studies offers substantial opportunities to improve our understanding of climate change, enhance predictive capabilities, and support mitigation and adaptation efforts.

1.8 Challenges and Opportunities of AI and ML for Climate Studies

The integration of Artificial Intelligence (AI) and Machine Learning (ML) for climate studies presents a wide array of both challenges and opportunities. These technologies have the potential to revolutionize how we analyse climate patterns, predict future climate scenarios, and address the impacts of climate change. The challenges and opportunities of using AI and ML for climate studies are listed below:

1.8.1 Challenges of AI and ML for Climate Studies

(a) *Data Quality and Availability*
One of the biggest challenges AI and ML face in climate studies is the availability and quality of data. Climate data is often sparse, fragmented, or incomplete, especially in remote regions or developing countries. Data from weather stations, satellites, and sensors may have inconsistencies, errors, or gaps, which makes training AI and ML models difficult. AI and ML models require large datasets that are not always accessible, and without high-quality data, the results can be unreliable.

(b) *Complexity of Climate Systems*
The climate system is extremely complex with many variables interacting in non-linear ways. Climate models must account for factors such as atmospheric pressure, ocean currents, greenhouse gas emissions, and more. Modelling these dynamics with AI and ML is challenging due to unpredictable and chaotic behaviour, making it hard for algorithms to draw accurate conclusions or predict future outcomes with confidence.

(c) *Computational Costs*
AI and ML algorithms involving deep learning require substantial computational resources and processing power to handle vast datasets. The complexity of climate modelling at a global scale demands high-performance computing infrastructure that is expensive to maintain. During real-time climate studies or predictions (e.g., monitoring extreme weather events), the computational demands are even more intensive, potentially limiting the scalability of AI/ML-based systems.

(d) *Bias and Generalization*
AI and ML models are trained on historical data, and if this data contains biases or does not fully represent certain regions or climate scenarios, it can lead to inaccurate predictions. AI models trained predominantly on data from one geographical region may not generalize well to other regions with different climates, geographies, or vulnerabilities. This bias can lead to skewed results, affecting the effectiveness of climate-related decisions in areas that are underrepresented in the training data.

(e) *Interpretability and Trust*
AI and ML models, especially deep learning models, are often referred to as "black boxes", meaning it can be difficult to interpret how they make decisions or predictions. Researchers and policymakers need to understand how AI/ML algorithms arrive at specific predictions to make informed decisions. The lack of transparency can lead to mistrust and hesitation in applying AI-generated recommendations for climate action.

1.8.2 Opportunities of AI and ML for Climate Studies

(a) *Improved Climate Predictions*
AI and ML process vast amounts of historical climate data and identify patterns that may be difficult for traditional climate models to detect. AI improve climate forecasting and predictive models. This can lead to more accurate long-term climate predictions, such as projections of temperature changes, rainfall patterns, or the frequency of extreme weather events. AI and ML also enhance the accuracy of seasonal predictions, which are vital for agriculture, water management, and disaster preparedness.
(b) *Real-Time Monitoring and Early Warning Systems*
AI and ML are invaluable for real-time climate monitoring. The analysis of data from satellites, weather stations, and IoT sensors detects shifts in climate patterns like rising temperatures or changes in atmospheric pressure. This allows for the early detection of climate anomalies like hurricanes, heatwaves, or floods, providing early warning systems that help communities prepare in advance. Faster and more accurate predictions of extreme weather events can save lives, reduce economic losses, and improve disaster management.
(c) *Enhanced Climate Modelling*
AI and ML improve the accuracy of climate models by allowing for the integration of large datasets from various sources, such as satellite imagery, ocean data, and land use records. These technologies simulate complex interactions between the atmosphere, oceans, and ecosystems, producing more detailed and precise models.
(d) *Climate Change Mitigation and Adaptation*
AI and ML have a crucial role in developing climate change mitigation strategies by optimizing energy consumption, reducing emissions, and enabling the transition to renewable energy sources. AI improve energy efficiency in buildings, optimize renewable energy grid management, and enhance carbon capture technologies. AI/ML helps in designing climate-resilient infrastructure, optimizing water resource management, and improving agriculture practices for adaptation to minimize the effects of climate change on vulnerable communities.
(e) *Optimizing Resource Management*
AI and ML optimize natural resource management like agriculture, water management, and forestry. These technologies help farmers predict crop yields,

optimize irrigation, and identify areas requiring attention due to drought or pest infestation (Sahani et al. 2019). AI-driven systems help to manage water resources more effectively by forecasting water demand, identifying areas of leakage in distribution systems, and recommending sustainable usage strategies. This ensures climate resilience for resource management and reduces overexploitation.

(f) *Sustainability and Environmental Protection*

AI and ML monitor environmental changes and help assess biodiversity loss, deforestation, or pollution. It analyses satellite imagery and sensor data to track changes in ecosystems and identify areas where conservation efforts are most crucial. This supports sustainable development with informed decisions for conservation, land-use planning, and habitat restoration.

(g) *Cost Reduction and Efficiency*

AI and ML automate many aspects of climate research and climate-related decision-making, and it leads to cost savings and increased efficiency. AI systems process large datasets quickly, allowing researchers and policymakers to make data-driven decisions without any delay. This helps to allocate resources more effectively and streamline climate-related projects and interventions.

While there are significant challenges associated with the usage of AI and ML in climate studies such as data quality, model interpretability, and bias, the potential opportunities for balancing these hurdles. AI and ML have the influence to revolutionize climate research, prediction, and disaster response to offer more accurate forecasts, better resource management, and enhanced climate resilience. These technologies continue to evolve for solutions to the challenges in the global effort to mitigate and adapt to climate change.

1.9 Summary

Artificial Intelligence (AI) and Machine Learning (ML) have emerged as powerful tools in climate disaster management, offering the ability to process vast amounts of data and provide timely, actionable insights. AI and ML are characterized by their ability to analyse complex data patterns, learn from historical data, and predict future outcomes. This makes them highly effective in managing the unpredictable nature of climate disasters, such as hurricanes, floods, and wildfires. In the context of climate-resilient disaster management, AI and ML can enhance early warning systems, improve disaster response efficiency, and support recovery efforts by providing real-time predictions and tailored recommendations.

The significance of AI and ML lies in their capacity to optimize disaster preparedness, minimize losses, and reduce the impact of disasters. These technologies enable more accurate risk assessments, facilitate better decision-making, and enhance resource allocation in times of crisis. AI-driven models can predict extreme weather events and help policymakers prepare in advance. ML algorithms can analyse historical disaster data to identify patterns and trends that might otherwise go unnoticed, further improving response strategies. However, there are several issues associated with AI and ML in climate disaster management. Challenges include the need for high-quality and comprehensive data, the risk of bias in predictive models, and the lack of transparency in certain AI systems, which can undermine trust in their predictions. Additionally, the computational resources required for real-time processing may be out of reach for some regions, and ethical concerns around data privacy and surveillance may arise.

Despite these challenges, AI and ML present opportunities to significantly improve climate disaster management by enabling faster, more accurate predictions, real-time monitoring, and customized response plans. They can also help optimize resource management, improve climate modelling, and support sustainability efforts. The strategies for applying AI and ML effectively include leveraging high-quality data, ensuring local context is considered in models, fostering interdisciplinary collaboration, and investing in capacity building in under-resourced regions. The application of AI and ML in climate disaster management offers great potential to improve preparedness, response, and recovery, but it requires overcoming challenges related to data, resources, and trust to fully realize its benefits.

References

Abbas M, Khan F, Liou Y-A, Ullah H, Javed B, Ali S (2024a) Assessment of the impacts of climate change on the construction of homogeneous climatic regions and ensemble climate projections using CMIP6 data over Pakistan. Atmos Res 304. https://doi.org/10.1016/j.atmosres.2024.107359

Abbas S, Abu Talib M, Ahmed I, Belal O (2024b) Integration of UAVs and FANETs in disaster management: a review on applications, challenges and future directions. Trans Emerg Telecommun Technol 35(12). https://doi.org/10.1002/ett.70023

Abe N, Buckingham K, Chen Y, Dilkina B, Eftelioglu E, Ganguly AR, Gel YR, Hodson J, Kannan R, Lee H, Mao J, Yu R (2023) Fragile earth: AI for climate sustainability – from wildfire disaster management to public health and beyond. In: Proceedings of the ACM SIGKDD international conference on knowledge discovery and data mining, pp 5845–5846. https://doi.org/10.1145/3580305.3599217

Ahmad M, Ali A (2023) Mapping the future of sustainable development through cloud-based solutions: a case study of OpenStreetMap. In: Promoting sustainable management through technological innovation. IGI Global. https://doi.org/10.4018/978-1-6684-9979-5.ch011

Ahmed T, Rahman T, Roy BB, Uddin J (2021) Drone detection by neural network using GLCM and SURF features. J Inf Syst Telecommun 9(33):15–23. https://doi.org/10.52547/jist.9.33.15

Al-Hamdan MZ, Quattrochi DA, Bounoua L, Lachir A, Zhang P (2016) Using Landsat, MODIS, and a biophysical model to evaluate LST in urban centers. Remote Sens 8(11). https://doi.org/10.3390/rs8110952

Algarni M (2023) Deploying artificial intelligence for optimized flood forecasting and mitigation. In: Proceedings of IEEE/ACS international conference on computer systems and applications, AICCSA. https://doi.org/10.1109/AICCSA59173.2023.10479337

Almalbis AEY, Gochian ABB, Murillo RC, Uy ADM, Concepcion R, Valencia IJ, Bandala A (2023) Interplay of remote sensing and artificial intelligence in open agricultural land monitoring. In: 2023 IEEE 15th international conference on humanoid, nanotechnology, information technology, communication and control, environment, and management, HNICEM 2023. https://doi.org/10.1109/HNICEM60674.2023.10589040

Alotaibi E, Nassif N (2024) Artificial intelligence in environmental monitoring: in-depth analysis. Discov Artif Intell 4(1). https://doi.org/10.1007/s44163-024-00198-1

Amani M, Ghorbanian A, Ahmadi SA, Kakooei M, Moghimi A, Mirmazloumi SM, Moghaddam SHA, Mahdavi S, Ghahremanloo M, Parsian S, Wu Q, Brisco B (2020) Google earth engine cloud computing platform for remote sensing big data applications: A comprehensive review. IEEE J Sel Top Appl Earth Obs Remote Sens 13:5326–5350. https://doi.org/10.1109/JSTARS.2020.3021052

Anderson V, Gough WA, Zgela M, Milosevic D, Dunjic J (2022) Lowering the temperature to increase heat equity: a multi-scale evaluation of nature-based solutions in Toronto, Ontario, Canada. Atmosphere 13(7). https://doi.org/10.3390/atmos13071027

Aravinda K, Sravanthi J, Lakhanpal S, Shrivastava A, Kalra R, Mohammad Q, Kumar Mishra N (2024) Water and hydrology: challenges, sustainable strategies and future trends. E3S Web Conf 552. https://doi.org/10.1051/e3sconf/202455201135

Arundel ST, Li W, Campbell BB (2023) Reimagining standardization and geospatial interoperability in today's GeoAI culture. In: Newsam S, Yang L, Mai G, Martins B, Lunga D, Gao S (eds) GeoAI 2023 – proceedings of the 6th ACM SIGSPATIAL international workshop on AI for geographic knowledge discovery. Association for Computing Machinery, Inc, pp 83–84. https://doi.org/10.1145/3615886.3627744

Ayub S, Uday P, Vaibhav S, Udaykumar P (2024) Advancing AI models for drought forecasting, challenges and future directions. Disaster Adv 17(12):58–71. https://doi.org/10.25303/1712da058071

Bakirman T, Kulavuz B, Bayram B (2023) Use of artificial intelligence toward climate-neutral cultural heritage. Photogramm Eng Remote Sens 89(3):163–171. https://doi.org/10.14358/PERS.22-00118R2

Bâra A, Oprea S-V (2024) Machine learning algorithms for power system sign classification and a multivariate stacked LSTM model for predicting the electricity imbalance volume. Int J Comput Intell Syst 17(1). https://doi.org/10.1007/s44196-024-00464-1

Barbierato E, Bernetti I, Capecchi I, Saragosa C (2019) Quantifying the impact of trees on land surface temperature: a downscaling algorithm at city-scale. In: T. K., P. G., M. S., A. V., & H. D. G (ed) Proceedings of SPIE – The International Society for Optical Engineering, vol 11174. SPIE. https://doi.org/10.1117/12.2532066

Barnawi A, Chhikara P, Tekchandani R, Kumar N, Alzahrani B (2021) Artificial intelligence-enabled internet of things-based system for COVID-19 screening using aerial thermal imaging. Futur Gener Comput Syst 124:119–132. https://doi.org/10.1016/j.future.2021.05.019

Basheer Ahammed KK, Pandey AC (2019) Geoinformatics based assessment of coastal multi-hazard vulnerability along the East Coast of India. Spat Inf Res 27(3):295–307. https://doi.org/10.1007/s41324-018-00236-y

Beernink S, Bloemendal M, Kleinlugtenbelt R, Hartog N (2022) Maximizing the use of aquifer thermal energy storage systems in urban areas: effects on individual system primary energy use and overall GHG emissions. Appl Energy 311. https://doi.org/10.1016/j.apenergy.2022.118587

Benis K, Reinhart C, Ferrão P (2017) Development of a simulation-based decision support workflow for the implementation of building-integrated agriculture (BIA) in urban contexts. J Clean Prod 147:589–602. https://doi.org/10.1016/j.jclepro.2017.01.130

Benz UC, Hofmann P, Willhauck G, Lingenfelder I, Heynen M (2004) Multi-resolution, object-oriented fuzzy analysis of remote sensing data for GIS-ready information. ISPRS J Photogramm Remote Sens 58(3–4):239–258

Bernetti I, Barbierato E, Capecchi I, Saragosa C (2020) Climate change and urban wellbeing: a methodology based on Sen theory and imprecise probabilities. Aestimum 2020:57–80. https://doi.org/10.13128/aestim-8086

Bhatia U, Ganguly AR (2017) Network science perspectives on engineering adaptation to climate change and weather extremes. In: Large-scale machine learning in the earth sciences. CRC Press. https://doi.org/10.4324/9781315371740

Brown H (2018) Infrastructural ecology as a planning paradigm: two case studies. Int J Sustain Dev Plan 13(2):187–196. https://doi.org/10.2495/sdp-v13-n2-187-196

Buscail C, Upegui E, Viel J-F (2012) Mapping heatwave health risk at the community level for public health action. Int J Health Geogr 11. https://doi.org/10.1186/1476-072X-11-38

Ceccato P, Ramirez B, Manyangadze T, Gwakisa P, Thomson MC (2018) Data and tools to integrate climate and environmental information into public health. Infect Dis Poverty 7(1). https://doi.org/10.1186/s40249-018-0501-9

Chai KEK, Lines RLJ, Gucciardi DF, Ng L (2021) Research screener: a machine learning tool to semi-automate abstract screening for systematic reviews. Syst Rev 10(1). https://doi.org/10.1186/s13643-021-01635-3

Chang F-J, Guo S (2020) Advances in hydrologic forecasts and water resources management. Water (Switzerland) 12(6). https://doi.org/10.3390/w12061819

Chang F-J, Chang L-C, Chen J-F (2023) Artificial intelligence techniques in hydrology and water resources management. Water (Switzerland) 15(10). https://doi.org/10.3390/w15101846

Charlesworth SM, Warwick F (2020) Sustainable drainage, green and blue infrastructure in urban areas. In: Sustainable water engineering. Elsevier. https://doi.org/10.1016/B978-0-12-816120-3.00002-6

Chavan S, Mulay P (2023) Assessing the impact of GeoAI in the world of spatial data and energy revolution. In: Risk detection and cyber security for the success of contemporary computing. IGI Global. https://doi.org/10.4018/978-1-6684-9317-5.ch008

Chen L, Han B, Wang X, Zhao J, Yang W, Yang Z (2023) Machine learning methods in weather and climate applications: a survey. Appl Sci (Switzerland) 13(21). https://doi.org/10.3390/app132112019

Chu T, Guo X (2013) Remote sensing techniques in monitoring post-fire effects and patterns of forest recovery in boreal forest regions: a review. Remote Sen 6. https://doi.org/10.3390/rs6010470

Cilli R, Elia M, D'Este M, Giannico V, Amoroso N, Lombardi A, Pantaleo E, Monaco A, Sanesi G, Tangaro S, Bellotti R, Lafortezza R (2022) Explainable artificial intelligence (XAI) detects wildfire occurrence in the Mediterranean countries of southern Europe. Sci Rep 12(1). https://doi.org/10.1038/s41598-022-20347-9

Conde FC, De Mata Muñoz M (2019) Flood monitoring based on the study of Sentinel-1 SAR images: the Ebro River case study. Water (Switzerland) 11(12). https://doi.org/10.3390/w11122454

Cordeiro JV (2021) Digital technologies and data science as health enablers: an outline of appealing promises and compelling ethical, legal, and social challenges. Front Med 8. https://doi.org/10.3389/fmed.2021.647897

Costanzo A, Falcone S, D'alessandro A, Vitale G, Giovinazzi S, Morici M, Dall'asta A, Buongiorno MF (2021) A technological system for post-earthquake damage scenarios based on the monitoring by means of an urban seismic network. Sensors 21(23). https://doi.org/10.3390/s21237887

David AO, Ndambuki JM, Muloiwa M, Kupolati WK, Snyman J (2024) A review of the application of artificial intelligence in climate change-induced flooding—susceptibility and management techniques. Civ Eng 5(4):1185–1198. https://doi.org/10.3390/civileng5040058

Dong J, Peng J, He X, Corcoran J, Qiu S, Wang X (2020) Heatwave-induced human health risk assessment in megacities based on heat stress-social vulnerability-human exposure framework. Landsc Urban Plan 203. https://doi.org/10.1016/j.landurbplan.2020.103907

Dtissibe FY, Ari AAA, Abboubakar H, Njoya AN, Mohamadou A, Thiare O (2024) A comparative study of machine learning and deep learning methods for flood forecasting in the far-north region, Cameroon. Sci Afr 23. https://doi.org/10.1016/j.sciaf.2023.e02053

Edeh MO, Dalal S, Alhussein M, Aurangzeb K, Seth B, Kumar K (2024) A novel deep learning model for predicting marine pollution for sustainable ocean management. PeerJ Comput Sci 10: 1–39. https://doi.org/10.7717/peerj-cs.2482

Efremova N, Foley JC, Unagaev A, Karimi R (2023) AI for sustainable agriculture and rangeland monitoring. Philos Stud Ser 152:399–422. https://doi.org/10.1007/978-3-031-21147-8_22

Elhag M, Abdurahman SG (2020) Advanced remote sensing techniques in flash flood delineation in Tabuk City, Saudi Arabia. Nat Hazards 103(3):3401–3413. https://doi.org/10.1007/s11069-020-04135-0

Fadlullah ZM, Kato N (2021) On smart IoT remote sensing over integrated terrestrial-aerial-space networks: an asynchronous federated learning approach. IEEE Netw 35(5):129–135. https://doi.org/10.1109/MNET.101.2100125

Fan Y-C, Yuan Y-Q, Yuan Y-C, Duan W-J, Gao Z-Q (2025) Research progress on the impact of climate change on wheat production in China. PeerJ 13(1). https://doi.org/10.7717/peerj.18569

Fujihara Y, Okuyama K, Hoshikawa K, Fujii H, Nakamura T, Lun S, Siev S (2021) Characteristics of the annual maximum and minimum water temperatures in tonle sap lake, Cambodia from 2000 to 2019. Remote Sens 13(19). https://doi.org/10.3390/rs13193972

Gao J, Li C, Zhao P, Zhang H, Mao G, Wang Y (2019) Insights into water-energy cobenefits and trade-offs in water resource management. J Clean Prod 213:1188–1203. https://doi.org/10.1016/j.jclepro.2018.12.126

Gao T, Lim S, Wong DKH, Schroepfer T (2024) From walking to staying: the impact of spatial qualities on socio-spatial segregation in innovation districts. Sustain Cities Soc 113. https://doi.org/10.1016/j.scs.2024.105688

Garbulowski M, Diamanti K, Smolińska K, Baltzer N, Stoll P, Bornelöv S, Øhrn A, Feuk L, Komorowski J (2021) R.ROSETTA: an interpretable machine learning framework. BMC Bioinformatics 22(1). https://doi.org/10.1186/s12859-021-04049-z

Gerlak AK, House-Peters L, Varady RG, Albrecht T, Zúñiga-Terán A, de Grenade RR, Cook C, Scott CA (2018) Water security: a review of place-based research. Environ Sci Policy 82:79–89. https://doi.org/10.1016/j.envsci.2018.01.009

Giupponi C, Biscaro C (2015) Vulnerabilities – bibliometric analysis and literature review of evolving concepts. Environ Res Lett 10(12). https://doi.org/10.1088/1748-9326/10/12/123002

Goodman C (2016) Triaging deforestation alerts: clustering alerts for review. In: GHTC 2016 – IEEE global humanitarian technology conference: technology for the benefit of humanity, conference proceedings, pp 84–90. https://doi.org/10.1109/GHTC.2016.7857264

Grammalidis N, Stergioulas A, Avramidis A, Karystinakis K, Partozis A, Topaloudis A, Kalantzi G, Tananaki C, Kanelis D, Liolios V, Panagiotis M (2023) A smart beekeeping platform based on remote sensing and artificial intelligence. In: T. K., H. D.G., M. S., & P. G (ed) Proceedings of SPIE – The International Society for Optical Engineering, vol 12786. SPIE. https://doi.org/10.1117/12.2681866

Gupta K, Hall RP (2021) Exploring Smart City project implementation risks in the cities of Kakinada and Kanpur. J Urban Technol 28(1–2):155–173. https://doi.org/10.1080/10630732.2020.1796115

Gupta S, Rikhtehgar Berenji H, Shukla M, Murthy NN (2023) Opportunities in farming research from an operations management perspective. Prod Oper Manag 32(6):1577–1596. https://doi.org/10.1111/poms.13967

Gupta RK, Jain A, Wang J, Pateriya RK (2024) Reshaping environmental science through machine learning and IoT. In: Reshaping environmental science through machine learning and IoT. IGI Global. https://doi.org/10.4018/979-8-3693-2351-9

Hamed AA, Ayer AA, Clark EM, Irons EA, Taylor GT, Zia A (2015) Measuring climate change on Twitter using Google's algorithm: perception and events. Int J Web Inf Syst 11(4):527–544. https://doi.org/10.1108/IJWIS-08-2015-0025

He B-J (2022) Green building: a comprehensive solution to urban heat. Energ Build 271. https://doi.org/10.1016/j.enbuild.2022.112306

Hou W, Wu S, Yang L, Yin Y, Gao J, Deng H, Wu M, Li X, Liu L (2022) Production–living–ecological risk assessment and corresponding strategies in China's provinces under climate change scenario. Land 11(9). https://doi.org/10.3390/land11091424
Ian V-K, Tang S-K, Pau G (2023a) Assessing the risk of extreme storm surges from tropical cyclones under climate change using bidirectional attention-based LSTM for improved prediction. Atmosphere 14(12). https://doi.org/10.3390/atmos14121749
Ian V-K, Tang S-K, Pau G (2023b) Comparative analysis of BALSSA and conventional NWP methods: a case study in extreme storm surge prediction in Macao. Atmosphere 14(11). https://doi.org/10.3390/atmos14111597
Igun E, Williams M (2018) Impact of urban land cover change on land surface temperature. Glob J Environ Sci Manag 4(1):47–58. https://doi.org/10.22034/gjesm.2018.04.01.005
Imhoff ML, Tucker CJ, Lawrence WT, Stutzer DC (2000) The use of multisource satellite and geospatial data to study the effect of urbanization on primary productivity in the United States. IEEE Trans Geosci Remote Sens 38(6):2549–2556. https://doi.org/10.1109/36.885202
Inapakurthi RK, Miriyala SS, Mitra K (2021) Deep learning based dynamic behavior modelling and prediction of particulate matter in air. Chem Eng J 426. https://doi.org/10.1016/j.cej.2021.131221
Isaac S, Casals M, Carta S, Tejedor B (2024) SMART URBAN RESILIENCY CONCEPTS AND GOALS: how can cities become smart but also sustainable and resilient. In: The Routledge companion to smart design thinking in architecture & urbanism for a sustainable, living planet. Taylor and Francis. https://doi.org/10.4324/9781003384113-18
Isaev E, Yuave N, Inape K, Jones C, Dawa L, Sidle RC (2024) Agricultural drought-triggering for anticipatory action in Papua New Guinea. Water (Switzerland) 16(14). https://doi.org/10.3390/w16142009
Jian L, Yang X, Zhou Z, Zhou K, Liu K, Eichler M, May M, Wiesmueller M, Saake M, Heiss R, Uder M, Wuest W, Alves WAL, Gobber CF, Silva DJ, Morimitsu A, Hashimoto RF, Marcotegui B, Peng J et al (2020) Image segmentation based on ultimate levelings: from attribute filters to machine learning strategies. Remote Sens Environ 175(2019):163671. https://doi.org/10.1016/j.isprsjprs.2008.07.005
Jitkajornwanich K, Vijaranakul N, Jaiyen S, Srestasathiern P, Lawawirojwong S (2024) Enhancing risk communication and environmental crisis management through satellite imagery and AI for air quality index estimation. MethodsX 12. https://doi.org/10.1016/j.mex.2024.102611
Kapetas L, Fenner R (2020) Integrating blue-green and grey infrastructure through an adaptation pathways approach to surface water flooding. Philos Trans R Soc A Math Phys Eng Sci 378(2168). https://doi.org/10.1098/rsta.2019.0204
Kesavan H, Choudhary S, Khan S, Samhita R, Sharma A (2024) AI based urban resilience planning: opportunities and challenges. J Environ Earth Sci 6(2):200–214. https://doi.org/10.30564/jees.v6i2.6681
Kim EK (2016) Innovative strategy to produce disaster map according to CDM life cycle in South Korea. Spat Inf Res 24(4):401–408. https://doi.org/10.1007/s41324-016-0037-7
Kow P-Y, Liou J-Y, Yang M-T, Lee M-H, Chang L-C, Chang F-J (2024) Advancing climate-resilient flood mitigation: utilizing transformer-LSTM for water level forecasting at pumping stations. Sci Total Environ 927. https://doi.org/10.1016/j.scitotenv.2024.172246
Kumar D, Shekhar S (2014) Research prospects in spatial data infrastructure for handling energy linked disaster's. IOSR J Environ Sci Toxicol Food Technol 8(5):53–67. https://doi.org/10.9790/2402-08535367
Kumar R, Banga HK, Kaur H (2020) Internet of Things-supported Smart City platform. IOP Conf Ser Mater Sci Eng 955(1). https://doi.org/10.1088/1757-899X/955/1/012003
Kumar A, Srivastav AL, Dubey AK, Dutt V, Vyas N (2023) Innovations in machine learning and IoT for water management. In: Innovations in machine learning and IoT for water management. IGI Global. https://doi.org/10.4018/979-8-3693-1194-3

Kumar D, Shekhar S, Tewary T (2024) Data analytics and artificial intelligence in Earth resource management. In: Data analytics and artificial intelligence for Earth resource management. Elsevier. https://doi.org/10.1016/B978-0-443-23595-5.00001-2

Kumari J, Rana DP (2023) GeoAI-based Covid-19 prediction model. In: Advances in data-driven computing and intelligent systems, Lecture notes in networks and systems, 653 LNNS, pp 669–684. https://doi.org/10.1007/978-981-99-0981-0_51

Kumari S, Agarwal S, Agrawal NK, Agarwal A, Garg MC (2025) A comprehensive review of remote sensing technologies for improved geological disaster management. Geol J 60(1): 223–235. https://doi.org/10.1002/gj.5072

Kuo C-M, Chen Y-Z, Chen S-T (2024) Applying Budyko framework to explore the projected hydrologic changes in southern Taiwan. Terr Atmos Ocean Sci 35(1). https://doi.org/10.1007/s44195-024-00079-1

Leiva-Murillo JM, Gomez-Chova L, Camps-Valls G (2013) Multitask remote sensing data classification. IEEE Trans Geosci Remote Sens 51(1):151–161. https://doi.org/10.1109/TGRS.2012.2200043

Li Y, Zhang H, Xue X, Jiang Y, Shen Q (2018) Deep learning for remote sensing image classification: A survey. Wiley Interdiscip Rev Data Min Knowl Discovery 8(6):1–17. https://doi.org/10.1002/widm.1264

Li G, Luo T, Song Y (2022) Spatial equity analysis of urban public services for electric vehicle charging—implications of Chinese cities. Sustain Cities Soc 76. https://doi.org/10.1016/j.scs.2021.103519

Li X, Yue J, Wang S, Luo Y, Su C, Zhou J, Xu D, Lu H (2024) Development of geographic information system architecture feature analysis and evolution trend research. Sustainability (Switzerland) 16(1). https://doi.org/10.3390/su16010137

Liang J, Xie Y, Sha Z, Zhou A (2020) Modeling urban growth sustainability in the cloud by augmenting Google Earth Engine (GEE). Comput Environ Urban Syst 84. https://doi.org/10.1016/j.compenvurbsys.2020.101542

Liao Y, Liao W, Wu D, Zhang C, Xu J, Majumdar S (2024) Inspired applications of artificial intelligence in disaster-resilient regional construction for climate risk. Urban Sustain Part F 3945:305–315. https://doi.org/10.1007/978-981-97-9658-8_16

Linardos V, Drakaki M, Tzionas P, Karnavas YL (2022) Machine learning in disaster management: recent developments in methods and applications. Mach Learn Knowl Extr 4(2):446–473. https://doi.org/10.3390/make4020020

Liu Y, Yang Y, Jing W, Yao L, Yue X, Zhao X (2017) A new urban index for expressing inner-city patterns based on MODIS LST and EVI regulated DMSP/OLS NTL. Remote Sens 9(8). https://doi.org/10.3390/rs9080777

Liu D, Liao T, Sun H, Ren F (2021a) Research progress and development direction of Chinese remote sensing software: taking PIE as an example. J Image Graphics 26(5):1169–1178. https://doi.org/10.11834/jig.200125

Liu J, Cao S, Chen X, Yang H, Peng J (2021b) Energy planning of renewable applications in high-rise residential buildings integrating battery and hydrogen vehicle storage. Appl Energy 281. https://doi.org/10.1016/j.apenergy.2020.116038

Liu S, Zhou Z, Ding H, Zhong Y, Shi Q (2021c) Crop mapping using sentinel full-year dual-polarized SAR data and a CPU-optimized convolutional neural network with two sampling strategies. IEEE J Sel Top Appl Earth Obs Remote Sens 14:7017–7031. https://doi.org/10.1109/JSTARS.2021.3094973

Ma L-J, Yuan J-S, Huang L (2024) Prospect of climate risks management in China under the framework of UN early warning for all initiative. Clim Change Res 20(1):48–61. https://doi.org/10.12006/j.issn.1673-1719.2023.122

Mahalakshmi S, Latha R (2019) Artificial intelligence with the internet of things on healthcare systems: a survey. https://doi.org/10.30534/ijatcse/2019/27862019

Maheswari EAP, Mernisi FA, Sidharta S, Puspitasari C (2024) Harnessing artificial intelligence for effective coastal flood disaster management: a systematic literature review. In: Proceedings of

2024 international conference on information management and technology, ICIMTech 2024, pp 530–535. https://doi.org/10.1109/ICIMTech63123.2024.10780929
Maity R, Srivastava A, Sarkar S, Khan MI (2024) Revolutionizing the future of hydrological science: impact of machine learning and deep learning amidst emerging explainable AI and transfer learning. Appl Comput Geosci 24. https://doi.org/10.1016/j.acags.2024.100206
Maji S, Swargiary D, Senapati A, Sikdar U (2024) ML and AI-based prediction of precipitation trends. In: S. V., A. V.K., L. K.-C., & C. R.G (ed) Procedia computer science, vol 235. Elsevier, pp 2279–2287. https://doi.org/10.1016/j.procs.2024.04.216
Majidi B, Hemmati O, Baniardalan F, Farahmand H, Hajitabar A, Sharafi S, Aghajani K, Esmaeili A, Manzuri MT (2021) Geo-spatiotemporal intelligence for smart agricultural and environmental eco-cyber-physical systems. Stud Comput Intell 911:471–491. https://doi.org/10.1007/978-3-030-52067-0_21
Mamatha NP, Bagali MU, Thangadurai N (2022) Disasters detection from remote satellite images using ML and deep learning approaches. In: International interdisciplinary humanitarian conference for sustainability, IIHC 2022 – proceedings, pp 1608–1615. https://doi.org/10.1109/IIHC55949.2022.10060337
Manunta P, Viel M, Ceriola G, Daglis IA, De Ridder K, Giannaros T, Keramitsoglou I, Maiheu B, Melas D, Herrero EM, Palacios M, Radius A, Sapage T, Tamame M, Tambuyzer H (2010) Monitoring and forecasting of Urban heat island phenomenon in ten European cities the uhi project. In: GI4DM 2010 conference – geomatics for crisis management, pp 39–44. https://www.scopus.com/inward/record.uri?eid=2-s2.0-85076720389&partnerID=40&md5=8de35d528a462a1eb5279b48251bf98c
Marzhan K, Dina K, Roland B (2018) Developing high-resolution remote sensing technology into an advanced knowledge management system to assess small-scale hydropower potential in Kazakhstan. In: Exergy for a better environment and improved sustainability 2, Green energy and technology, pp 581–605. https://doi.org/10.1007/978-3-319-62575-1_41
Masoumi Z, van Genderen JL, Maleki J (2019) Fire risk assessment in dense urban areas using information fusion techniques. ISPRS Int J Geo Inf 8(12). https://doi.org/10.3390/ijgi8120579
Mawgoud AA, Eltabey MM, Abu-Talleb A (2021) A distributed artificial intelligence framework to evolve infrastructure resilience in telecommunications sector. In: Advanced machine learning technologies and applications. AMLTA 2021, Advances in intelligent systems and computing, vol 1339, pp 774–784. https://doi.org/10.1007/978-3-030-69717-4_71
Millar M-A, Yu Z, Burnside N, Jones G, Elrick B (2021) Identification of key performance indicators and complimentary load profiles for 5th generation district energy networks. Appl Energy 291. https://doi.org/10.1016/j.apenergy.2021.116672
Mohajane M, Costache R, Karimi F, Bao Pham Q, Essahlaoui A, Nguyen H, Laneve G, Oudija F (2021) Application of remote sensing and machine learning algorithms for forest fire mapping in a Mediterranean area. Ecol Indic 129. https://doi.org/10.1016/j.ecolind.2021.107869
Mohamed SK, Abdel Samee BE (2022) Social distancing model utilizing machine learning techniques. In: Advances in data science and intelligent data communication technologies for COVID-19, Studies in systems, decision and control, vol 378, pp 41–53. https://doi.org/10.1007/978-3-030-77302-1_3
Mondol SC, Adhikary SK, Nath H, Shuvo SP (2025) A deep learning approach using long short-term memory networks for enhanced prediction of rainfall in the northeastern region of Bangladesh. In: R. M., A. M.S., & M. M.H (ed) AIP conference proceedings, vol 3262, Issue 1. American Institute of Physics. https://doi.org/10.1063/5.0247626
Mosavi A, Ozturk P, Chau K-W (2018) Flood prediction using machine learning models: literature review. Water (Switzerland) 10(11). https://doi.org/10.3390/w10111536
Motwani A, Shukla PK, Pawar M (2021) Novel framework based on deep learning and cloud analytics for smart patient monitoring and recommendation (SPMR). https://doi.org/10.1007/s12652-020-02790-6

Mujetahid A, Nursaputra M, Soma AS (2023) Monitoring illegal logging using Google Earth Engine in Sulawesi Selatan tropical forest, Indonesia. Forests 14(3). https://doi.org/10.3390/f14030652

Murray AT, Baik J, Echeverri Figueroa V, Rini D, Moritz MA, Roberts DA, Sweeney SH, Carvalho LMV, Jones C (2023) Developing effective wildfire risk mitigation plans for the wildland urban interface. Int J Appl Earth Obs Geoinf 124. https://doi.org/10.1016/j.jag.2023.103531

Murugadoss R, Nesamani SL, Banushri A, Monica KM, Vairavel M, Rajini SNS, Gopi P (2024) A review of using deep learning from an ecology perspective to address climate change and air pollution. Global NEST J 26(2). https://doi.org/10.30955/gnj.005697

Navarro A, Carbonell M (2008) Assessment of groundwater contamination caused by uncontrolled dumping in old gravel quarries in the Besòs aquifers (Barcelona, Spain). Environ Geochem Health 30(3):273–289. https://doi.org/10.1007/s10653-007-9123-z

Nayak P, Das B, Kumar Panigrahi J (2018) Intent of technology, innovation and value creation for start-up entrepreneurs. Int J Mech Eng Technol 9(2):629–636. https://www.scopus.com/inward/record.uri?eid=2-s2.0-85044638747&partnerID=40&md5=ba703f48af05959fd1cd8d5374f95fb0

Nkwunonwo UC (2013) Land use/land cover mapping of the Lagos Metropolis of Nigeria using 2012 SLC-off Landsata ETM+ satellite images. Int J Sci Eng Res 4(11):1217–1223

O'Donnell EC, Lamond JE, Thorne CR (2017) Recognising barriers to implementation of blue-green infrastructure: a Newcastle case study. Urban Water J 14(9):964–971. https://doi.org/10.1080/1573062X.2017.1279190

Oliveira A, Leal V, Galamba M, Cunha R (2022) AI for urban climate: an EO-based approach for high-resolution mapping of human exposure to heatwaves. In: International geoscience and remote sensing symposium (IGARSS), 2022 July, pp 4475–4478. https://doi.org/10.1109/IGARSS46834.2022.9883071

Orimoloye IR, Olusola AO, Belle JA, Pande CB, Ololade OO (2022) Drought disaster monitoring and land use dynamics: identification of drought drivers using regression-based algorithms. Nat Hazards 112(2):1085–1106. https://doi.org/10.1007/s11069-022-05219-9

Othman AK, Hamzah MI, Abu Hassan LF (2020) Modeling the contingent role of technological optimism on customer satisfaction with self-service technologies: a case of cash-recycling ATMs. J Enterp Inf Manag 33(3):559–578. https://doi.org/10.1108/JEIM-09-2019-0295

Ouafiq EM, Saadane R, Chehri A (2023) Data-driven agriculture and role of AI in smart farming. In: Data-driven intelligence in wireless networks: concepts, solutions, and applications. CRC Press. https://doi.org/10.1201/9781003216971-6

Oyounalsoud MS, Yilmaz AG, Abdallah M, Abdeljaber A (2024) Drought prediction using artificial intelligence models based on climate data and soil moisture. Sci Rep 14(1). https://doi.org/10.1038/s41598-024-70406-6

Pandey PC, Pandey M, Sharma RK, Kumar R, Srivastava PK, Giuliani G (2024) Concepts of disasters and research themes: editorial message. In: Geo-information for disaster monitoring and management. Springer. https://doi.org/10.1007/978-3-031-51053-3_1

Paswan AK, Pathan SK, Agarwal A, Verma V (2024) Artificial intelligence in efficient management of water resources. In: Data analytics and artificial intelligence for Earth resource management. Elsevier. https://doi.org/10.1016/B978-0-443-23595-5.00010-3

Pranto AM, Aziz UI, Das LC, Ghosh S, Islam A (2024) Forecasting precipitation using a Markov chain model in the coastal region in Bangladesh. Nat Environ Pollut Technol 23(4):2201–2209. https://doi.org/10.46488/NEPT.2024.v23i04.024

Puno GR, Amper RAL, Opiso EM, Cipriano JAB (2020) Mapping and analysis of flood scenarios using numerical models and GIS techniques. Spat Inf Res 28(2):215–226. https://doi.org/10.1007/s41324-019-00280-2

Quan SJ, Dutt F, Woodworth E, Yamagata Y, Yang PP-J (2017) Local climate zone mapping for energy resilience: a fine-grained and 3D approach. In: L. H., Y. J., S. F., D. U., & C. S.K (ed) Energy procedia, vol 105. Elsevier, pp 3777–3783. https://doi.org/10.1016/j.egypro.2017.03.883

Quesada-Román A, Torres-Bernhard L, Hernández K, Martínez-Rojas N (2024) Historical trends and future implications of disasters in Honduras. Nat Hazards 120(13):12313–12339. https://doi.org/10.1007/s11069-024-06686-y

Rad A, Naderi B, Soltani M (2011) Clustering and ranking university majors using data mining and AHP algorithms: a case study in Iran. Expert Syst Appl 38(1):755–763. https://doi.org/10.1016/j.eswa.2010.07.029

Rahman MH, Zaman IU, Ahmed K, Bushra F, Khan NZ, Rahman RM (2022) Machine learning techniques for river discharge prediction using ERA5 and GloFAS data. In: C. S. & P. R (ed) 2022 IEEE 13th annual ubiquitous computing, electronics and Mobile communication conference, UEMCON 2022. Institute of Electrical and Electronics Engineers Inc, pp 554–560. https://doi.org/10.1109/UEMCON54665.2022.9965629

Rajkumar N, Viji C, Latha PM, Vennila VB, Shanmugam SK, Pillai NB (2024) The power of AI, IoT, and advanced quantum based optical systems in smart cities. Opt Quant Electron 56(3). https://doi.org/10.1007/s11082-023-06065-0

Ramírez SAH (2019) The new global governance in the aerospace industry: new technological capabilities in Brazil, China and Canada. Proceedings of the international astronautical congress, IAC, 2019, October. https://www.scopus.com/inward/record.uri?eid=2-s2.0-85079165092&partnerID=40&md5=0b39c9fd50b9379e240264d5061dd26f

Ravanelli R, Nascetti A, Cirigliano RV, Di Rico C, Leuzzi G, Monti P, Crespi M (2018) Monitoring the impact of land cover change on surface urban heat Island through Google Earth Engine: proposal of a global methodology, first applications and problems. Remote Sens 10(9). https://doi.org/10.3390/rs10091488

Riadi B, Barus B, Widiatmaka, Yanuar M, Pramudya B (2018) Identification of flood area in the coastal region using remote sensing in Karawang Regency, West Java. In: S. R. D. Susanto R.D (ed) IOP conference series: earth and environmental science (vol 162, issue 1). Institute of Physics Publishing. https://doi.org/10.1088/1755-1315/162/1/012042

Rijal M, Luo P, Mishra BK, Zhou M, Wang X (2024) Global systematical and comprehensive overview of mountainous flood risk under climate change and human activities. Sci Total Environ 941. https://doi.org/10.1016/j.scitotenv.2024.173672

Rolnick D, Donti PL, Kaack LH, Kochanski K, Lacoste A, Sankaran K, Ross AS, Milojevic-Dupont N, Jaques N, Waldman-Brown A, Luccioni AS, Maharaj T, Sherwin ED, Mukkavilli SK, Kording KP, Gomes CP, Ng AY, Hassabis D, Platt JC et al (2023) Tackling climate change with machine learning. ACM Comput Surv 55(2). https://doi.org/10.1145/3485128

Saha S, Roy J, Pradhan B, Hembram TK (2021) Hybrid ensemble machine learning approaches for landslide susceptibility mapping using different sampling ratios at East Sikkim Himalayan, India. Adv Space Res 68(7):2819–2840. https://doi.org/10.1016/j.asr.2021.05.018

Sahani J, Kumar P, Debele S, Spyrou C, Loupis M, Aragão L, Porcù F, Shah MAR, Di Sabatino S (2019) Hydro-meteorological risk assessment methods and management by nature-based solutions. Sci Total Environ 696. https://doi.org/10.1016/j.scitotenv.2019.133936

Samantray P, Gouda KC (2024) A review on the extreme rainfall studies in India. Nat Hazards Res 4(3):347–356. https://doi.org/10.1016/j.nhres.2023.08.005

Samarinas N, Spiliotopoulos M, Tziolas N, Loukas A (2023) Synergistic use of earth observation driven techniques to support the implementation of water framework directive in Europe: a review. Remote Sens 15(8). https://doi.org/10.3390/rs15081983

Schaefer M, Ebrahimi Salari H, Köckler H, Thinh NX (2021) Assessing local heat stress and air quality with the use of remote sensing and pedestrian perception in urban microclimate simulations. Sci Total Environ 794. https://doi.org/10.1016/j.scitotenv.2021.148709

Selvam AP, Al-Humairi SNS (2025) Environmental impact evaluation using smart real-time weather monitoring systems: a systematic review. Innov Infrastruct Solutions 10(1). https://doi.org/10.1007/s41062-024-01817-7

Senanayake S, Pradhan B, Wedathanthirige H (2024) Spatial intelligence in AI applications for assessing soil health to monitor farming systems and associated ESG risk. In: Intelligence

systems for earth, environmental and planetary sciences: methods, models and applications. Elsevier. https://doi.org/10.1016/B978-0-443-13293-3.00001-4
Shafiei Shiva J, Chandler DG, Kunkel KE (2019) Localized changes in heat wave properties across the United States. Earth's Future 7(3):300–319. https://doi.org/10.1029/2018EF001085
Shahvar MP, Scuderi D, Valenti D, Collura A, Miccichè S, Farina V, Marsella G (2025) Climate change multi-risk assessment for mango cultivation in Sicily, Italy, by using Bayesian network. Acta Horticulturae 1(1415). https://doi.org/10.17660/ActaHortic.2025.1415.15
Shaikh TA, Mir WA, Rasool T, Sofi S (2022) Machine learning for smart agriculture and precision farming: towards making the fields talk. Arch Comput Methods Eng 29(7):4557–4597. https://doi.org/10.1007/s11831-022-09761-4
Sharma M, Luthra S, Joshi S, Kumar A (2021) Implementing challenges of artificial intelligence: evidence from public manufacturing sector of an emerging economy. Gov Inf Q 39. https://doi.org/10.1016/j.giq.2021.101624
Sharmila K, Shanthi C, Devi R, Kamala Kannan T (2019) A comprehensive study on novel hybrid approach for decision support system in disease diagnosis. Indian J Public Health Res Dev 10(3):200–204. https://doi.org/10.5958/0976-5506.2019.00487.X
Shi Y, Katzschner L, Ng E (2018) Modelling the fine-scale spatiotemporal pattern of urban heat island effect using land use regression approach in a megacity. 618:891–904. https://doi.org/10.1016/j.scitotenv.2017.08.252
Simarmata N, Wikantika K, Tarigan TA, Aldyansyah M, Tohir RK (2021) Utilization of multitemporal imagery for analysis of changes in mangrove cover by using cloud computing method in the East Coast region of Lampung Province. In: Y. H. T. S. E. Nurfani E. Prasetyo B. A (ed) IOP conference series: earth and environmental science (vol 830, issue 1). IOP Publishing Ltd. https://doi.org/10.1088/1755-1315/830/1/012013
Singh K, Malhotra J (2019) IoT and cloud computing based automatic epileptic seizure detection using HOS features based random forest classification. J Ambient Intell Humaniz Comput 14. https://doi.org/10.1007/s12652-019-01613-7
Singh CR, Manoharan G (2024) Strengthening resilience: AI and machine learning in emergency decision-making for natural disasters. In: Internet of Things and AI for natural disaster management and prediction. IGI Global. https://doi.org/10.4018/9798369342848.ch012
Sit M, Demiray BZ, Xiang Z, Ewing GJ, Sermet Y, Demir I (2020) A comprehensive review of deep learning applications in hydrology and water resources. Water Sci Technol 82(12): 2635–2670. https://doi.org/10.2166/wst.2020.369
Song X, Xu N, Zhang J, He R (2024) Urban flooding in China: current status, causes and challenges. Shuikexue Jinzhan/Adv Water Sci 35(3):357–373. https://doi.org/10.14042/j.cnki.32.1309.2024.03.001
Sonthi VK, Nagarajan S, Murali Krishna MMVB, Giridhar K, Lalitha VL, Mohan VM (2022) Imminent threat with authentication methods for AI data using blockchain security. In: EAI/Springer innovations in communication and computing. https://doi.org/10.1007/978-3-030-70501-5_14
Sreekeshava KS, Kolathayar S, Vinod Chandra Menon N, Bhargavi C (2024) Civil engineering for multi-hazard risk reduction – an introduction. In: Civil engineering for multi-hazard risk reduction, Lecture notes in civil engineering, vol 457, pp 1–10. https://doi.org/10.1007/978-981-99-9610-0_1
Sreelakshmi S, Vinod Chandra SS, Shaji E (2022) Landslide identification using machine learning techniques: review, motivation, and future prospects. Earth Sci Inf 15(4):2063–2090. https://doi.org/10.1007/s12145-022-00889-2
Story C (2012) Geospatial data processing in distributed computing environment. January
Sun Y, Huang S-K, Zhao X (2024) Predicting hurricane evacuation decisions with interpretable machine learning methods. Int J Disaster Risk Sci 15(1):134–148. https://doi.org/10.1007/s13753-024-00541-1
Supriya Y, Gadekallu TR (2023) Particle swarm-based federated learning approach for early detection of Forest fires. Sustainability (Switzerland) 15(2). https://doi.org/10.3390/su15020964

Talaviya T, Shah D, Patel N, Yagnik H, Shah M (2020) Implementation of artificial intelligence in agriculture for optimisation of irrigation and application of pesticides and herbicides. Artif Intell Agric 4:58–73. https://doi.org/10.1016/j.aiia.2020.04.002
Talha M, Nejadhashemi AP, Moller K (2025) Soft computing paradigm for climate change adaptation and mitigation in Iran, Pakistan, and Turkey: a systematic review. Heliyon 11(2). https://doi.org/10.1016/j.heliyon.2025.e41974
Tanniru S, Ramsankaran R (2023) Machine learning-based estimation of high-resolution snow depth in Alaska using passive microwave remote sensing data. IEEE J Sel Top Appl Earth Obs Remote Sens 16:6007–6025. https://doi.org/10.1109/JSTARS.2023.3287410
Thangavel K, Spiller D, Sabatini R, Amici S, Sasidharan ST, Fayek H, Marzocca P (2023) Autonomous satellite wildfire detection using hyperspectral imagery and neural networks: a case study on Australian wildfire. Remote Sens 15(3). https://doi.org/10.3390/rs15030720
Theodoraki C, Messeghem K, Rice MP (2018) A social capital approach to the development of sustainable entrepreneurial ecosystems: an explorative study. Small Bus Econ 51(1):153–170. https://doi.org/10.1007/s11187-017-9924-0
Tran T-N-D, Tapas MR, Do SK, Etheridge R, Lakshmi V (2024) Investigating the impacts of climate change on hydroclimatic extremes in the Tar-Pamlico River basin, North Carolina. J Environ Manag 363. https://doi.org/10.1016/j.jenvman.2024.121375
Tshimanga RM, Tshitenge JM, Kabuya P, Alsdorf D, Mahe G, Kibukusa G, Lukanda V (2024) A regional perceptive of flood forecasting and disaster management systems for The Congo River Basin. In: Flood forecasting: a global perspective. https://doi.org/10.1016/B978-0-443-14009-9.00027-4
Velterop E, Sirorattanakul K, Novellino A, Muhire D, Amrutkar R, Patial S, Vargas-Sanabria D, Ibitolu H, Pyne B, de Prados AV, Charles-Aimé NM, Towashiraporn P (2019) Current and near-future state of space technology for disaster situations. Proceedings of the international astronautical congress, IAC, 2019, October
Venkatesan R, Srinivasan B, Rajendiran P (2019) Tiger hash based AdaBoost machine learning classifier for secured multicasting in mobile healthcare system. https://doi.org/10.1007/s10586-018-2241-9
Venter ZS, Brousse O, Esau I, Meier F (2020) Hyperlocal mapping of urban air temperature using remote sensing and crowdsourced weather data. Remote Sens Environ 242. https://doi.org/10.1016/j.rse.2020.111791
Wang J (2023) Towards GeoAI as a containerized microservice. In: D. M.L., R. M., E. A., K. P., & N. M.A (ed) GIS: proceedings of the ACM international symposium on advances in geographic information systems. Association for Computing Machinery. https://doi.org/10.1145/3589132.3628365
Wang W, Zhou W, Ng EYY, Xu Y (2016) Urban heat islands in Hong Kong: statistical modeling and trend detection. 83(2):885–907. https://doi.org/10.1007/s11069-016-2353-6
Wang Y, Wang H, Cui P, Chen D, Tang J, Ou T, Hao J, Wang J, Zhang G, Lei Y, Lei Y, Wu C (2024) Disaster effects of climate change and the associated scientific challenges. Kexue Tongbao/Chin Sci Bull 69(2):286–300. https://doi.org/10.1360/TB-2023-0325
Xu R, Xia Y, Chen P (2024) A multi-objective particle swarm optimization algorithm for drone path planning in Forest firefighting. In: 2024 11th international conference on electrical and electronics engineering, ICEEE 2024, pp 523–527. https://doi.org/10.1109/ICEEE62185.2024.10779280
Yadav A (2023) Leveraging artificial intelligence for sustainable development and environmental resilience. In: Exploring ethical dimensions of environmental sustainability and use of AI. https://doi.org/10.4018/9798369308929.ch008
Yadav AK, Chandel SS, Chrysoulakis N, Habib SM, El-Raie Emam Suliman A, Al Nahry AH, Abd El Rahman EN (2015) Spatial modeling for the optimum site selection of solar photovoltaics power plant in the northwest coast of Egypt. J Indian Soc Remote Sens 30(4):100313. https://doi.org/10.1016/j.renene.2014.10.046

Yeşilköy S, Baydaroğlu I, Singh N, Sermet Y, Demir I (2024) A contemporary systematic review of cyberinfrastructure systems and applications for flood and drought data analytics and communication. Environ Res Commun 6(10). https://doi.org/10.1088/2515-7620/ad85c4

Zabihi O, Siamaki M, Gheibi M, Akrami M, Hajiaghaei-Keshteli M (2023) A smart sustainable system for flood damage management with the application of artificial intelligence and multi-criteria decision-making computations. Int J Disaster Risk Reduct 84. https://doi.org/10.1016/j.ijdrr.2022.103470

Zeng Y-F, Chang M-J, Lin G-F (2024) A novel AI-based model for real-time flooding image recognition using super-resolution generative adversarial network. J Hydrol 638. https://doi.org/10.1016/j.jhydrol.2024.131475

Zhang B, Di L, Yu G, Shao Y, Shrestha R, Kang L (2013) A web service based application serving vegetation condition indices for flood crop loss assessment. In: 2013 2nd international conference on agro-geoinformatics: information for sustainable agriculture, agro-geoinformatics 2013, pp 215–220. https://doi.org/10.1109/ARGO-GEOINFORMATICS.2013.6621910

Zhou G, Zhou L (2021) Ecological meteorology: origin, concept, and prospects; [生态气象: 起源, 概念和展望]. Kexue Tongbao/Chin Sci Bull 66(2):210–218. https://doi.org/10.1360/TB-2020-1274

Zhou G, Zhou L, Ji Y, Lü X, Zhou M (2021) Basin integrity and temporal-spatial connectivity of the water ecological carrying capacity of the Yellow River. Kexue Tongbao/Chin Sci Bull 66(22): 2785–2792. https://doi.org/10.1360/TB-2021-0039

Zhou B, Zou Q, Jiang H, Yang T, Zhou W, Chen S, Yao H (2024) A novel framework for predicting glacial lake outburst debris flows in the Himalayas amidst climate change. Sci Total Environ 946. https://doi.org/10.1016/j.scitotenv.2024.174435

Zhuang H, Zhang J, Sivaparthipan CB, Muthu BA (2020) Sustainable smart city building construction methods. Sustainability (Switzerland) 12(12). https://doi.org/10.3390/SU12124947

Chapter 2
The Climate Science, Designing, and Implementation of Climate Policy

Syed Shahid Mazhar, Farhina Sardar Khan, Prateek Srivastava, and Ambrina Sardar Khan

Abstract This research explores the complex relationships between climate science, policy-making efforts, and the development of effective adaptation to change as well as resilience strategies based on scientific literature and documents related to policymaking. It breaks down the intricate issues caused by climate change, such as increased temperatures, extreme weather events, sea-level surges, and changes in rainfall patterns. This study highlights the importance of incorporating scientific knowledge into policy structures to strengthen adaptive capabilities and enable resilience at local, regional, and global scales (Adger et al. 2007).

A thorough review of critical aspects of climate science, such as climate modelling and risk evaluations, along with vulnerabilities, is presented. This study ventures to discover ways by which such scientific understanding can be used to drive the development and implementation of policies (Jones and Preston 2011). The research also emphasizes cases and examples of successful development climate adaptation resilience and provides insight into the vital role of interdisciplinary cooperation and stakeholder involvement, as well as long-term strategies (Pelling 2011).

By interconnecting the disciplines of climate science and policy, this study strives to close gaps in present knowledge with the overall goal of significantly adding value toward building sound strategies ready for efficiently addressing the various effects

S. S. Mazhar
Integral Business School, Integral University, Lucknow, Uttar Pradesh, India

F. S. Khan
Integral Business School, Integral University, Lucknow, Uttar Pradesh, India

Department of Commerce, Integral University, Lucknow, Uttar Pradesh, India

P. Srivastava
Department of Botany, University of Allahabad, Prayagraj, Uttar Pradesh, India

A. S. Khan (✉)
Department of Environmental Sciences, Integral University, Lucknow, Uttar Pradesh, India
e-mail: ambrina@iul.ac.in

D. Kumar, N. P. Bassill (eds.), *Artificial Intelligence and Machine Learning for Climate Disaster Management*, Disaster Resilience and Green Growth,
https://doi.org/10.1007/978-981-95-1588-2_2

stemming from a changing geophysical environment. Its broad aim is to increase societal and ecological strength in the face of these threats.

Keywords Climate science · Climate change · Policy · Climate adaptation · Resilience · Adaptive capacity · Climate modelling · Vulnerability assessment · Risk analysis · Cross-sectoral collaboration · Stakeholder engagement

2.1 Introduction: Climate Science and Policy

Climate change is a global challenge that occurs so rapidly that the complexity of this process requires urgent attention. Climate science plays a key role in delivering the appropriate knowledge and understanding to facilitate effective policy responses through its structured analysis of climate systems and causes of climatic changes (Webster et al. 2003). However, climate policy is a broad term that involves various strategies involving governments and organizations to reduce greenhouse gas emissions, adapt to the impacts of climate change, and build resilience.

Climatic science and policy need to converge to develop coherent approaches based on evidence-based dealing with climate change. As new challenges emerge due to climate change and the strive towards a sustainable future, policymakers can address such issues by introducing scientific findings and knowledge into their policy frameworks.

Climate science is a fundamental foundation for formulating policies on climate change. Using advanced climate models and data analysis, scientists can evaluate the reasons for and impact of climate change, and come up with more realistic scenarios as well whilst identifying which areas are at most risk. Understanding science's cognizance of climate dynamics, which also involves interactions between the atmosphere and sea surface processes as well as biological processes, aids policymakers in comprehending the complex mechanisms behind how climate change occurs, and its scale of consequences.

Policy responses to climate change include mitigation and adaptation of efforts. Policies that concentrate on mitigation aim to reduce greenhouse gas emissions, transition into a low-carbon economy, and promote sustainable practices. These policies are informed by scientific knowledge on how to identify sources of emissions, determine whether mitigation strategies work or not, and look at additional benefits such as better health in humanity, improved ecologies, and strong economies.

At the same time, policies on climate change adaptation aim at enhancing resilience and adaptive capacity as a response to impacts caused by climatic changes (Webster et al. 2003).

On the other side, adaptation climate change policy focuses on increasing resilience and adaptive capacity in response to impacts of fragments or effects that are caused by global warming. Their formulation can be seen in research findings of vulnerability, risk, and impacts based on regions, sectors and communities highlighted. Developing a climate adaptation plan will allow policymakers to determine and focus on the measures that are more pressing or affecting populations at risk and longer term, as well as promoting sustainability by integrating climate science into planning.

Effective governance, stakeholder engagement and collaboration across sectors are crucial determinants of successfully implementing climate policies. Webster et al. (2003) stated that "The policy frameworks should integrate the perspectives and knowingness of various stakeholders; ranging from researchers – scientists through policymakers, communities (in this instance represented by community women), marginalized groups such as refugees and also nongovernment organizations involved in service provision within these refugee camps". Policymakers can drive an inclusive and participatory approach to policies if they leverage the depth of knowledge and diversity in perspectives required for developing endurable, fair climate policy.

The massive destruction caused by climate change is acknowledged all over the world, a sentence that usually illustrates how seriously they are taken. At different administrative levels, many governments have considered and adopted strategies to reduce the impact of climate change on their environments. The mitigation approaches that a government may consider are laying down policies such as reducing aggregate emissions priorities and encouraging systems for energy efficiency. As such, a deeper understanding of climate policies plays an important role in shaping decisions across the public and private spheres regarding what authorities need to do or influence on this issue. The sphere of policy options is wide and intricate. For instance, the Intergovernmental Panel on Climate Change (IPCC) has outlined six primary types of policy tweaks designed to directly prevent greenhouse gas GHG emissions alongside 53 specific ideas about how to structure international agreements that regulate climate change.

Adoption of such policies is far less advanced as compared to climate mitigation. On the other hand, these are more formalized at a global level and have taken shape in the form of agreements like the United Nations Framework Convention on Climate Change (UNFCCC 2009), Kyoto Protocol etc. The necessary effectiveness of the previously implemented policies or how potential future ones might operate will be assessed through scientific research, looking not only at what is currently in place but also what may come in due course and policy related to climate.

This analysis aims to focus on the intersection between climate science and politics with a view towards either climate adaptation and resilience, or more broadly speaking humanitarian issues associated with solar energy home generator system deployment. For instance, it looks at how scientific knowledge impacts policy frameworks in addressing the challenges of climate change effectively by reviewing the Intergovernmental Panel on Climate Change assessment reports, United Nations Framework Convention on Climate Change Agreements, scientific evidences and models, implementation of climate policies etc. The purpose is to anlyse the scientific relationship with the development and the implementation of the climatic action (Webster et al. (2003).

Table 2.1 Overview of climate laws in European countries

Nation	Legislative framework	Initial enactment	Key updates and developments
United Kingdom	Climate change act	November 2008	Revised 2050 emissions target in July 2019
Denmark	Law on the climate council, policy declarations, and National Targets	June 2014	New legislation agreed upon in Dec 2019; draft law proposed in Jan 2020
Finland	National climate legislation	June 2015	Strengthened targets adopted in 2019
France	Act for energy transition and green growth	August 2015	Institutional and target changes in Sept 2019
Ireland	Climate action and low carbon development framework	December 2015	Major reform proposed under climate action (amendment) bill
Sweden	Climate legislation	June 2017	–
Netherlands	National climate act	July 2019	–
Germany	Federal law on climate protection	December 2019	–
Spain	Draft law on climate change and energy transition	February 2019	Final version anticipated in 2020; updated draft shared in June 2019
Norway	Climate change act	June 2017	Legal duty to achieve carbon neutrality by 2050
Austria	Climate protection act	2011	New climate targets and carbon pricing in 2022
Portugal	Framework climate law	November 2021	Climate neutrality by 2050; carbon budgeting introduced
Italy	National integrated energy and climate plan (NECP)	2019 (EU compliance)	Revised NECP submitted in 2023
Belgium	Federal and regional climate plans (no unified national law)	Various	Coordination between regions (Flemish, Walloon, Brussels-capital)
Switzerland	CO_2 act	2000	2021 revision rejected in referendum; updated version proposed
Greece	National climate law	May 2022	Net-zero goal by 2050 and coal phase-out by 2028
Czech Republic	Climate protection policy (strategic framework)	2017 (policy adopted)	Implementation aligned with EU green Deal

In the intricate relationship between climate science and policy, policymakers, scientists, and stakeholders can collaborate to develop effective and equitable approaches that mitigate greenhouse gas emissions, adapt to climate change impacts, and foster resilience for a sustainable and climate-resilient future. Many European countries have adopted climate laws and policies as shown in Table 2.1.

2.2 Importance of Climate Science

The study of climate science has become an integral part of policy documentation as climate and climate change are drastically affecting all the primary sectors of economic, social, and environmental development. Some of the major factors that make it all the more important are documented as thc following:

(a) *Comprehending Climate Dynamics*: The fundamental goal of climate science is to elevate our comprehension of climate dynamics, encompassing the catalysts and mechanisms underlying climate fluctuations. Through the examination of these intricate mechanisms, researchers strive to discern configurations, tendencies, and reciprocal mechanisms that underlie shifts in temperature, precipitation, occurrences of extreme weather, and other pivotal climatic factors. This profound comprehension stands as an indispensable pillar in the formulation of impactful climate policies (Hasselmann 1997).

(b) *Evaluating Climate Effects and Vulnerabilities:* Climate Science is essentially concerned with the evaluation of the effects, from an ecological as well as human point of reference against any predisposition brought about by change in climate. It involves the broader sense of analysis on the perils brought about by increasing temperatures, sea-level rise, and changing patterns in precipitation trends as well as other factors that are attached to climate dynamics. By identifying areas, fields, and groups likely to be affected by fragility, policy designers can develop iterative solutions for adaptation, preparing themselves against threats that they are bound to face. (IPCC 2014).

(c) *Guiding Mitigation Approaches*: Climate science plays a crucial role in constructing and determining the effectiveness of strategies for mitigating climate change. For researchers studying greenhouse gas (GHG) emissions, their sources from human activities like burning fossil fuels, and their effects on global temperatures and extreme weather occurrences is one way that the researchers contribute. The main goal of research is to create and assess mitigation solutions, such as switching to renewable energy, improving energy efficiency, rehabilitating ecosystems, and putting sustainable agriculture into practice. In order to assess the efficacy of policies, forecast future events under various policy options, and establish goals such as the Paris Agreement's 1.5°C global warming limit, scientists employ scenario planning and climate modeling. (IPCC 2018).

(d) *Enhancing Adaptation and Resilience*: Climate science forms the basis of cognizance for taking precautions and overcoming obstacles posed by climate change, as well as serves to guide policies based on contexts related to adapting to these changes so that people can be resilient against unforeseen disruptions. Scientists offer major recommendations to decisionmakers designed to facilitate adaptation to climate change through vulnerability estimation, projections of plausible climate scenarios, and detection of social-ecological drivers of impacts. Such recommendations facilitate shifts in attitudes and actions across different sectors through formulation of resilience-based frameworks, risk management, and mainstreaming adaptation actions like crop development that can tolerate drought and upgrade of facilities for water storage (Chan and Tam 2020).

(e) *Facilitating Evidence-Driven Decision-Making*: There is a complementary common achievement between policy-forming institutions such as the UNFCCC and climate science institutions such as the IPCC, which helps them provide policymakers with the most reliable scientific evidence. This opportunity helps informed and effective decisions be made regarding how to mitigate and adapt to climate change based on sound evidence instead of speculation or opinion. This is precisely the kind of feeling that aims at bridging the de facto gap between scientific understanding and practical application in policies. (Cash et al. 2003).

(f) *Promoting Collaboration and Stakeholder Engagement*: Enhancing Collaboration and Stakeholder Engagement encouraging the integration of climate science with policy initiatives aimed at fostering collaboration among all stakeholders, from scientists to policymakers to local communities' business corporations, a wide array of stakeholders, like scientists, policymakers, local communities, enterprises, and civil society. Through interdisciplinary research advancement, facilitating knowledge dispersion, and setting up participatory mechanisms this objective seeks to make sure that climate policies are inclusive and relevant in all contexts. In this way, the aim is to form policies that are comprehensive and specific for particular conditions by aligning with the perspectives or ideas of all relevant stakeholders (Cash et al. 2003; Reed et al. 2009).

(g) *Advancing Climate Policy Monitoring and Evaluation*: Climate science directly participates in the development of monitoring and evaluation frameworks related to climate policies. Researchers contribute to the substantiation of policy adjustments and enhancements based on scientific evidence, through the delineation of indicators' continuity in tracking the advancement of comprehensive evaluations over efficacy and measures of policies (IPCC 2019). This specific ambition aspires to speed up the implementation of adaptive management techniques, and learning gained from policy delivery experiences.

(h) *Promoting Equity and Just Transitions*: Addressing the multi-faceted consequences of climate change, from societal to economic and environmental perspectives in line with principles of fairness and justice, is where both climate science and policy come together (IPCC 2014). This endeavour centres on guaranteeing that climate policies conscientiously account for the uneven effects of climate change on marginalized populations, fostering transitions toward societies that are low-carbon and resilient in a manner that is just, inclusive, and equitable.

These ideas serve as guiding principles for the integration of climate science and policy and highlight the importance of evidence-based decision-making, stakeholder engagement, and transformative action to address the challenges of climate change as shown in Fig. 2.1.

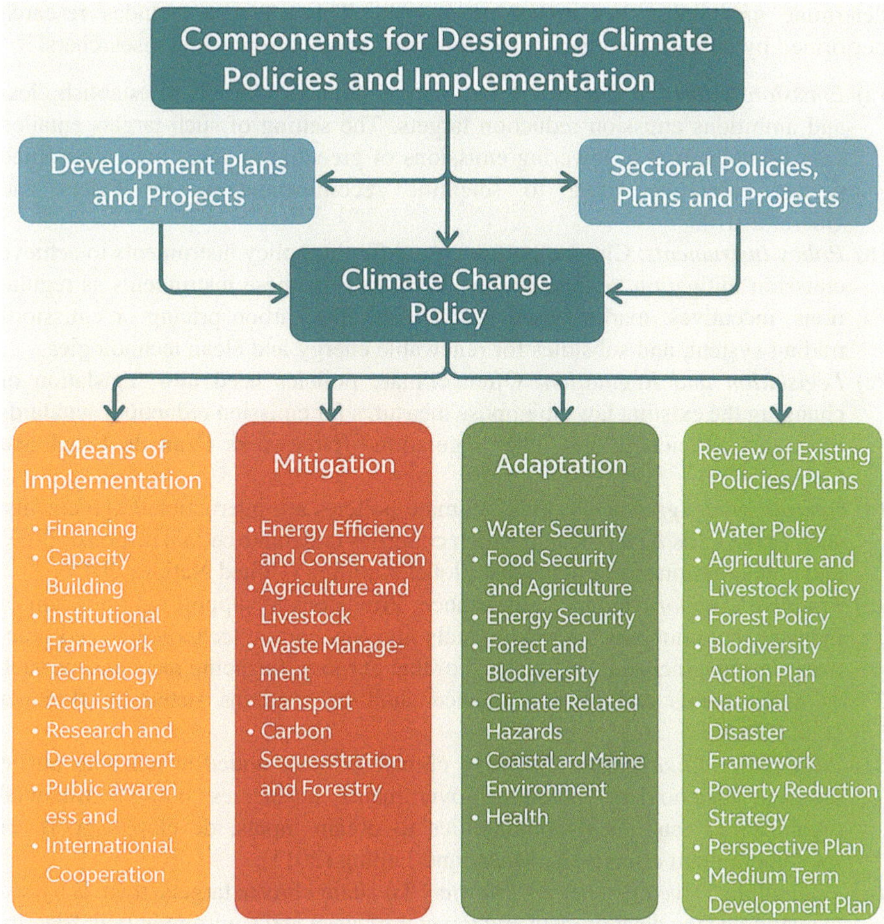

Fig. 2.1 Components are used for designing climate policies and implementation

2.3 Components and Methodologies for Climate Studies

When studying the intersection of climate science and policy, researchers employ various methodologies to examine the relationship between scientific knowledge and policy development. The methodology to be utilized will depend on the objectives of the research, available resources, and the nature of the questionnaire. Here are some common elements that researchers may consider when studying climate science and policy:

Science and Data Analysis: According to the Climate Change 2014 Synthesis Report, climate policies should be based on scientific studies and research to understand the present state of the climate and predict its future changes as well as determine greenhouse gas emissions sources. This element includes research performed by climate scientists, meteorologists, and environmental researchers.

(a) *Emission Reduction Targets*: A key part of climate policy is to establish clear and ambitious emission reduction targets. The setting of such targets entailed specific objectives in lowering emissions of greenhouse gases over a specified period, often according to scientific recommendations by Peters and Quéré (2016).
(b) *Policy Instruments*: Climate policies use different policy instruments to achieve emission mitigation targets. Stavins 2019, can list these instruments as regulations, incentives, market-based mechanisms like carbon pricing or emissions trading system, and subsidies for renewable energy and clean technologies.
(c) *Legislation and Regulation*: Often climate policies need new legislation or changing the existing laws to impose measures for emission reduction, standards of energy efficiency, and other regulations frameworks (Van de Graaf and Colgan 2017).
(d) *International Agreements*: Many climate policies are international agreements, such as the Paris Agreement in which countries undertake certain responsibilities and collaborations to fight against global warming (United Nations 2015).
(e) *Financial Support:* Financial assistance: Provisions to support the transition of industry, communities, and individuals into low-carbon technologies and practices are often included in policies. For that purpose, financing mechanisms such as green bonds and climate finance shall be used, as stated by Buchner et al. (2015).
(f) *Stakeholder Engagement*: Effective climate policies aimed at reaching out to different stakeholders, including governments, businesses, non-governmental organisations, and local communities to obtain inputs, develop support, or implement them effectively, Moser and Dilling (2011).
(g) *Technology Development and Transfer:* To attain climate targets, there is a need to encourage the development and transfer of clean and environmentally friendly technologies, according to Newell and Paterson 2010, incentives for R&D or measures facilitating the transfer of technology in developing countries can be part of the policy.
(h) *Monitoring and Reporting*: Sufficient monitoring and reporting systems are needed to monitor the progress towards these emission reduction targets and evaluate the effectiveness of climate policy, as suggested in the 2009 UN Framework Convention on Climate Change report. In this regard, transparency and accountability are of paramount importance.
(i) *Adaptation Strategies*: Climate policies should also deal with adaptation to climate change's effects, which includes plans for protecting those at risk of harmful climatic conditions such as infrastructure, ecosystems, and the environment. Education and public awareness: compliance with climate policy can be encouraged, supported by the increase in public awareness of climate change and its importance as well as citizens' education on this topic, as documented by O'Neill and Nicholson Colein (2009).

(j) *Evaluation and Adjustment*: Climate policies need to be regularly assessed in order to evaluate their effectiveness and make necessary adjustments as the challenges change with time (Piggot and Cooper 2020).
(k) *International Collaboration*: Climate change is a worldwide problem, and the world must cooperate. Policies aimed at combating the effects of climate change through collaborations using platforms such as the United Nations and bilateral agreements will strengthen their impacts.

2.4 Outcomes of Climatic Study in Policy Matters

Since there is a very strong interlinking factor between climate study and policy development, the study becomes all the more important in a comprehensive understanding of mitigation strategies from climate change and its impact. The documentation of the climate change policies needs extensive baseline research and data before implantation as climate science and climate change vary demographically, economically, socially, and environmentally. Some of the benefits and outcomes of such studies help in the following:

(a) *Policy Integration*: Climate science and policy research is a confluence that elucidates the importance of assimilating scientific insights into policy structures. Numerous studies represent cases in which scientific evidence significantly influenced policy options resulting in the development of robust strategies for combating climate change mitigation and adaptation (Cash et al. 2003). This fusion may include the integration of climate models, vulnerability assessments, or risk analyses into policy frameworks making them more powerful.
(b) *Policy Gaps and Challenges*: Research often reveals gaps and challenges in mainstreaming climate science with policy. These may include obstacles to the successful implementation of scientific insights into decision-making processes, inadequacies in communication between scientist groups and policymakers, or limited capability for policymakers to interpret from understand this data effectively (Cash et al. 2003; Weible and Sabatier 2007). The outcomes might provide insights into areas that require improvements to better address the gap between climate science and policy.
(c) *Science-Policy Interfaces*: Scholarship often explores the relationships and dynamics between climate scientists and policymakers. Findings can highlight instances of effective collaboration and communication between these factions, leading to the creation of policies that are based on substantiated evidence (Guston 2001; Pielke Jr 2007). Also, they may identify ways to enhance the science-policy interface that would include building stronger partnerships and systems for interdisciplinary studies as well as facilitating knowledge transfer between scientists and these policymakers.
(d) *Science-Policy Effectiveness in Climate Change and Mitigation*: Investigation studies often measure the effectiveness of scientific-based climate policies. The results can shed light on the effects and outcomes of policy actions such as

reducing emissions, applying adaptive strategies, or implementing renewable energy agendas (IPCC 2014; Ostrom 2005). These assessments can provide information regarding the effectiveness of different policy strategies and guide ensuing policy frameworks.

(e) *Science-Stakeholder Engagement and Public Perception*: Investigations are often focused on the influence or effect of stakeholder engagement and public perception in shaping a climate policy (Huang et al. 2015; Maibach et al. 2010). The results can help illuminate the impact of public opinion, community participation, and inclusion of diverse interests in policy processes. Understanding these dynamics can greatly empower the growth of more inclusive and socially accepted blueprints of policy.

(f) *Policy Innovations and Best Practices*: Scholarly investigation might discover creative policy mechanisms and case studies that integrate climate science into the frameworks of policymaking (Jordan et al. 2018; Meadowcroft 2009). The outlooks may underscore successful case study analyses, emphasize identifiable lessons learned, and provide directions for policymakers and stakeholders seeking to develop pragmatic measures in responding to climate change.

2.5 Data Assimilation and Utilization

(a) *Climate Data:* Climatologists use datasets from different sources to investigate past climate patterns, evaluate current trends, and predict future climatic scenarios. Organizations that provide climate data are the National Oceanic and Atmospheric Administration (NOAA), the National Aeronautics and Space Administration (NASA), as well as World Meteorological Organization (WMO).

(b) *Policy Documents and Reports*: Government agencies, international organizations, and research institutions can offer statistics on policies. These include domestic and international climate policies, and plans for dealing with the problem of climate change including reports from intergovernmental organizations such as UNFCCC, IPPC, or World Bank. These documents contain information that can help us to know policy objectives, strategies, and progress towards targets of climate change.

(c) *Emissions Inventories*: Frequently, it is possible to publish an inventory of domestic greenhouse gas emissions that gives data about sources and levels of GHGs emitted from every industry. Such inventories may be accessed by government bodies responsible for the protection of the environment, climate change, or energy such as the United States Environmental Protection Agency U.S., European Union, European Environment Authority, etc.

(d) *Climate Models and Simulations*: Different emission trajectories, policy scenarios, and climate models generate data for the simulation of the next climate scenarios. Research institutions and international climate modelling initiatives, such as the Coupled Model Intercomparison Project CMIP develop these models. Climate models provide forecasts for temperature and precipitation changes, as well as other variables related to them that assist in policy analysis and decision-making.

(e) *Socioeconomic Data*: It is also crucial to gather statistics on social perspectives about population demography, employment indicators, and land use or energy consumption to grasp the context and consequences of climate policies. This information may be gathered from national statistics agencies, international databases such as the World Bank's development indicators, or research institutions.

(f) *Surveys and Interviews*: Researchers often collect primary information using surveys, interviews, or focus groups to understand stakeholder's views perceptions, and attitudes regarding policy issues. These data can help understand how effective the use of policies, stakeholders, and communication is in delivering climate change information to policymakers as well as citizens.

It is necessary to make sure that the data sources used in studying climate science and policy are reliable, credible, and properly cited. Access to certain types of data and databases may be based on institutional subscriptions, data exchange agreements, or open data policies. Climate science and policy, when applied through strong research collaboration with stakeholder participation, can aid development that is informed by evidence towards sustainability. A framework was proposed for the development of an action plan based on the climate change policy of Pakistan and its integration into the planning process by Khan et al. (2016) well depicted in Fig. 2.2.

The purpose of climate science and policy encompasses an extensive array of domains, which comprise comprehending climate dynamics, evaluating impacts and vulnerabilities, guiding mitigation and adaptation strategies, facilitating evidence-based decision-making, fostering collaboration and stakeholder involvement, overseeing monitoring and evaluation, and advancing equitable and just transitions (IPCC 2014; Meadowcroft 2009). These goals highlight the urgency and complexity of the climate crisis, signalling that interdisciplinary approaches and future-oriented perspectives are needed. Effective climate policies require the smooth integration of scientific knowledge, data, and models into policy formation, implementation, and assessment (IPCC 2014; Ostrom 2005). Such policy frameworks should be flexible and include relativeness or responsiveness to the new scientific understanding of evolving climate dynamics.

Moreover, an essential aspect includes the development of comprehensive decision-making processes that incorporate different stakeholders; this should be complemented by a focus on justice and equity in favour of disadvantaged communities (Cash et al. 2003; O'Brien et al. 2009). This inclusive approach adheres to the nature of complexity in climate challenges and aims to come up with policies that are strong, fair, and resilient. To handle the effects of climate change successfully, coordination and partnership between scientists, policymakers in government sectors, business people and even civil society as well can play a vital role. To bridge the gap between climate science and policy, there should be effective communication pathways and platforms for knowledge sharing as well as perpetuating its

Fig. 2.2 Framework for the development of an action plan based on climate change policy and its integration into the planning process

understanding of either side's perspectives or constraints. By promoting a robust science–policy interface, we can increase the design and implementation of evidence-based climate policies that meet social requirements while contributing to a sustainable and stable future (Cash et al. 2003; O'Brien et al. 2009). Climate science and policy are critical components of the global response to climate change. Scientific knowledge informs policymaking and enhances evidence-based decision-making, innovation, and the welfare of current individuals as well as those in future generations. If we make collaboration, equity, and adaptive management our main goals; that will help us find a way to a more sustainable and climate-resilient world.

References

Adger WN, Agrawala S, Mirza MMQ, Conde C, O'Brien K, Pulhin J et al (2007) Assessment of adaptation practices, options, constraints and capacity. Climate change 2007: impacts, adaptation and vulnerability. In: Contribution of working group II to the fourth assessment report of the intergovernmental panel on climate change, pp 717–743

Buchner BK, Falconer A, Herve-Mignucci M (2015) Global landscape of climate finance 2015. Climate Policy Initiative

Cash DW, Clark WC, Alcock F, Dickson NM, Eckley N, Guston DH et al (2003) Knowledge systems for sustainable development. Proc Natl Acad Sci 100(14):8086–8091

Chan H-W, Tam K-P (2020) Exploring the association between climate change concern and mitigation behaviour between societies: a person-context interaction approach. Asian J Soc Psychol 24. https://doi.org/10.1111/ajsp.12430

Guston DH (2001) Boundary organizations in environmental policy and science: an introduction. Sci Technol Hum Values 26(4):399–408

Hasselmann K (1997) Multi-pattern fingerprint method for detection and attribution of climate change. Clim Dyn 13(9):601–611

Huang Y, Ding Y, Zhao D (2015) Public perceptions of climate change in China: results from the questionnaire survey. Nat Hazards 77(3):1849–1863

IPCC (2014) Climate change 2014: impacts, adaptation, and vulnerability. In: Contribution of working group II to the fifth assessment report of the intergovernmental panel on climate change. Cambridge University Press

IPCC (2018) Global warming of 1.5°C. An IPCC special report on the impacts of global warming of 1.5°C above pre-industrial levels and related global greenhouse gas emission pathways, in the context of strengthening the global response to the threat of climate change. World Meteorological Organization

IPCC (2019) IPCC guidelines for national greenhouse gas inventories. Prepared by the National Greenhouse Gas Inventories Programme. Accessed from: https://www.ipcc-nggip.iges.or.jp/public/2006gl/vol1.html

Jones L, Preston F (2011) Adaptive capacity in the public sector: a comparative study of municipal adaptation planning in England and Sweden. Glob Environ Chang 21(2):392–400

Jordan A, Huitema D, Hildén M, van Asselt H, Rayner T, Schoenefeld J (2018) Emergence of polycentric climate governance and its future prospects. Nat Clim Chang 8(6):1–7

Khan M, Khan J, Ali Z, Ahmad I, Ahmad M (2016) The challenge of climate change and policy response in Pakistan. Environ Earth Sci 75. https://doi.org/10.1007/s12665-015-5127-7

Maibach E, Roser-Renouf C, Leiserowitz A (2010) Global warming's six Americas 2009: an audience segmentation analysis. Yale University and George Mason University

Meadowcroft J (2009) What about the politics? Sustainable development, transition management, and long-term energy transitions. Policy Sci 42(4):323–340

Moser SC, Dilling L (2011) Communicating climate change: closing the science-action gap. In: The Oxford handbook of climate change and society, pp 161–174

Newell P, Paterson M (2010) Climate capitalism: global warming and the transformation of the global economy. Cambridge University Press

O'Brien KL, Sygna L, Haugen JE (2009) Vulnerable or resilient? A multi-scale assessment of climate impacts and vulnerability in Norway. Clim Chang 97(3–4):509–543

O'Neill S, Nicholson-Cole S (2009) "Fear won't do it": promoting positive engagement with climate change through visual and iconic representations. Sci Commun 30(3):355–379

Ostrom E (2005) Understanding institutional diversity. Princeton University Press

Pelling M (2011) Adaptation to climate change: from resilience to transformation. Routledge

Peters GP, Le Quéré C (2016) GCP carbon budget 2016: global carbon budget 2016. Earth System Science Data 8(2):605–649

Pielke RA Jr (2007) The honest broker: making sense of science in policy and politics. Cambridge University Press

Piggot G, Cooper A (2020) Policy feedbacks and climate governance: the lessons of failed emissions trading reforms in Australia. Environ Politics 29(1):55–76

Reed MS, Graves A, Dandy N, Posthumus H, Hubacek K, Morris J et al (2009) Who's in and why? A typology of stakeholder analysis methods for natural resource management. J Environ Manag 90(5):1933–1949

Stavins RN (2019) The future of US carbon pricing policy. Rev Environ Econ Policy 13(1):87–107

UNFCCC (2009) Reporting and review of information: guidelines for national communications from parties not included in annex I to the convention, part I: UNFCCC reporting guidelines on annual inventories of anthropogenic emissions by sources and removals by sinks of greenhouse gases

United Nations (2015) Paris agreement. Retrieved from https://unfccc.int/sites/default/files/english_paris_agreement.pdf

Van de Graaf T, Colgan JD (2017) The global energy challenge: environment. Palgrave Macmillan, development and security

Webster MD, Forest CE, Reilly JM, Babiker MH, Kicklighter DW, Mayer M, Prinn RG, Sarofim MC, Sokolov AP, Stone PH, Wang C (2003) Uncertainty analysis of climate change and policy response. Clim Chang 61:295–320

Weible CM, Sabatier PA (2007) A guide to the advocacy coalition framework. In: Theories of the policy process, vol 2, pp 123–156

Chapter 3
Strategic Integration of Green Innovation and Information Systems for Sustainable Development

Sonal Jain and Deepak Kumar

Abstract Green innovation focuses on developing eco-friendly technologies and practices that minimize environmental impact while fostering economic growth. The strategic integration of green innovation and information systems articulates a new approach to sustainable development. These innovations can be seamlessly implemented, monitored, and optimized with advanced information systems. Information systems enable the collection and analysis of real-time data, enhancing decision-making and resource management to support sustainability. The alignment of green innovation with digital technologies fosters collaboration across industries and sectors to encourage the extensive adoption of sustainable practices. It drives transparency and accountability of stakeholders with accessible, data-driven insights. This integration also facilitates the creation of smart systems for energy efficiency, waste reduction, and the promotion of circular economies. This work highlights the effect of green innovation on business performance. This presents the primary quantitative methods with the survey and analysis of the data collected from the participants. The entire analysis has been conducted with the SPSS tool. This presents the role of green innovation in explaining the impact of business performance within the business. The synergy between green innovation and information systems shows a powerful framework for addressing global environmental challenges and advancing economic development to pave the way for a sustainable future.

S. Jain
KIIT School of Economics & Commerce (KSEC), Kalinga Institute of Industrial Technology (KIIT) Deemed to be University, Bhubaneswar, Odisha, India

D. Kumar (✉)
Atmospheric Sciences Group, Department of Geosciences, College of Arts & Sciences, Texas Tech University, Lubbock, Texas, USA

Center of Excellence in Weather & Climate Analytics, Atmospheric Sciences Research Center (ASRC), University at Albany, State University of New York (SUNY), Albany, New York, USA

D. Kumar, N. P. Bassill (eds.), *Artificial Intelligence and Machine Learning for Climate Disaster Management*, Disaster Resilience and Green Growth,
https://doi.org/10.1007/978-981-95-1588-2_3

Keywords Green innovation · Organizational performance · Information system · Sustainable business

3.1 Introduction

The persistent global challenges of environmental degradation and climate change need innovative approaches to sustainable development (Pelorosso et al. 2017). Green innovation is described by eco-friendly technologies and has emerged as a vital component in addressing these issues. It has the potential to drive sustainability, which is significantly amplified with the strategic integration with the advanced information systems (Saha et al. 2021). Information systems provide robust frameworks for data collection, analysis, and implementation, enhancing the efficiency and scalability of green solutions (Dwevedi et al. 2018; Ludwig et al. 2023). This synergy promotes resource optimization and environmental stewardship to support economic growth (Falavigna and Ippoliti 2020; Wu et al. 2021). This work explores the transformative potential of combining green innovation with information systems to achieve sustainable development goals.

Green innovation minimizes the damage created within the environment. The improvement of the economic performance of the organization is determined. The environmental performance of the company is being highlighted with the help of this green innovation (Dwevedi et al. 2018; Jirandehi and Sajadi 2024).

This work determines the effect of green innovation on business performance (Ahmad et al. 2019; Dantas et al. 2020; Xie et al. 2022). The negative environmental effect is being accelerated in this research portion. Improvement of the service about the business structure is explained within this portion. The green funds is being used for the environmental and economic businesses to bring innovation within the performance of the business sector.

Figure 3.1 highlights the impact of green innovation on sustainable business performance. The economic and social performance of the business with the support of green innovation is identified for the study (Chang and Cheng 2019; Cheng et al. 2021a). It is noticed that the 35.5 billion USD revenue of the business increased with the support of green innovation. Therefore, the annual growth of the business is developed by 21.6% in the year 2022. Environmental management policies play an important role in improving the productivity of the business; moreover, different opportunities for the business are being developed (Cao et al. 2021; Xue et al. 2018). An information system is another factor that has an impact on sustainable business performance; hence, the growth of the business performance must be addressed. Therefore, the financial structure of the business must be increased. The work tries to understand the strategic integration of green innovation to improve business performance (Becker et al. 2021; Schwecke et al. 2008; Wahab et al. 2024). The core research objectives are (a) identification of the role of green innovation in business performance, (b) assessment of different factors to improve the green innovation process, (c) analysis of the indicators of sustainable business performance, and (d) describe the impact of information systems on business performance. The key

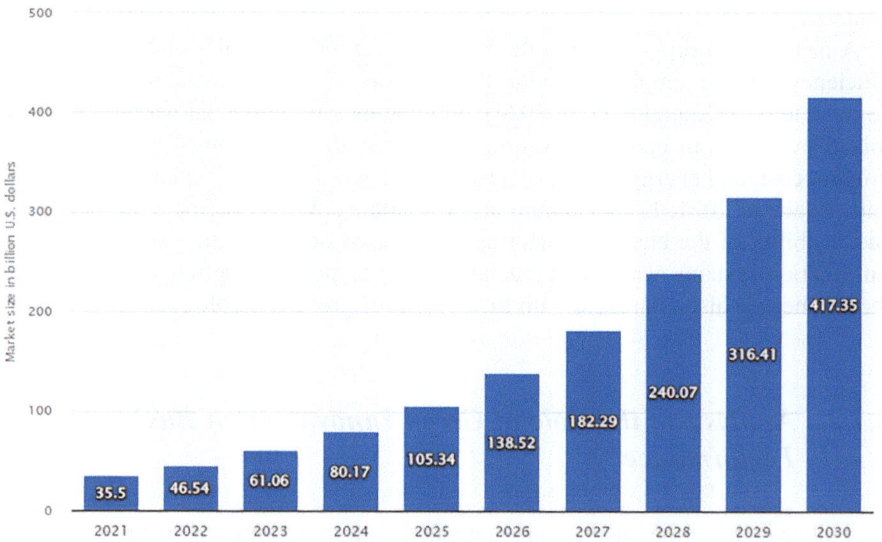

Fig. 3.1 Impact of green innovation to improve business performance. (Source: www.statista.com)

questions include (a) what is the role of green innovation in business performance? (b) What are the different factors that improve the green innovation process? (c) Which indicators improve sustainable business performance? and (d) What are the ways to assess the impact of the information system on business performance?

The overall analysis of the research findings is allowed to be highlighted in this literature review portion. Analysis of the main features with the support of the research objectives is highlighted in this section. The economic and environmental effects of green innovation are highlighted in this article section.

3.1.1 Theoretical Implications~Behavioural Management Theory

Individuals' behaviour is focused on with the support of the behavioural management theory. The performance of the business depends on the role of the employees (García-Rodríguez et al. 2015; Tseng et al. 2019). Moreover, positive working environments are created; hence, it improves the growth of the organizational sector. The cost reduction process is being generated, and it develops the motivationprocess of the employees. The physiological environment of the business is determined, and it improves the growth of the business (Pan et al. 2018; Saritha and Santhosh Kumar

2019). The trust of the employees must be highlighted, and it helps to accelerate business growth.

A healthier workplace within the business section accelerates and improves the efficiency of the employees with the support of the innovation process. As commented by (Xianqing et al. 2024), eliminating environmental wastage is determined by access to green innovation. Moreover, the reduction of the population, product cost, and energy deduction process is determined. As emphasized by (Bhat and Alqahtani 2021; Kamalapuram and Choudhury 2024; Sharma et al. 2024) the sustainability of the business performance is based on green innovation; therefore, information systems also play a crucial role in this process. Optimizing the time of the business is also addressed with the support of green technology.

3.1.2 Analysis of the Role of Green Innovation on Business Performance

Several growths of the business are generated with the support of green innovation. The performance of the business is evaluated with the assistance of green performance. This study is based on the importance of green innovation for sustainable business performance (Mohammadian et al. 2020; Tai et al. 2020). Therefore, all factors are designed with the support of green products. As commented by (Shang et al. 2024; Ye et al. 2018), the negative impact on the environment has to be eliminated with access to green innovation within the business structure.

Figure 3.2 explains the impact of green innovation on the performance of the business. Reduction of energy use supports green innovation; moreover, waste recycling, resource sustainability, and population control are the three major factors that improve the productivity of the business (Cheng et al. 2021b; Miller et al. 2022; Zhou et al. 2015). On the other hand, the job satisfaction of the employees is highlighted; hence, it helps to improve productivity. A positive impact on the performance of the organization is determined.

Fig. 3.2 Role of green innovation on business performance. (Source: Influenced by Pyszczynski et al. 2020)

3.1.3 Identification of Different Factors to Improve the Green Innovation Process

Technology readiness is the key factor to improve the structure of the business. The internal structure of the organization plays an impactful role in improving the green innovation process. As highlighted by (Anthwal et al. 2024; Hong and Xiao 2024), administrative regulation of the environmental rate is to be addressed and affect the reduction of the carbon emission process generated. The environmental population is reduced with the support of adopting green technology; moreover, sustainable growth of the business must be highlighted Fig. 3.3 identifies the different factors to improve green innovation to increase business performance.

The economic construction of the business plays a potential role in this business performance. The technical progress of the business is being accelerated. On the other hand, as argued by (Idoko et al. 2020; Rena et al. 2022), minimizing the product cost and, therefore, time is highlighted with the support of the green innovation process. Changing the environment of the business accelerates green innovation technology.

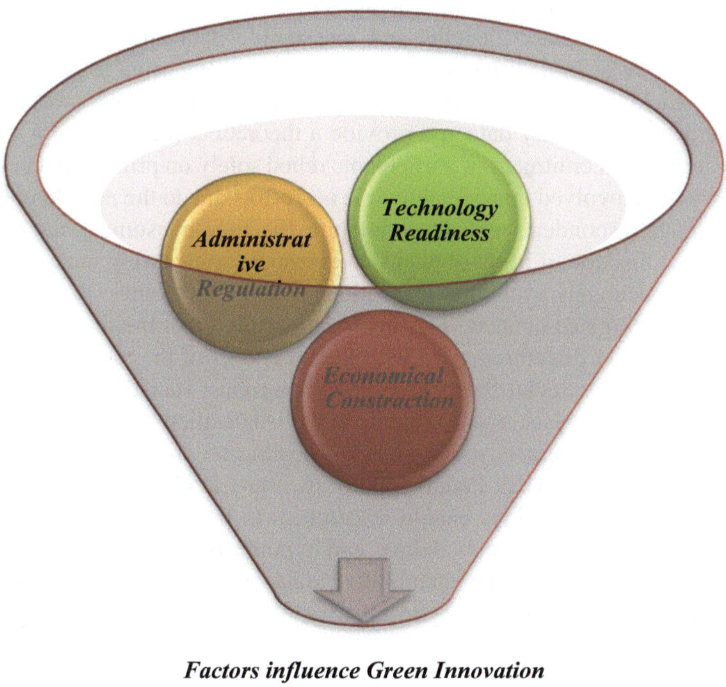

Fig. 3.3 Different factors to improve green innovation. (Source: Influenced by Przychodzen et al. 2020)

3.1.4 Impact of Indicators of Sustainable Business Performance

Decrease in the operating process improves the sustainability of the business performance, and corporate reputation is based on the business performance (Haleem et al. 2020; Kumar 2021). Positive working environments must be generated and increase the growth of the organization. (Chen et al. 2024; Viegas et al. 2014) reveal that innovation within the organization helps to improve creativity. The assessment of employee job satisfaction includes consideration of sustainable business practices. The performance of the business reduces the negative impact on the business. The environmental sustainability is being assessed to check the development of green market sustainability as indicator of the business performance. The process of repairing environmental damage is evaluated using a range of indicators that help measure the extent of degradation and the effectiveness of restoration efforts.

3.2 Methodology

The primary data for this quantitative research was collected through a structured survey, which received 75 valid responses from participants. This method was chosen to obtain firsthand insights and measurable data directly relevant to the research objectives. In addition to the survey, a comprehensive review of existing literature—including scholarly articles and peer-reviewed journals—was conducted to supplement the primary data and provide a theoretical foundation for the study. The selection and recruitment of participants relied solely on primary data collection methods, which involved considerable time and effort due to the need to ensure data reliability and respondent relevance. Despite the time-consuming nature of this process, the collection of primary data was essential for capturing nuanced perspectives and for accurately evaluating the influence and effectiveness of green innovation within the targeted context. This approach ensured that the findings were both evidence-based and directly aligned with the core aims of the research. The Statistical Package for the Social Sciences (SPSS) is a robust statistical software widely utilized for data analysis, management, and visualization across various fields, including academic research, social sciences, healthcare, and business (Enwereuzor & Ugwu, 2021; Srivastava & Pathak, 2020). Its user-friendly interface and comprehensive analytical capabilities enable researchers to perform descriptive statistics, inferential analysis, predictive modeling, and hypothesis testing. In this study, SPSS was employed to conduct the quantitative data analysis, ensuring accurate and reliable interpretation of the survey results.

3.3 Results and Finding

3.3.1 Demographic Data ~ Gender

Demographic data provides statistical information to describe the characteristics of a population. It includes key variables such as age, gender, ethnicity, education level, income, occupation, marital status, and geographic location. This type of data is essential for understanding population trends, behaviours, and needs, making it invaluable across various fields, including marketing, public policy, healthcare, and urban planning (Mouratidis 2021). Through analysis of demographic data, organizations and governments identify the patterns and disparities to design targeted strategies and allocate resources efficiently (Olya et al. 2022; Sindhwani et al. 2021). Businesses use demographic insights to tailor products or services to specific customer segments for policymakers to address social inequalities and improve community development initiatives. Demographic data provides a foundation for informed decision-making to help anticipate future trends in a rapidly changing world.

Table 3.1 shows that out of 75 total respondents, 52 were identified as female, accounting for 69.3% of the group. A total of 17 were identified as male, making up 22.7%. About 6 respondents preferred not to specify their gender, representing 8.0% of the total. The cumulative percentage shows how the responses add up sequentially, reaching 100% when all categories are included. Figure 3.4 indicates that 69.3% of the respondents identify as female, while 22.67% identify as male, and 6% as other genders. This gender distribution suggests a significant female majority among participants, which may influence the interpretation of the findings, particularly if gender plays a role in shaping perceptions or behaviors related to the research topic. Therefore, the results has been analyzed with consideration of this demographic imbalance, as it could affect the generalizability of the study's conclusions.

Gender distribution provides a proportion of individuals identifying as different genders within a given population (Ahmed et al. 2017; Cheza et al. 2021; Sánchez-Pujalte et al. 2021). It is commonly expressed as percentages or ratios of males, females, and other gender identities. Understanding gender distribution is crucial for analyzing population dynamics and addressing societal needs (Riyas et al. 2021; Shams Esfandabadi et al. 2020). This data serves various purposes across disciplines. Beyond binary classifications, modern analyses increasingly recognize non-binary and gender-diverse individuals, ensuring more comprehensive and

Table 3.1 Demographic data on gender

What is your gender?					
		Frequency	Percent	Valid percent	Cumulative percent
Valid	Female	52	69.3	69.3	**69.3**
	Male	17	22.7	22.7	**92.0**
	Prefer not to say	6	8.0	8.0	**100.0**
	Total	**75**	**100.0**	**100.0**	

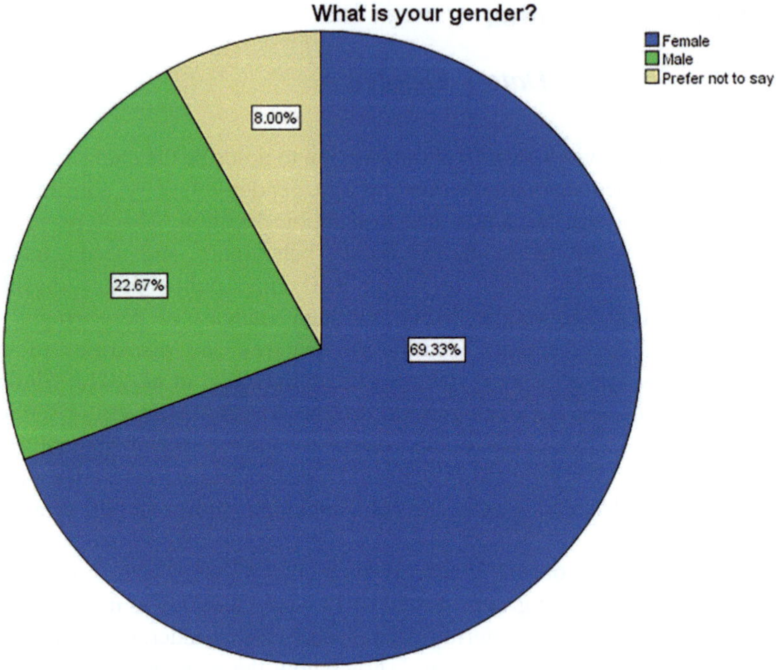

Fig. 3.4 Gender distribution

inclusive representation. This evolving understanding highlights the importance of considering cultural, social, and contextual factors when analyzing gender distribution to address the unique needs of all groups and foster equitable growth and development. In public policy, it helps address gender-based disparities in education, employment, and healthcare. In business, it informs market segmentation and consumer behaviour analysis. Gender distribution also plays a significant role in identifying gaps in representation and promoting equity, inclusivity, and diversity.

3.3.2 Demographic Data ~ Age Group

Demographic data concerning age groups provides essential insights into the composition of a population based on age. This data is often segmented into specific ranges, such as children (0–14 years), youth (15–24 years), adults (25–64 years), and seniors (65+ years). Analysing age groups helps identify trends related to population growth, ageing, and generational shifts. It also informs policies and strategies in sectors like education, healthcare, employment, and marketing, ensuring resources are targeted effectively to meet the unique needs of each age group.

Table 3.2 Demographic data on age group

What is your age?						
		Frequency	Percent	Valid percent	Cumulative percent
Valid	Between 20 and 25 years	19	25.3	25.3	**25.3**
	Between 25 and 30 years	17	22.7	22.7	**48.0**
	Between 35 and 40 years	29	38.7	38.7	**86.7**
	Between 40 and 50 years	10	13.3	13.3	**100.0**
	Total	**75**	**100.0**	**100.0**	

Understanding these dynamics is crucial for sustainable development and planning. Table 3.2 shows how the data is spread across different age categories and their corresponding percentages. It shows that 19 respondents are aged between 20 and 25 years, which accounts for 25.3% of the total, 17 respondents fall between the ages of 25–30 years, making up 22.7%, 29 respondents, the largest group, are aged between 35 and 40 years, representing 38.7%, 10 respondents belong to the 40–50 years age bracket, contributing 13.3% to the total.

The Cumulative Percent column shows how these percentages add up sequentially. The cumulative percentage reaches 86.7% for the respondents of the age group between 35 and 40 years, and the addition of the 40–50 years group sums to 100%. This provides a clear view of the age distribution, highlighting that the majority (over 60%) fall between 35 and 40 years or younger. Age group distribution provides a classification of a population into distinct age brackets to understand demographic patterns and trends. This segmentation typically organizes individuals into categories such as children, adolescents, adults, and seniors, or into specific age ranges like 0–20, 20–25, 25–30, 30–35, and 35–40 and so on. It provides valuable insights into the population structure and plays a crucial role in various fields, including economics, healthcare, education, and urban planning.

Age group distribution also helps in identifying potential challenges, such as an ageing population leading to increased dependency ratios. Figure 3.5 exhibits the age distribution, and it shows that 38.67% of respondents are in the age category of 35–40, 22.67% of respondents are in the age group of age 25–30, and 13.3% of respondents are from the age of 40–50 category. The analysis of age group distribution supports informed policymaking and strategic planning, ensuring resources are allocated effectively to meet the specific needs of each age segment.

3.3.3 Demographic Data ~ Income Level

Table 3.3 illustrates the distribution of respondents based on their monthly income range. The explanation and analysis show that there are 13 respondents (17.3%) fall into this income bracket above Rs. 60,000, with 17 respondents, representing 22.7% of the total, making it the second-largest group below Rs. 25,000, 8 respondents (10.7%) fall in this range, forming the smallest income group between Rs. 25,000 to

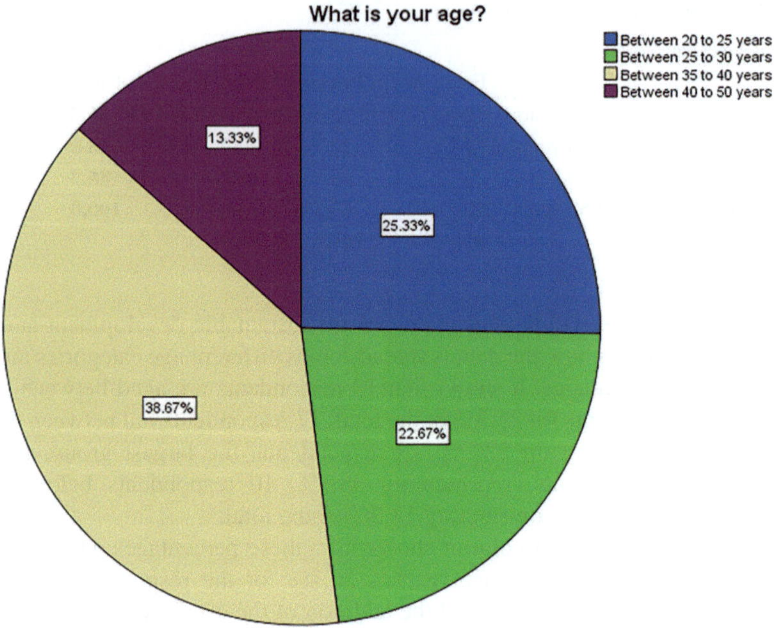

Fig. 3.5 Age group distribution

Table 3.3 Demographic data on income level

What is your monthly income range?					
		Frequency	Percent	Valid percent	Cumulative percent
Valid	Above Rs. 60,000	13	17.3	17.3	**17.3**
	Below Rs. 25,000	17	22.7	22.7	**40.0**
	Between Rs. 25,000 and Rs. 35,000	8	10.7	10.7	**50.7**
	Between Rs. 35,000 and Rs. 45,000	14	18.7	18.7	**69.3**
	Between Rs. 45,000 and Rs. 60,000	23	30.7	30.7	**100.0**
	Total	**75**	**100.0**	**100.0**	

Rs. 35,000, 14 respondents are in this bracket, constituting 18.7% are between Rs. 35,000 to Rs. 45,000, and the highest number of respondents (23), accounting for 30.7% are between Rs. 45,000 to Rs. 60,000. The data suggests a diverse income distribution among respondents. It shows that the largest portion of the group (30.7%) earns Rs. 45,000 to Rs. 60,000, which could indicate this income range is common or ideal within the sampled population. Those earning below Rs. 25,000 form a significant 22.7%, highlighting the presence of lower-income individuals. A

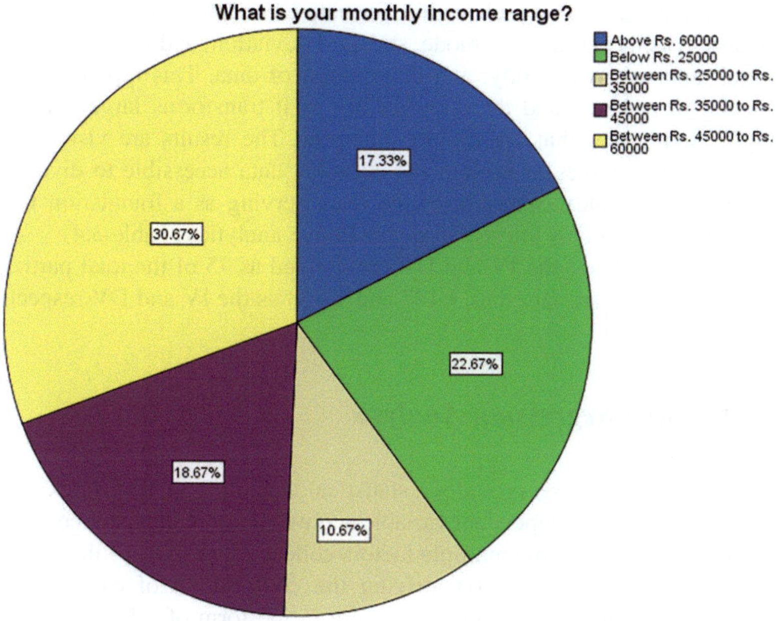

Fig. 3.6 Distribution of income level

smaller proportion (10.7%) earns between Rs. 25,000 and Rs. 35,000, suggesting that fewer respondents fall in this middle-income category. The high-income category (above Rs. 60,000) constitutes 17.3%, indicating a reasonably significant presence of higher earners. The data showcases a balanced income distribution, with a tilt towards the higher income range of Rs. 45,000 to Rs. 60,000. This shows the economic diversity of the data being collected.

The distribution of income levels represents the way income is spread across individuals or households within a population. It is a key indicator of economic equality, societal well-being, and living standards. Income distribution can vary significantly across regions, industries, and demographics, reflecting broader socio-economic dynamics such as education, employment opportunities, and government policies. Figure 3.6 shows that there are 30.67% of respondents who have an income in the range of Rs. 45,000 to Rs. 60,000, and there are 22.47% of people who are from below the 25,000 income range. Moreover, 10.67% of the respondents have income above 60,000.

3.3.4 Descriptive Data Analysis

Descriptive data analysis summarizes and interprets the raw data to uncover patterns, trends, and insights. It involves the use of statistical tools and techniques to organize and present data in a comprehensible format. Descriptive analysis focuses on measures such as mean, median, mode, standard deviation, and range to understand the central tendency, variability, and distribution of data. This type of analysis is fundamental in research and decision-making as it transforms large datasets into meaningful summaries that are easy to interpret. The results are visualized with charts, graphs, and tables to explain the complex data accessible to diverse audiences. It provides a clear snapshot of the data, serving as a foundation for more advanced analyses such as inferential or predictive analytics (Table 3.4).

In the above analysis, the IV and DV are defined as 75 of the total participants. The statistical values are raised as 3.477 and 1.468 as the IV and DV respectively.

3.3.5 Multiple Regression Analysis

Multiple regression analysis provides a statistical technique analysis to examine the relationship between one dependent variable and two or more independent variables. It helps in understanding how multiple factors collectively influence the outcome of the dependent variable and in quantifying the contribution of each factor. The method creates a mathematical model, usually in the form of a linear equation, to predict the dependent variable based on the independent variables (Nguyen and Do 2020; Sabrin et al. 2020). The general equation is $\mathbf{Y} = \boldsymbol{\beta}_0 + \boldsymbol{\beta}_1\mathbf{X}_1 + \boldsymbol{\beta}_2\mathbf{X}_2 + \ldots + \boldsymbol{\beta}_n\mathbf{X}_n + \boldsymbol{\varepsilon}$, where $\mathbf{Y}$ is the dependent variable, $\mathbf{X}_1, \mathbf{X}_2, \ldots \mathbf{X}_n$ are the independent variables, $\boldsymbol{\beta}_0$ is the intercept, $\boldsymbol{\beta}_1, \boldsymbol{\beta}_2, \ldots \boldsymbol{\beta}_n$ are the coefficients, and $\boldsymbol{\varepsilon}$ represents the error term. Multiple regression is widely used in fields such as economics, marketing, healthcare, and social sciences to analyse relationships, make predictions, and inform decision-making processes (Tables 3.5, 3.6, 3.7, and 3.8).

The above analysis reveals a strong positive correlation between the dependent variable (DV) and the independent variable (IV), with a correlation coefficient of 0.858. This high value indicates a statistically significant and direct relationship between the two variables, suggesting that changes in the independent variable are closely associated with corresponding changes in the dependent variable. Such a strong correlation highlights the relevance and impact of the IV on the DV within the context of this study. The survey collected responses from a total of 75 participants. This sample size provides the dataset for the quantitative analysis conducted in the study, offering sufficient data to explore the research questions and draw meaningful conclusions within the scope of the investigation. Research on green innovation plays a critical role in shaping and developing organizational structures (Anthwal et al., 2024; Mina et al., 2021; Perez-Uribe et al., 2024). It emphasizes the formulation of new business strategies that are integral to achieving sustainability within the corporate sector. The efficiency of green innovation serves as a key indicator for assessing organizational growth and understanding the influence of various internal and external factors. Furthermore, examining the impact of these factors on the green innovation process provides valuable insights into how organizations can enhance their economic structures while promoting sustainable development.

Table 3.4 Descriptive statistics

Descriptive statistics													
	N	Range	Minimum	Maximum	Sum	Mean		Std. deviation	Variance	Skewness		Kurtosis	
	Statistics	Statistic	Statistics	Statistic	Statistic	Statistic	Std. error	Statistic	Statistic	Statistic	Std. error	Statistic	Std. error
IV	75	20.00	5.00	25.00	799.00	10.65	0.40	3.47	12.01	1.41	0.28	3.77	0.55
DV	75	8.00	2.00	10.00	318.00	4.24	0.17	1.47	2.16	1.33	0.28	3.66	0.55
Valid N (listwise)	75												

Table 3.5 Model summary

				Std.	Change statistics					
Model	R	R square	Adjusted R square	error of the estimate	R square change	F change	df1	df2	Sig. F change	Durbin-watson
1	0.58[a]	0.34	0.32	1.21	0.34	36.99	1	73	0.00	2.17

The R square value is defined as 0.336, where the F change value is also raised as 36.995 in the model summary analysis. Fundamentally, the estimated value is raised as 1.20486 in this analysis
There is a beta value of 0.858 as the coefficient of this survey analysis. Also, there is 12.168 as the IV significance value in this analysis in an appropriate way

Table 3.6 ANOVA

Model		Sum of squares	df	Mean square	F	Sig.
1	Regression	53.71	1	53.71	36.99	0.00
	Residual	105.97	73	1.45		
	Total	159.68	74			

There are regression value is raised as 53.706 where the df value is only 1. Also, there are 105.974 sums of squares as the df value of 74 comes from the mean value of 1.452 in this survey analysis

Table 3.7 Coefficients

KMO and Bartlett's test		
Kaiser-Meyer-Olkin measure of sampling adequacy		0.500
Bartlett's test of Sphericity	Approx. Chi-Square	29.723
	df	1
	Sig.	0.000

In the validity test, there is KMO and Bartlett's value is defined as 0.500. Also, there are 29.723 is defined as the chi-square value in this survey analysis

Table 3.8 Correlation test

Correlations

		IV	DV
IV	Pearson correlation	1	0.858**
	Sig. (2-tailed)		0.000
	N	75	75
DV	Pearson correlation	0.858**	1
	Sig. (2-tailed)	0.000	
	N	75	75

Coefficients

		Unstandardized coefficients		Standardized coefficients		
Model		B	Std. error	Beta	*t*	Sig.
1	(constant)	−0.045	0.357		−0.125	0.901
	IV	0.203	0.017	0.858	12.168	0.000

There is coefficient value of −0.045 as the IV of 0.203 in this survey analysis

3.4 Conclusion

The strategic integration of green innovation and information systems offers a transformative pathway toward sustainable development. The combination of eco-friendly technologies with advanced data-driven systems, organizations, and societies will achieve more efficient resource utilization to reduce environmental impacts and enhance decision-making processes. This synergy not only fosters environmental stewardship but also drives economic growth and social equity, aligning with the broader goals of sustainable development. This chapter highlights the impact of green innovation on business performance. The work highlights the positive impact of the information process. The research highlights the importance of collaboration across industries and stakeholders in adopting and implementing these integrated strategies. As global challenges like climate change and resource depletion intensify, the role of green innovation and information systems will become increasingly critical in shaping a sustainable and resilient future.

References

Ahmad M, Zhao Z-Y, Li H (2019) Revealing stylized empirical interactions among construction sector, urbanization, energy consumption, economic growth and CO2 emissions in China. Sci Total Environ 657:1085–1098. https://doi.org/10.1016/j.scitotenv.2018.12.112

Ahmed T, Chandran VGR, Klobas J (2017) Specialized entrepreneurship education: does it really matter? Fresh evidence from Pakistan. Int J Entrep Behav Res 23(1):4–19. https://doi.org/10.1108/IJEBR-01-2016-0005

Anthwal A, Uniyal A, Gairolla J, Singh R, Gehlot A, Abbas M, Akram SV (2024) Industry 4.0 technologies adoption for digital transition in drug discovery and development: a review. J Ind Inf Integr 38. https://doi.org/10.1016/j.jii.2024.100562

Becker C, Barboza EG, Martins EM (2021) State of the political-administrative and morphological view of balneary Esplanada and Campo Bom in the municipality of Jaguaruna, Santa Catarina [Uma visão política-administrativa e morfológica dos balneários Esplanada e Campo Bom do Município de Jaguaru]. Revista Brasileira de Geomorfologia 22(4):747–768. https://doi.org/10.20502/RBG.V22I4.1806

Bhat JR, Alqahtani SA (2021) 6G ecosystem: current status and future perspective. IEEE Access 9: 43134–43167. https://doi.org/10.1109/ACCESS.2021.3054833

Cao W, Zhou Y, Li R, Li X, Zhang H (2021) Monitoring long-term annual urban expansion (1986–2017) in the largest archipelago of China. Sci Total Environ 776. https://doi.org/10.1016/j.scitotenv.2021.146015

Chang A-Y, Cheng Y-T (2019) Analysis model of the sustainability development of manufacturing small and medium- sized enterprises in Taiwan. J Clean Prod 207:458–473. https://doi.org/10.1016/j.jclepro.2018.10.025

Chen Q, Gou Y, Wang T, Liu P, Zhu J (2024) The evolutionary path and emerging trends of circulating fluidized bed technology: an integrated analysis through bibliometric assessment and data visualization. Energies 17(14). https://doi.org/10.3390/en17143514

Cheng T-M, Chang S-Y, Lien W-H (2021a) Work-leisure balance: perceived organizational leisure support. J Leis Res 52(2):202–226. https://doi.org/10.1080/00222216.2020.1762518

Cheng X, Zhang H, Cao J, Ma Y (2021b) The effect of mindfulness-based programs on psychological distress and burnout in kindergarten teachers: a pilot study. Early Childhood Educ J 50. https://doi.org/10.1007/s10643-021-01254-6

Cheza A, Tlou B, Zhou DT (2021) Incidence of non-communicable diseases (NCDs) in HIV patients on ART in a developing country: case of Zimbabwe's Chitungwiza central hospital – a retrospective cohort study (2010–2019). PLoS One 16(5). https://doi.org/10.1371/journal.pone.0252180

Dantas NCFM, Duarte OS, Ferreira WC, Ayala AP, Rezende CF, Feitosa CV (2020) Plastic intake does not depend on fish eating habits: identification of microplastics in the stomach contents of fish on an urban beach in Brazil. Mar Pollut Bull 153. https://doi.org/10.1016/j.marpolbul.2020.110959

Dwevedi R, Krishna V, Kumar A (2018) Environment and big data: role in smart cities of India. Resources 7(4). https://doi.org/10.3390/resources7040064

Enwereuzor IK, Ugwu MC (2021) Nexus between emotional intelligence and altruism as a matter of compassion for others. Psychol Stud 66(1):97–104. https://doi.org/10.1007/s12646-020-00592-6

Falavigna G, Ippoliti R (2020) The impact of institutional performance on payment dynamics: evidence from the italian manufacturing industry. J Bus Econ Manag 21(5):1285–1306. https://doi.org/10.3846/jbem.2020.13195

García-Rodríguez FJ, Gil-Soto E, Ruiz-Rosa I, Sene PM (2015) Entrepreneurial intentions in diverse development contexts: a cross-cultural comparison between Senegal and Spain. Int Entrep Manag J 11(3):511–527. https://doi.org/10.1007/s11365-013-0291-2

Haleem A, Khan MI, Khan S (2020) Conceptualising a framework linking halal supply chain management with sustainability: an India centric study. J Islamic Market 12. https://doi.org/10.1108/JIMA-07-2019-0149

Hong Z, Xiao K (2024) Digital economy structuring for sustainable development: the role of blockchain and artificial intelligence in improving supply chain and reducing negative environmental impacts. Sci Rep 14(1). https://doi.org/10.1038/s41598-024-53760-3

Idoko EC, Nebo GN, Ukenna SI (2020) Determinants of field salespersons' sales performance in deposit money banks: does organizational commitment mediate? Banks Bank Syst 15(4): 204–220. https://doi.org/10.21511/bbs.15(4).2020.17

Jirandehi AB, Sajadi SM (2024) Co-citation analysis and co-occurrence analysis of resilience keywords in StartUps. Sci Res J 10(2):139–162. https://doi.org/10.22070/rsci.2024.17648.1667

Kamalapuram SK, Choudhury D (2024) Industry 4.0 technologies for cultivated meat manufacturing. Food Bioeng 3(1):14–28. https://doi.org/10.1002/fbe2.12080

Kumar A (2021) Transition management theory-based policy framework for analyzing environmentally responsible freight transport practices. J Clean Prod 294. https://doi.org/10.1016/j.jclepro.2021.126209

Ludwig B, Donabauer G, Ramsauer D, Subari K, al. (2023) URWalking: indoor navigation for research and daily use. Kunstliche Intell 37(1):83–90. https://doi.org/10.1007/s13218-022-00795-1

Miller K, Krivtsov V, Cohen L (2022) Inventory of green roofs within Edinburgh, Scotland. Lecture Notes Civil Eng 203:1345–1353. https://doi.org/10.1007/978-981-16-7160-9_136

Mina H, Kannan D, Gholami-Zanjani SM, Biuki M (2021) Transition towards circular supplier selection in petrochemical industry: a hybrid approach to achieve sustainable development goals. J Clean Prod 286. https://doi.org/10.1016/j.jclepro.2020.125273

Mohammadian HD, Wittberg V, Castro M, Bolandian G (2020) The 5thWave and i-sustainability plus theories as solutions for SocioEdu consequences of Covid-19. In: Proceedings of 2020 IEEE learning with MOOCS LWMOOCS, pp 118–123. https://doi.org/10.1109/LWMOOCS50143.2020.9234360

Mouratidis K (2021) How COVID-19 reshaped quality of life in cities: a synthesis and implications for urban planning. Land Use Policy 111. https://doi.org/10.1016/j.landusepol.2021.105772

Nguyen HC, Do DT (2020) Factors influencing the perception of the selling price of luxury apartments. J Asian Finance Econ Bus 7(5):185–194. https://doi.org/10.13106/JAFEB.2020.VOL7.NO5.185

Olya MH, Badri H, Teimoori S, Yang K (2022) An integrated deep learning and stochastic optimization approach for resource management in team-based healthcare systems. Expert Syst Appl 187. https://doi.org/10.1016/j.eswa.2021.115924

Pan X, Zhang J, Song M, Ai B (2018) Innovation resources integration pattern in high-tech entrepreneurial enterprises. Int Entrep Manag J 14(1):51–66. https://doi.org/10.1007/s11365-017-0464-5

Pelorosso R, Gobattoni F, Leone A (2017) The low-entropy city: a thermodynamic approach to reconnect urban systems with nature. Landsc Urban Plan 168:22–30. https://doi.org/10.1016/j.landurbplan.2017.10.002

Perez-Uribe R, Ocampo-Guzman D, Salcedo-Perez C, Carvajal-Contreras A (2024) Organizational management sustainability in VUCA contexts. In: Organizational management sustainability in VUCA contexts. IGI Global. https://doi.org/10.4018/979-8-3693-0720-5

Rena Y et al (2022) Eco-innovations and sustainability in solid waste management: an indian upfront in technological, organizational, start-ups and financial framework. J Environ Manag 302. https://doi.org/10.1016/j.jenvman.2021.113953

Riyas MJ, Syed TH, Kumar H, Kuenzer C (2021) Detecting and analyzing the evolution of subsidence due to coal fires in Jharia coalfield, India using sentinel-1 SAR data. Remote Sens 13(8). https://doi.org/10.3390/rs13081521

Sabrin S, Karimi M, Fahad MGR, Nazari R (2020) Quantifying environmental and social vulnerability: role of urban Heat Island and air quality, a case study of Camden, NJ. Urban Clim 34. https://doi.org/10.1016/j.uclim.2020.100699

Saha S, Roy J, Pradhan B, Hembram TK (2021) Hybrid ensemble machine learning approaches for landslide susceptibility mapping using different sampling ratios at East Sikkim Himalayan, India. Adv Space Res 68(7):2819–2840. https://doi.org/10.1016/j.asr.2021.05.018

Sánchez-Pujalte L, Mateu DN, Etchezahar E, Gómez Yepes T (2021) Teachers' burnout during covid-19 pandemic in Spain: trait emotional intelligence and socioemotional competencies. Sustainability (Switzerland) 13(13). https://doi.org/10.3390/su13137259

Saritha S, Santhosh Kumar G (2019) Change detection in urban landscapes: a tensor factorization approach. Spat Inf Res 27(5):587–600. https://doi.org/10.1007/s41324-019-00255-3

Schwecke M, Simmons B, Maheshwari B, Ramsay G (2008) Integrating alternative water sources in urbanised environments. WIT Trans Ecol Environ 112:351–359. https://doi.org/10.2495/SI080341

Selvam AP, Al-Humairi SNS (2025) Environmental impact evaluation using smart real-time weather monitoring systems: a systematic review. Innov Infrastruct Solut 10(1). https://doi.org/10.1007/s41062-024-01817-7

Shams Esfandabadi Z, Ravina M, Diana M, Zanetti MC (2020) Conceptualizing environmental effects of carsharing services: a system thinking approach. Sci Total Environ 745. https://doi.org/10.1016/j.scitotenv.2020.141169

Shang Y, Zhou S, Zhuang D, Żywiołek J, Dincer H (2024) The impact of artificial intelligence application on enterprise environmental performance: evidence from microenterprises. Gondwana Res 131:181–195. https://doi.org/10.1016/j.gr.2024.02.012

Sharma R, Mehta K, Yu P (2024) Fostering innovation in venture capital and startup ecosystems. In: Fostering innovation in venture capital and startup ecosystems. IGI Global. https://doi.org/10.4018/979-8-3693-1326-8

Sindhwani R, Kumar R, Behl A, Singh PL, Kumar A, Gupta T (2021) Modelling enablers of efficiency and sustainability of healthcare: a m-TISM approach. Benchmarking 29. https://doi.org/10.1108/BIJ-03-2021-0132

Srivastava S, Pathak D (2020) Impact of emotional intelligence on narcissism-satisfaction with life relationship: a study on Indian managers. Int J Manag Pract 13(2):200–215. https://doi.org/10.1504/ijmp.2020.105667

Tai X, Xiao W, Tang Y (2020) A quantitative assessment of vulnerability using social-economic-natural compound ecosystem framework in coal mining cities. J Clean Prod 258. https://doi.org/10.1016/j.jclepro.2020.120969

Talha M, Nejadhashemi AP, Moller K (2025) Soft computing paradigm for climate change adaptation and mitigation in Iran, Pakistan, and Turkey: a systematic review. Heliyon 11(2). https://doi.org/10.1016/j.heliyon.2025.e41974

Tseng M-L, Lin C, Remen Lin C-W, Wu K-J, Sriphon T (2019) Ecotourism development in Thailand: community participation leads to the value of attractions using linguistic preferences. J Clean Prod 231:1319–1329. https://doi.org/10.1016/j.jclepro.2019.05.305

Viegas CV, Varvakis G, Selig PM (2014) Strategic management accounting (SMA) and green supply chain management (GSCM): advances and challenges for integration in recent academic research. In: Process control: theory, applications and challenges. Nova Science Publishers, Inc. https://www.scopus.com/inward/record.uri?eid=2-s2.0-84959080198&partnerID=40&md5=fbcff811cabe6561c56ee7be19d5659b

Wahab A, Muhammad M, Ullah S, Abdi G, Shah GM, Zaman W, Ayaz A (2024) Agriculture and environmental management through nanotechnology: eco-friendly nanomaterial synthesis for soil-plant systems, food safety, and sustainability. Sci Total Environ 926. https://doi.org/10.1016/j.scitotenv.2024.171862

Wu M, Wu J, Zang C (2021) A comprehensive evaluation of the eco-carrying capacity and green economy in the Guangdong-Hong Kong-Macao Greater Bay Area, China. J Clean Prod 281. https://doi.org/10.1016/j.jclepro.2020.124945

Xianqing X, Xinyi Y, Mei Z, Guokun W (2024) Intelligent manufacturing enabling technologies system construction for high-quality development of furniture. Linye Kexue/Scientia Silvae Sinicae 60(11):177–189. https://doi.org/10.11707/j.1001-7488.LYKX20230427

Xie R, Yu D, Zhang X, Yang Z, Yang J, Ye J (2022) Analysis of regional carbon emission decoupling coupling in China based on ArcGIS analysis-empirical evidence from urban-rural integration in Fujian Province. Front Energy Res 10. https://doi.org/10.3389/fenrg.2022.910565

Xue J, Liu G, Casazza M, Ulgiati S (2018) Development of an urban FEW nexus online analyzer to support urban circular economy strategy planning. Energy 164:475–495. https://doi.org/10.1016/j.energy.2018.08.198

Ye Y, Wang C, Zhang Y, Wu K, Wu Q, Su Y (2018) Low-carbon transportation oriented urban spatial structure: theory, model and case study. Sustainability (Switzerland) 10(1). https://doi.org/10.3390/su10010019

Zhou W, Vredenburgh D, Rogoff EG (2015) Informational diversity and entrepreneurial team performance: moderating effect of shared leadership. Int Entrep Manag J 11(1):39–55. https://doi.org/10.1007/s11365-013-0274-3

Chapter 4
Artificial Intelligence and Machine Learning to Assess the Effects of COVID-19 on Land Surface Temperature (LST) and Air Quality Index

Anugrah Rohini Lall, Ashish Mani, and Deepak Kumar

Abstract This research work is an empirical study that brings out the impact of COVID-19 on land surface temperature and its relationship with environmental quality. The study emphasizes one of the important remote sensing parameters, land surface temperature (LST), to identify the positive effect of the lockdown on the environment due to COVID-19 pandemic in Dehradun city, an interim capital of Uttarakhand. The findings of this study bring out that there is a significant effect on LST during lockdown. The trends analysis of land surface temperature which has been performed for the last 12 years reflects dramatic variations. Data analysis was done by using satellite imagery Landsat 8 OLI and Landsat 5 TM for April in the years 2009, 2019, 2020, and 2021. The study also focuses on the effect of variation in LST on environmental quality and estimates the relationship between land surface temperatures and air quality. To justify the study, air quality data of the Dehradun city from Uttarakhand Pollution Control Board has been verified with LST which shows a positive relation. Correlation is used to find out the relationship between LST and Air quality. The result shows a sudden decrease in air quality with an unexpected decrease in surface temperature in April 2020 due to the lockdown in

A. R. Lall (✉)
Department of Humanities and Social Sciences, Graphic Era Deemed to be University, Dehradun, Uttarakhand, India
e-mail: anugrah.lall@geu.ac.in

A. Mani (✉)
Amity School of Natural Resources & Sustainable Development (ASNRSD), Amity University Uttar Pradesh (AUUP), Noida, Uttar Pradesh, India

D. Kumar
Atmospheric Sciences Group, Department of Geosciences, College of Arts & Sciences, Texas Tech University, Lubbock, Texas, USA

Center of Excellence in Weather & Climate Analytics, Atmospheric Sciences Research Center (ASRC), University at Albany, State University of New York (SUNY), Albany, New York, USA

D. Kumar, N. P. Bassill (eds.), *Artificial Intelligence and Machine Learning for Climate Disaster Management*, Disaster Resilience and Green Growth,
https://doi.org/10.1007/978-981-95-1588-2_4

Dehradun city and again an increase in March 2021 after COVID-19 lockdown. LST range decreased from (26.67 °C min to 42.95 °C max) in April 2009 to (28.39 °C min to 43.35 °C max) in April 2019 then (17.30 °C min to 31.97 °C max) in April 2020 and again increased by (21.25 °C min to 32.68 °C max) in April 2021. Similarly, air quality levels in Dehradun showed a marked improvement during the COVID-19 lockdown period. Data indicate a shift from 'moderate' Air Quality Index (AQI) levels (100–200) in April 2009 and April 2019 to 'satisfactory' levels (51–100) in April 2020. However, post-lockdown observations reveal a return to the earlier, 'moderate' AQI range. This temporary improvement highlights the significant impact of reduced human activities on environmental quality. The findings demonstrate the potential for targeted interventions in improving urban air quality and offer valuable insights for decision-makers involved in urban planning, climate change analysis, and forest fire risk management. By informing more sustainable environmental policies and practices, the study contributes to the broader goals of sustainable development in urban regions.

Keywords GIS · Remote Ssensing · Land surface temperature · Air quality index · COVID- 19

4.1 Introduction

With the Outbreak of COVID-19 in December 2019, mobility of people was restricted worldwide. The pandemic has affected the world so badly that the outbreak was declared a Public Health Emergency of International Concern on 30th January 2020 (WHO 2020) Globally till the 23rd of March, more than 14,000 people have died and more than 334,000 have been infected by COVID-19 (WHO 2020). Therefore, a nationwide lockdown was imposed in India from the 24th of March for 3 weeks up to the 14th of April and later extended up to the 3rd of May (Chen et al. 2020). Furthermore, it is extended till June 30.

Though COVID-19 has brought mankind to a standstill, it has made significant changes in environmental quality. Several major cities have witnessed positive effects on air quality and land surface temperature (LST). The study of LST plays an important role in global climate change studies. The study of surface temperature is important to a range of issues and themes in earth sciences climate changes, global environmental change, and human–environment interactions. This research work tried to study variations in land surface temperatures in Dehradun city. After Dehradun became the interim capital of Uttarakhand, its population, construction, and transportation activity are growing rapidly in the region. This eventually increased pollution and deteriorated the environmental quality of the city. According to a WHO study, Dehradun is the 30th most polluted city. The geographical and meteorological conditions do not permit easy dispersal of pollution.

There have been significant changes in land cover as well as land use after Dehradun became the interim capital of Uttarakhand in 2001. The heat released from urban houses, transportation, and industry is the main cause of Urban Heat

Island (UHI) (Chakraborty et al. 2015). The rising urban temperatures can be partly attributed to the widespread use of air conditioning systems in densely populated city areas. While these systems provide thermal comfort, their extensive use significantly increases energy consumption, often relying on fossil fuel-based electricity. This, in turn, contributes to higher greenhouse gas emissions, exacerbating the urban heat island (UHI) effect and further elevating ambient temperatures. The feedback loop between increased cooling demand and rising urban temperatures presents a critical challenge for sustainable urban development and energy policy. The establishment of power plants and the burning of fossil fuels is also affecting energy consumption and pollution (Weng and Yang 2004). The intensity of UHI is determined by the interaction of several factors, including the degree of urbanization (total built-up area, population size, industrial development, and transportation), the geometry of a city, the physical environment, and human activities (Yang et al. 1984). Whether a city is a metropolis or a county seat, the UHI effect is linked to the composition of the underlying surface, i.e. the pattern and structure of land use and land cover. Urban development has often intensified the degree of UHI (Weng and Yang 2004). Some researchers have found that natural and socio-economic factors simultaneously create certain effects on LST patterns (Buyantuyev and Wu 2010; Jenerette et al. 2007; Kuang et al. 2015).

But after the lockdown was imposed on 24th March 2020, there have been significant positive effects noticed in the environment. It is evident through the study that there are beneficial impacts of lockdown due to COVID-19 in terms of global restoration of air quality and temperature. (Anderson et al. 2020). Air quality on the other hand affects land surface temperature. Major sources of NO_3 (NO_3 = NO+NO_2) include fossil fuel combustion, mediated biomass burning, industrial and transportation emissions, and lightning (Krotkov et al. 2016; Xue et al. 2020). NO_2 affects atmospheric oxidation rates participating in the formation of surface ozone (Krotkov et al. 2016; Meng et al. 2008). It is also found through some studies that there is a correlation. Feizizadeh and Blaschke (2013) found a correlation between land cover, land surface temperature, and air pollutants in the case of Tabriz city in Iran. It is also found that traffic and the increasing number of private cars due to urbanization have a significant effect on air quality. In recent decades, however, the increasing traffic has switched the attention to nitrogen oxides, organic compounds, and small particles. (Fenger 1999). The study reveals that the worst polluted traffic circle doubles the city's average pollutants during busy hours (Bhaumik 2007). Thus, it is found through various studies that there is a rise in surface temperature due to air pollution and rapid urbanization. The research papers state that land surface temperature is affected by air pollution. Therefore, the present study attempts to bring out the changes in the LST of the city before and after the lockdown. Besides, the research will find out the significant relationship between pollution level through (AQI) and LST in the city. The study will help the decisions/policymakers of various organizations in urban planning, climate change analysis, and forest fire risk management study for the growth of sustainable development. Also, it will help the government to understand the necessary action to meet the environmental challenge, which is of great concern if there is no lockdown.

4.2 Material and Methods

4.2.1 Study Area

The present study region Dehradun City is the capital of Uttarakhand state and lies between 30° 10′ and 30° 24′ north latitude and 78° 10′ and 78° 54′ east longitudes (Fig. 4.1). The elevation is between 540 m and 1048 m with the total area of about 183.70 km^2 (Mani et al. 2023a, b, c). Dehradun lies at the foot slopes of the Himalayas. It is encircled by stream Song on the east, river Asan on the west, Mussoorie on the north, and Shivalik range in the south (Mani et al. 2023a, b, c). In the late spring months, the temperature ranges from 44 °C to 23 °C.

In winter the maximum and minimum temperatures range from 23.4 °C to 5.2 °C. Dehradun encounters heavy to mild showers from June to mid-August. The greater part of the yearly precipitation (roughly 2000 mm) is obtained during the months from June to September, July and August being the rainiest months of the period. The Dehradun city is known for its schools. It is also called a hub for education. Furthermore, Dehradun city is the trading and business place for the whole hill entity of the Uttarakhand state. Amidst the hills, Dehradun Valley has always been a tourist attraction. Because of its peaceful and serene environment, it is a gateway for tourists

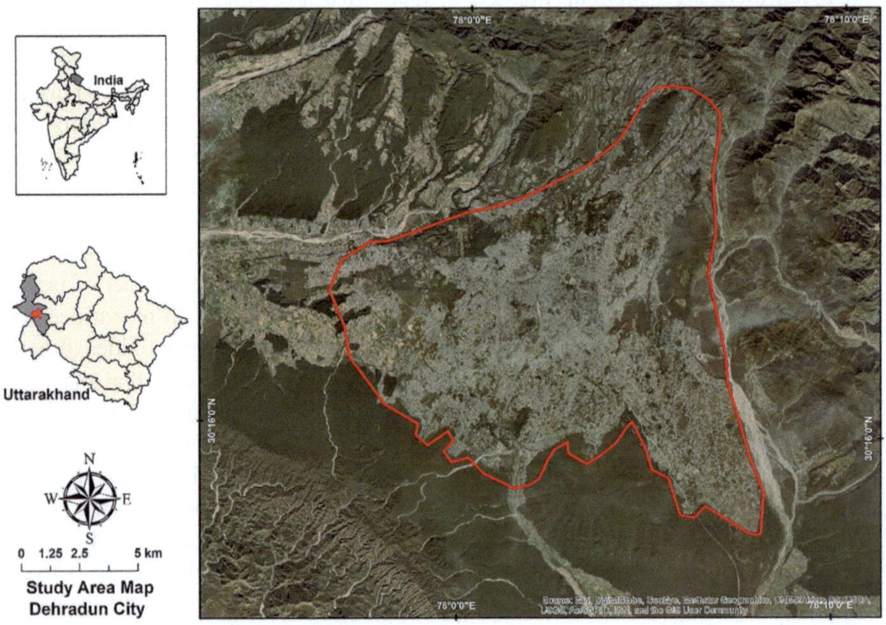

Fig. 4.1 Study area map of Dehradun city

who visit Mussoorie and Dhanaulti, Rishikesh, or Haridwar. After Uttarakhand came into existence on November ninth, 2000, as the 27th State of the Indian Union, Dehradun has become the interim capital of Uttarakhand. There have been explicit changes in the environment and economic activities in Dehradun. There are significant effects on the land surface temperature of Dehradun (Mani et al. 2023a, b, c). After the outbreak of the COVID-19 lockdown, LST and air quality have shown decline.

4.2.2 Methods

The present study is based on data collected through GIS. Satellite imagery of Landsat 8 (OLI-Operational Land Imager) for years 2009, 2019, and 2020 is of April and 2021 imagery is of March. These imageries have been downloaded from the source USGS Earth Explorer (https://earthexplorer.usgs.gov/). For image processing and preparing maps, Arc GIS Software (10.3) has been used. To find out the effect of lockdown on environmental quality, LST for 4 years is taken. Besides, the Air Quality Index is also taken to calculate its relationship with LST.

4.2.2.1 LST

LST implies the radiative skin temperature of the land surface (Mani et al. 2021). It is a measure of how hot or cold the Earth's surface would feel to the touch. In this study, LST values of four different years of data have been calculated using a spatial analysis tool in Arc GIS software. A thermal band for the Landsat satellite has been used. For Landsat 5 TM thermal band is the 6th band and for Landsat 8 OLI thermal bands are the 10th band & 11th band. The maps have been prepared in Arc GIS Software.

The Steps Used for LST Calculation of Landsat 8 OLI

- Convert (DN) Digital Numbers to (TOA) Top of Atmospheric(L).

$$\text{TOA(L)} = \text{ML}^{*}\text{Qcal} + \text{AL}$$

where

TOA (L) represents Spectral Radiance.
ML represents the Radiance of MultiBand.
Qcal represents Band 10 or Band 11 from downloaded satellite imagery.
AL represents the Radiance of AddBand

- Convert TOA to (BT) Brightness Temperature of satellite.

$$BT = (K2/(\ln(K1/L) + 1)) - 273.15$$

where

BT represents the Brightness Temperature of Satellite.
K_1 represents the Band-specific thermal conversion constant from the metadata or MTL file (K_1_CONSTANT_BAND).
K_2 represents Band-specific thermal conversion constant from the metadata or MTL file (K_2_CONSTANT_BAND).
L represents Spectral Radiance

- Calculate the NDVI.

$$\text{NDVI} = (\text{NIR Band–RED Band})/(\text{NIR Band} + \text{RED Band})$$

where

NDVI represents Normalized Differential Vegetation Index.
NIR Band = 5th Band & RED Band = 4th Band.

- Formula of (Pv) proportion of vegetation.

$$\text{Pv} = \text{Square}((\text{NDVI–NDVImin})/(\text{NDVImax–NDVImin}))$$

- Calculate emissivity (ε).

$$\varepsilon = 0.004^{*}\text{Pv} + 0.986$$

- Finally, calculate the LST.

$$\text{LST} = (\text{BT}/(1 + (\text{Qcal}^{*}\text{BT}/1.4388) * \text{Ln}(\varepsilon)))$$

The Steps Used for LST Calculation of Landsat 5 TM

The calculations for LST of Landsat 5 and Landsat 8 OLI are almost similar except for the first step, which is given below

- Convert (DN) Digital Numbers to (L_λ) Spectral Radiance.

$$L_\lambda = L_{\min} + (L_{\max} – L_{\min}) * \text{DN}/255$$

where

L_λ represents Spectral radiance,
L_{min} represents the Spectral radiance of DN value 1.
L_{max} represents the Spectral radiance of DN value 255.
DN represents Digital Number.

Finally, after all the values of LST have been calculated, the maps have been prepared for the years 2009, 2019, 2020, and 2021. The decade difference is shown to find out the changes that have taken placc from 2009 to 2020.

The comparison between 2019 and 2020 provides valuable insights into changes in Land Surface Temperature (LST) leading up to the COVID-19 lockdown, capturing the thermal conditions immediately before widespread restrictions were implemented. Furthermore, the LST maps for 2020 and 2021 offer a clear visual representation of pre-lockdown and post-lockdown scenarios. These temporal comparisons highlight the significant impact of reduced human activities during the lockdown period and the subsequent rebound in urban thermal patterns, thereby emphasizing the influence of anthropogenic factors on urban heat dynamics.

The study reveals the changes in LST in the given year and the decline during lockdown which led to lower pollution levels. The pollution level is studied from AQI collected from the Uttarakhand Pollution Control Board.

4.2.2.2 Air Quality Index

An air quality index (AQI) is used by government organizations to communicate to their people the level of pollution in the air. In India, Central Pollution Control Board, a central government agency, has broken the AQI into different categories as shown in Fig. 4.2.

To justify the current study, the air quality data has been downloaded from the Uttarakhand Pollution Control Board and the AQI maps representing different air pollutants PM_{10}, $PM_{2.5}$, SO_2 and NO_2 values over a different period. These values have been calculated from three air quality stations in Dehradun city:

AIR QUALITY INDEX (AQI)	CATEGORY
0-50	Good
51-100	Satisfactory
101-200	Moderate
201-300	Poor
301-400	Very Poor
401-500	Severe

Fig. 4.2 Indian air quality index and its categories. (Source: Central Pollution Control Board)

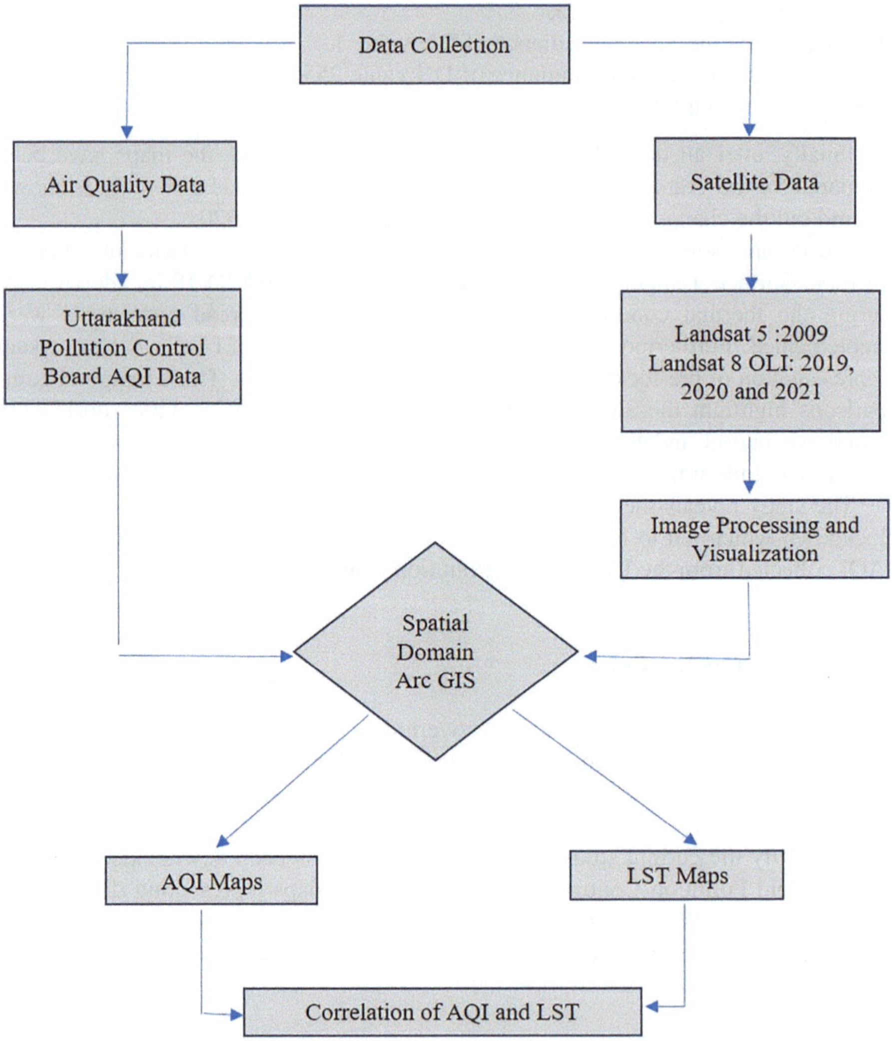

Fig. 4.3 Flow chart of the methodology

(a) Clock Tower, Dehradun—Commercial.
(b) Near Parag Dairy, Raipur Road, Dehradun–Commercial/Residential.
(c) Himalaya Drug Company—Commercial Industrial.

The flow chart of the fmethodology is shown in Fig. 4.3.

4.2.3 Objectives of the Study

Hence, the study conducted has the following objectives:

(a) To find out the effect of the lockdown on LST of Dehradun City.
(b) To find out the effect of variation in LST on the air quality of the city.
(c) To find out the relationship between land surface temperatures and air quality.
(d) To help the decision/policymaker to achieve sustainable development.

4.2.4 Hypothesis

H_0 There is no significant relationship between LST and Air Quality.
H_1 There is a significant relationship between LST and Air Quality.
H_0 There is a significant effect of lockdown on LST.
H_1 There is a significant effect of lockdown on LST.

4.3 Result and Discussion

4.3.1 LST

The present study is conducted in Dehradun city of Uttarakhand to find out the effect of lockdown on LST. Hence, the land surface temperature map for four different years has been shown in Fig. 4.4a–d.

The data has been collected for 4 years i.e. 2009, 2019, 2020, and 2021. In the LST maps, there is an increase in surface temperature, which is from 26.67 °C min–42.95 °C max in April 2009 to 28.39 °C min–43.35 °C max in April, 2019, and during the COVID-19 lockdown, there is a sudden drop of temperature (17.30 °C min–31.97 °C max) in April 2020 and again in the following year 2021 the land surface temperature rises (21.25 °C min–32.68 °C max) in March 2021. This reflects the impact of the lockdown on LST. The high density of residential, transportation, and commercial activities is the reason for the high LST (Sahani et al. 2021). These temporal changes took place due to a decline in transportation activities during lockdown.

The trend analysis shown in Fig. 4.5 reveals a gradual increase in Land Surface Temperature (LST) in the city from 2009 to 2019, indicating a consistent warming trend likely driven by rapid urbanisation, increased built-up areas, and reduced vegetation cover. However, a notable deviation from this trend was observed during the COVID-19 lockdown in 2020, when there was a significant decline in LST. This temporary reduction is attributed to decreased human activities, reduced vehicular emissions, and lower industrial output, highlighting the strong influence of anthropogenic factors on urban thermal environments. However, during the lockdown in

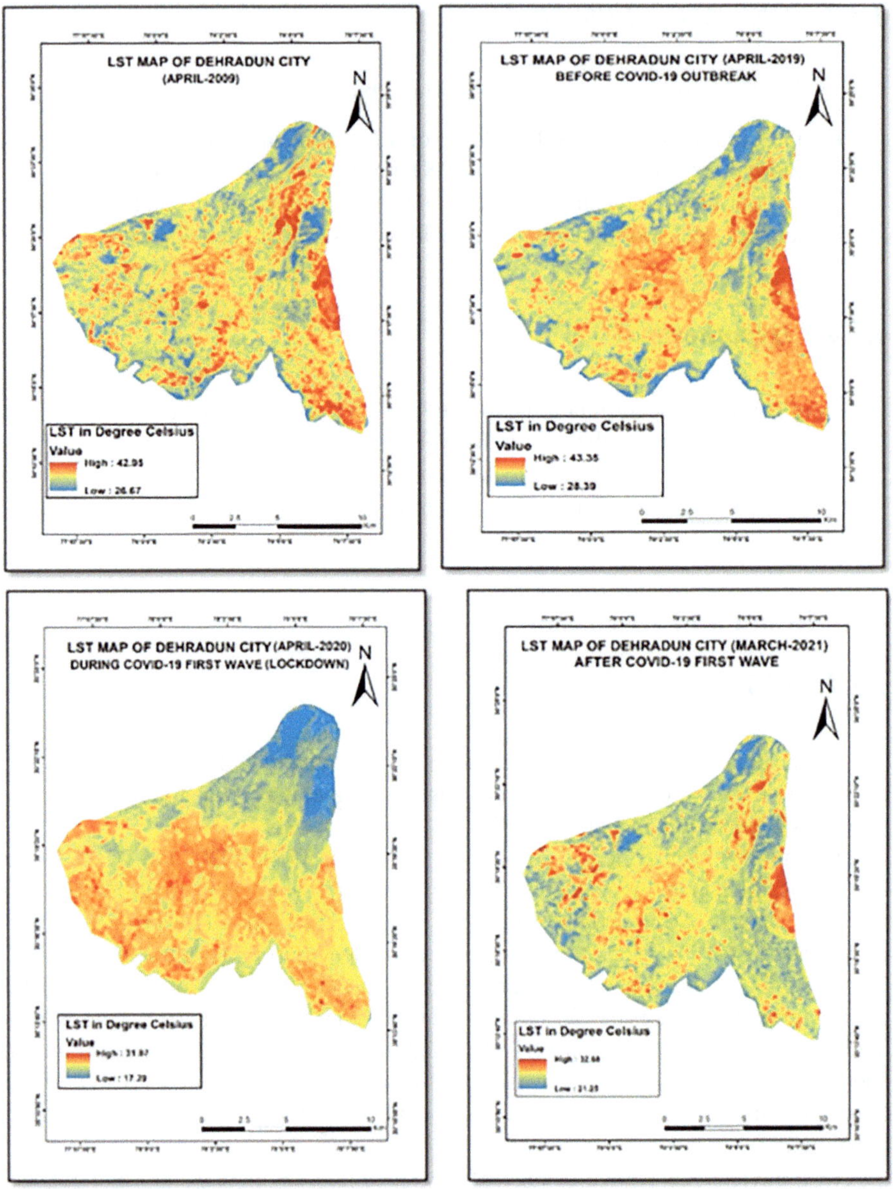

Fig. 4.4 (**a**) LST map of Dehradun in 2009 (**b**) LST map of Dehradun before COVID-19, (**c**) LST map of Dehradun during first wave of COVID-19 and (**d**) LST map of Dehradun after first wave of COVID-19

Fig. 4.5 LST trend over the years in Dehradun City before, during, and after COVID-19

the year 2020, there was a significant drop in LST. However, there is again a rise in LST post-lockdown. With the increase in commercial activities in 2021, the LST started rising. This also reflects that LST needs to be moderated with increased economic activities.

4.3.2 Air Quality Index

The Air Quality Index Map for four different years has been shown in Fig. 4.6a–d.

Major sources of NO_3 (NO_3 = NO+NO_2) include fossil fuel combustion, mediated biomass burning, industrial and transportation emissions, and lightning (Krotkov et al. 2016; Xue et al. 2020). There has been a dramatic downfall in pollution levels. Due to the lockdown, it can be observed that NO2 declined significantly. Similarly, SO2 had declined. Consequently, significant variations in air pollutant concentrations, specifically PM_{10}, $PM_{2.5}$, SO_2, and NO_2 are clearly depicted in the Air Quality Index (AQI) maps across different time periods. These variations indicate a substantial reduction in pollutant levels during the COVID-19 lockdown, leading to a noticeable improvement in overall environmental quality. The sharp decline in emissions can be attributed to the temporary halt of industrial activities, reduced vehicular traffic, and limited construction work. This period serves as a critical reference point, illustrating the extent to which anthropogenic activities contribute to urban air pollution and emphasising the potential benefits of implementing long-term sustainable environmental policies. The AQI IS calculated using https://www.airnow.gov/aqi/aqi-calculator-concentration. The data shows that AQI in 2019 was 345, in 2020 it was 209, and in 2021 it again rose to 367.

Figure 4.7 shows a graph representing AQI trends over the years; the graph and maps show concentrations of PM_{10}, $PM_{2.5}$, SO_2, and NO_2.

The impact of LST implies that COVID-19 pandemic has a positive effect on the environment. However, the pandemic drastically affects the human being but it manages to bring the environment into a stable condition. Fig. 4.7 shows that

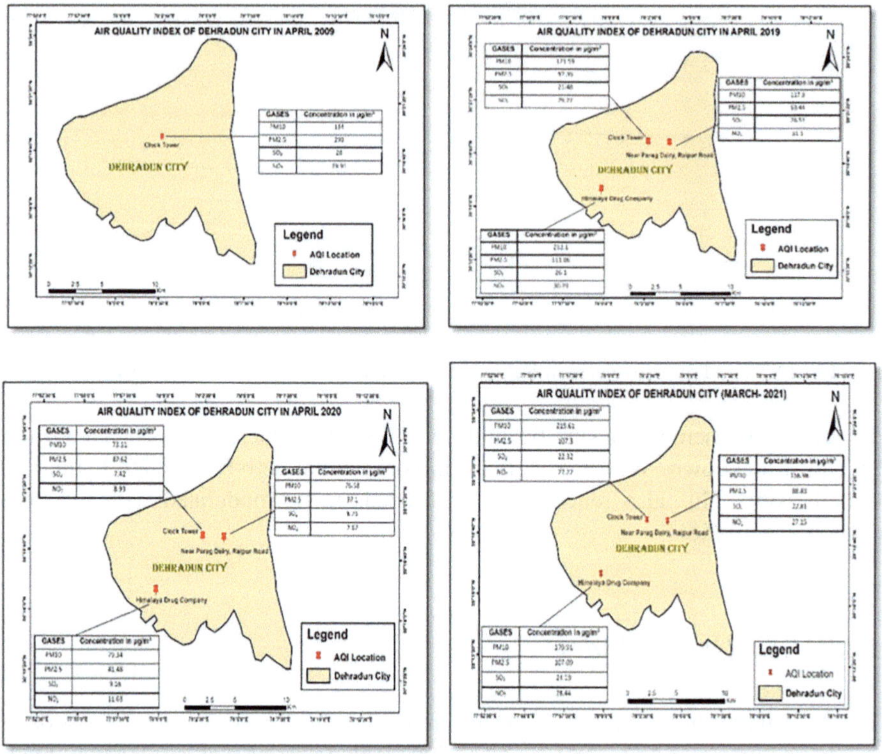

Fig. 4.6 (**a**) Air Quality Map of Dehradun in 2009 (**b**) Air Quality Map of Dehradun before COVID-19, (**c**) Air Quality Map of Dehradun During First Wave COVID-19 and (**d**) Air Quality Map of Dehradun after First Wave of COVID-19

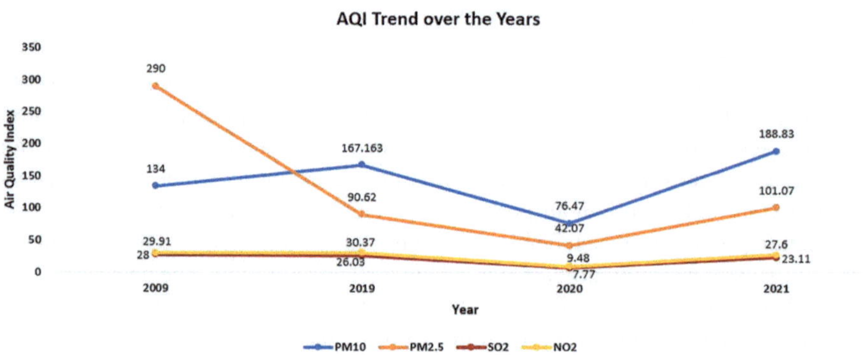

Fig. 4.7 AQI trend over the years of Dehradun City before, during, and after COVID-19 19

there is a significant decline in AQI $PM_{2.5}$ as well as PM_{10} from 2019 to 2020. SO_2 and NO_2 have also shown a drastic decline from 2019 to 2020.

Furthermore, The LST and AQI relation is justifying our map's result. The trend shows the conclusive result that LST and AQI are positively correlated.

4.3.3 Correlation

4.3.3.1 Correlation of AQI and LST

The data collected is calculated through EXCEL CORREL Formula to find out the relationship between AQI and LST for 3 years i.e. 2019, 2020, and 2021. The value estimated is 0.56, which reflects that the two variables are strongly correlated. Here we assume that the significant p-value is <0.05. Hence, the Null Hypothesis is rejected and results reflect that there is a positive relationship between LST and AQI. If there is an increase in LST, it will give rise to AQI. This indicates that when the surface temperature increases, the air quality of the particular surrounding is polluted, and on the other hand if the surface temperature decreases, air quality will be clean. Since there were low economic activities as well as traffic during the lockdown, LST decreased eventually air quality index also declined.

4.4 Conclusion

Although COVID-19, a newly emerged dangerous infectious disease, poses some serious threat to humans and even has affected economic activities to a great extent, but has also provided an opportunity to heal our planet and thus be considered a 'Blessing in Disguise'. The study reveals that environmental quality has significantly improved during lockdown. Our study has found a significant decline in land surface temperature along with positive impacts on the air quality index of the region during lockdown. This positive impact on the environment may be short but has provided us the opportunity to learn to reduce our carbon footprints along with reducing pollution levels in the long term which is crucial for sustainable development. Due to the lockdown, anthropogenic activities have already been controlled which has proved to be beneficial for both our environment and biodiversity. This study brings out the changes in LST, which is an important factor in climate change. Policymakers and environmentalists should work on policies to maintain LST. The adverse effect of climate change is evident therefore reduction in carbon emissions and pollution control is required. Sustainable practices are needed to be adopted to reduce LST. Various sustainable practices should be identified and employed at all levels. Higher management authorities and policymakers should promote such activities and recognize the efforts of citizens and employers. A behavioural approach should be encouraged. The mindset needs to be changed so that people adopt a sustainable approach voluntarily and with responsibility. Awareness campaigns should be organized and conducted to promote sustainability at all levels. Discouraging usage of vehicles is required so that less number of vehicles should be on roads. Individuals alone cannot change the pattern but collective action is needed to address the issue of climate change. The lockdown has proved that there is a trade-off between environmental quality and economic activity. The economic activities were hampered significantly but environmental quality has shown improvement in terms of air quality and LST. The decline in economic activities has affected economic growth subsequently and the country cannot put pause to the activities. Hence, we need to understand that economic development should not deteriorate environmental quality. So economic tools are required to address the environmental issues. Policymakers should adopt a sustainable approach and devise policies to meet these negative externalities. The present study brings out the differences in air quality. Concerned agencies should devise schemes or plans to control the threat of increasing pollution levels after the lockdown is lifted. Dehradun is a beautiful city known for its tourism but climate change is detrimental to the development of the city and the state as a whole. Some examples prove how destructive could be the deterioration of the environment. Increasing LST and air quality can be damaging for the city in all ways. An appropriate action plan is needed to be prepared to protect the environment of Dehradun and to meet the demands of the increasing population and footfalls in the city. The current study would help the policymakers to identify the problem and to bring out appropriate solutions.

References

Anderson RM, Heesterbeek H, Klinkenberg D, Hollingsworth TD (2020) How will country-based mitigation measures influence the course of the COVID-19 epidemic? Lancet 395:931–934

Bhaumik S (2007) Air pollution suffocates Calcutta. BBC News. Available online: https://news.bbc.co.uk/2/hi/south Asia/ 6614561.stm. Accessed on 3 May 2007

Buyantuyev A, Wu J (2010) Urban heat islands and landscape heterogeneity: linking spatiotemporal variations in surface temperatures to land-cover and socioeconomic patterns. Landsc Ecol 25:17–33

Chakraborty SD, Kant Y, Mitra D (2015) Assessment of land surface temperature and heat uxes over Delhi using remote sensing data. J Environ Manag 148:143–152

Chen B, Liang H, Yuan X, Hu Y, Xu M, Zhao Y et al (2020) Roles of meteorological conditions in COVID-19 transmission on a worldwide scale. medRxiv 2020.

Feizizadeh B, Blaschke T (2013) Examining urban Heat Island relations to land use and air pollution: multiple end member spectral mixture analysis for thermal remote sensing. IEEE J Selected Top 6(3):1749–1756. https://doi.org/10.1109/jstars.2013.2263425

Fenger J (1999) Urban air quality. Atmos Environ 33(29):4877–4900. https://doi.org/10.1016/S1352-2310(99)00290-3

Jenerette GD et al (2007) Regional relationships between surface temperature, vegetation, and human settlement in a rapidly urbanizing ecosystem. Landsc Ecol 22:353–365

Krotkov NA et al (2016) Aura OMI observations of regional SO2 and NO2 pollution changes from 2005 to 2015. Atmos Chem Phys 16(7):4605–4629. https://doi.org/10.5194/acp-16-4605-2016

Kuang W, Liu Y, Dou Y, Chi W, Chen G, Gao C et al (2015) What are hot and what are not in an urban landscape: quantifying and explaining the land surface temperature pattern in Beijing, China. Landsc Ecol 30:357–373

Mani A, Kumar D, Kumar D (2021) Evaluating decadal change in green cover of Dehradun City. In: Sharma P (ed) Geospatial technology and smart cities. The urban book series. Springer, Cham. https://doi.org/10.1007/978-3-030-71945-6_26

Mani A, Bansal D et al (2023a) Assessment of COVID-19 impact on the vegetation and urbanisation of Dehradun City using geospatial techniques. In: Geoinformatics for sustainable urban development. Taylor & Francis Group, CRC Press, p 22

Mani A, Bansal D, Kumari M, Kumar D (2023b) Land use land cover changes and climate change impact on the water resources: a study of Uttarakhand state. In: River conservation and water resource management. Springer Nature Singapore, Singapore, pp 1–16

Mani A, Kumari M, Badola R (2023c) Evaluating urban topography and land use changes for Urban River management using geospatial techniques. Eng Proc 58:12. https://doi.org/10.3390/ecsa-10-16004

Meng ZY et al (2008) Vertical distributions of SO2 and NO2 in the lower atmosphere in Beijing urban areas, China. Sci Total Environ 390(2):456–465. https://doi.org/10.1016/j.scitotenv.2007.10.012

Sahani N, Goswami SK, Saha A (2021) The impact of COVID-19 induced lockdown on the changes of air quality and land surface temperature in Kolkata city, India. Spat Inf Res 29(4): 519–534

Weng Q, Yang S (2004) Managing the adverse thermal effects of urban development in a densely populated Chinese city. J Environ Manag 70:145–156

World Health Organisation (2020) Summary table of SARS cases by country. 1 Nov 2002–2020. country2003_08_15.pdf

Chapter 5
Trend Analysis of Long-Term Temperature Data for Prediction of Heat Waves Through Statistical Analysis Using Extreme Value Theory for Climate Disaster Management

Sagar Tomar and Rakesh Kadaverugu

Abstract The term "heat wave" refers to long periods of unusually hot weather. Evidence shows that heat waves happen more frequently than any other catastrophic event. It has profound effects on human productivity, industrial development, agriculture yield, environment quality, natural resources, and importantly on health and biodiversity. And heatwaves can potentially affect the national economy. Central India is more prone to heat waves and has adverse effects. With rising global warming, it is expected that heat waves will become more intense, frequent, and prolonged. Heat waves may be harmful, especially for more susceptible individuals, such as the elderly, children, and people who have specific medical issues. As a result, India experiences more heat waves throughout the summer (April to June), more often, and for longer periods overall. Future heatwaves are expected to occur more frequently, endure longer, and be stronger in almost all inhabited places. The proposed statistical framework applies the extreme value theory (EVT) and allows for the estimation of the probability of uncommon occurrences that have not yet been seen and have not been recorded in the past. The present chapter discusses the possibility of determining the locations and chances of future heat waves of varying sizes. The result shows that the return levels of extreme temperature (or heat wave) are 47.96 °C and 44.59 °C for Nagpur and Hyderabad for a return period of 10 years. The frequency and severity of severe temperatures are shown by the return values and agreement between the observed data and the predicted distributions. The results help in predictive maintenance to mitigate and adapt to extremely hot conditions and also provide support in the near real-time prediction of heat events, thus helping in resource optimization and providing better decision-making to achieve sustainable developmental goals.

S. Tomar · R. Kadaverugu (✉)
CSIR- National Environmental Engineering Research Institute, Nagpur, Maharashtra, India

Academy of Scientific and Innovative Research (AcSIR), Ghaziabad, Uttar Pradesh, India
e-mail: r_kadaverugu@neeri.res.in

D. Kumar, N. P. Bassill (eds.), *Artificial Intelligence and Machine Learning for Climate Disaster Management*, Disaster Resilience and Green Growth,
https://doi.org/10.1007/978-981-95-1588-2_5

Keywords Heat waves · Extreme value theory · Temperature · Statistical analysis · Decision-making · Resource optimization · Sustainability

5.1 Introduction

Extreme weather conditions are unanticipated, uncommon, severe, or out-of-season weather occurrences that fall beyond the historical range for an area or region. Heat waves are extended periods of exceptionally high temperatures that can have a big effect on the environment and civilization (Van Der Wiel et al. 2020). Although they can happen at any time of the year, in areas with pronounced seasonality, they are more frequently related to the summertime. Although there isn't a single definition for what constitutes a heat wave, they are commonly distinguished by a minimum severity (temperatures over a particular threshold) and length (often 3 days or more in a row) (Barriopedro et al. 2023).

Heat waves have wide-ranging effects on society and can have serious repercussions, especially for weaker members of society or groups. Elderly people, those without access to air conditioning, and people who work outside are more likely to get sick or die from heat-related conditions (Adams n.d.). Heat waves can cause dehydration, heat exhaustion, and heatstroke as well as worsen pre-existing medical illnesses such as respiratory and cardiovascular diseases. Heat waves have the potential to severely strain medical infrastructure and increase mortality (Murari et al. 2015). Additionally, heat waves have a big influence on the economy. Especially in industries like agriculture and construction, heat waves frequently result in decreased worker productivity (Li and Amatus 2020). Because crops and cattle can be stressed beyond their capacity to withstand heat and drought, the agriculture sector, which supports a sizable portion of the global workforce, is particularly sensitive to the effects of heat waves. As a result, harvests are reduced and consumer prices are increased. This may lead to lost labour and adverse effects on livelihoods. Heat waves can also interfere with energy infrastructure, resulting in power outages and higher cooling demand (Burillo et al. 2017). To better understand the features, effects, and drivers of heat waves, it is important to conduct research at several regional and temporal dimensions. Studying heat waves at the local level, such as in metropolitan areas, can assist in identifying mitigation methods for their effects on infrastructure, infrastructure demand, labour productivity, and health (Zschenderlein et al. 2020; Dhole et al. 2023).

Climate models are useful tools for predicting changes in heat waves in the future (Diffenbaugh and Scherer 2011). The majority of land regions are expected to experience longer, more intense, widespread, and more frequent heat waves as a result of continued global warming (Nori-Sarma et al. 2019). These heat waves would surpass previous record-breaking occurrences (Zhang et al. 2020). Although expected increases in heat waves vary by geography and other factors, they typically rise linearly as global warming progresses (Wang et al. 2017). The risk of hot days occurring just once per several decades can greatly rise with even a minor increase in global temperature. For preparedness and adaptation measures, accurate heat wave prediction is important because it allows for risk assessment, early warning systems, health protection, infrastructure planning, urban design, and effect reduction on the economy and society (Zhang et al. 2014a). Authorities may better plan for and respond to heat waves, minimizing their detrimental impacts on populations, by taking into account climatic mechanisms, monitoring and forecasting, vulnerability assessment, and adaptation methods (Knowlton et al. 2009).

Two methods may be used to characterize extreme weather events: the impact-related threshold approach, which employs thresholds to evaluate if an event is extreme, and the probabilistic approach, which looks at the likelihood of an occurrence of a particular size occurring during a given reference period (USDA n.d.). An analytical and predictive statistical approach is called extreme value theory (Pandey 2001). Extreme Value Theory (EVT) assists in comprehending the likelihood and severity of high heat episodes while analysing heat wave data (Furrer et al. 2010). The chance of severe temperatures may be calculated, and their possible effects can be evaluated, using the GEV distribution and other models (Sillmann et al. 2011). When analysing heat wave data within the framework of climate change, it is essential to take into account the assumptions of stationarity and investigate non-stationary techniques (Dreesman and Tutz 2001).

Regression analysis is a statistical approach that is used in statistical models to study the link between meteorological variables and the incidence of heat waves (Arismendi et al. 2014). The Auto-Regressive Integrated Moving Average (ARIMA) model, which is frequently used for time series forecasting, is an illustration of a statistical model (Lal et al. 2012). To learn trends and make forecasts, machine learning approaches require training models using past weather data. As a result of these models' ability to handle huge and complicated information, they can capture more intricate correlations between meteorological variables and heat waves (Sandeep and Prasad 2018). Decision trees, random forests, and neural networks are a few popular machine-learning techniques for forecasting the weather (Iyengar 1999; Thibault and Grandjean 1991; Zhang et al. 2014a, b). It is recognized that certain characteristics or patterns in the weather might predict the development of heat waves. These indicators can be included as predictors of statistical or machine-learning models to improve the prediction power of those models (Mulder et al. 2012).

The most effective technique is determined by the precise requirements and information available. It is possible to make better predictions of upcoming heat waves by integrating different approaches and taking into consideration a range of factors. Conventional heat wave prediction methods, such as statistical models and machine learning techniques, have their drawbacks and difficulties such as limited accuracy and lack of understanding of underlying physical processes (Li et al. 2020). The correlations between meteorological indicators and heat waves are frequently assumed to be linear and stationary by conventional approaches (Chen and Ramachandra Rao 2003). Extreme heat episodes, however, can display non-stationary and non-linear behaviour, making it difficult to effectively record and predict them using these techniques (Buzan et al. 2015). Deep neural networks are one example of a machine learning technique used to forecast heat waves that can be computationally expensive and demand a lot of computing power (Benalcazar and Kamiński 2019). This may reduce their usefulness for operational usage and real-time prediction.

The sequential Mann-Kendall test (Seq-MK) statistic was used in this chapter to evaluate any qualitative changes in the trend of temperature with alternative methods of predicting heat waves are required, EVT is one such strategy. EVT is a statistical theory that concentrates on the probability of extreme occurrences (Roberts 2000). By putting a lot of emphasis on the extreme tails of the distribution, it offers a framework for modelling and forecasting extreme phenomena, including heat waves (Lee et al. 2013). Extreme events' occurrence in the tail of the distribution is explicitly modelled by EVT. This makes it possible to estimate the likelihood of rare events like heat waves with greater accuracy. EVT is more flexible and adaptive to various data distributions since it does not rely on certain distributional assumptions. This is especially helpful for severe occurrences, which might not adhere to conventional distributional assumptions. A more specialized and narrowly focused investigation of heat waves is possible because of the unique framework for analysing severe occurrences provided by EVT. To capture the persistence and clustering of heat waves throughout time, EVT can add temporal dependency in severe event modelling. EVT is more efficient than traditional approaches in handling small amounts of data (Xiaoyi Zhou et al. 2012). It concentrates on the distribution's extreme tails, where there are few data points, and offers reliable predictions even with limited observations (Neftci 2000).

5.2 Data and Methodology

5.2.1 Study Area

Nagpur is the third-largest city in the central Indian state of Maharashtra. Geographically, Nagpur is located in the Nagpur district in the country of India's geographic centre. It is bounded by the Satpura Range in the north and the Balaghat Range in the south and is located on the Deccan Plateau. Smart city in central India, Nagpur has experienced a drop in plant cover over time. Between the latitudes of 21.05 and 21.23 N and the longitudes of 79.00 and 79.18 E, the city spans a 222.8 km^2 area. The climate of Nagpur is tropical savannah, with scorching summers, moderate winters, and a wide range of seasonal temperatures. The area is 308 m above mean sea level and is located in a tropical dry climate zone with temperatures ranging from 5 to 47 °C throughout the year (Kadaverugu et al. 2021) (Fig. 5.1).

Geographically, Hyderabad is located along the Musi River's banks in the Deccan Plateau. Its average elevation above sea level is about 536 m. The Eastern Ghats in the east and the fertile Telangana Plateau in the west surround the city, which is located in central-southern India (Nalla n.d.). In a tropical savanna climate, Hyderabad has hot, humid summers, a monsoon season, and mild winters. Hyderabad has a hot, dry summer from March to June with frequent highs of 40 °C (104 °

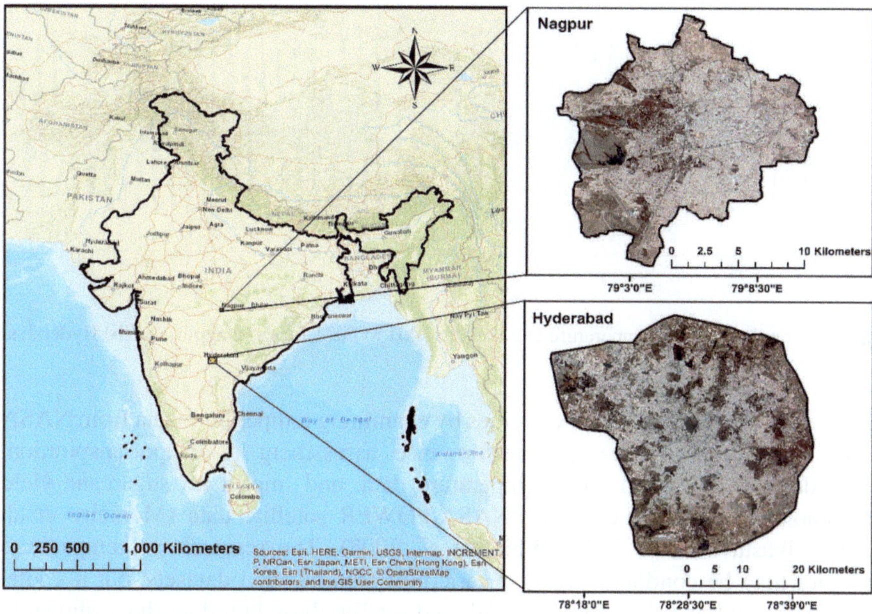

Fig. 5.1 Location map of Nagpur and Hyderabad (Central India's cities)

F). The city experiences a typical dry and rainy climate on top of a hot, semi-desert environment. The annual average precipitation is around 725.8 mm. The annual average temperature is 26.6 °C. May is the hottest month, with temperatures ranging from 15 to 40 °C.

Due to the urban heat island effect, temperatures in urban areas are often higher than in rural ones (Liao et al. 2017). Due to their huge density, asphalt, concrete, and buildings both absorb and radiate heat, causing this effect (Kim 1992; Kadaverugu et al. 2022). Urban heat islands and the efficacy of heat mitigation measures used in such regions can be better understood by studying heat waves in urban areas like Nagpur and Hyderabad. The combined study of Hyderabad and Nagpur helps to find potential heatwave vulnerability, economic importance, and rapid urban expansion. Insights into the difficulties of urbanization and the efficiency of mitigation techniques in reducing the effects of heatwaves on infrastructure, the environment, and human health may be gained by monitoring heatwaves in these fast-growing cities.

5.2.2 *Data Required*

The importance of NASA- Prediction of Worldwide Energy Resource (POWER) data for the temperature parameter in extreme value theory is its capacity to deliver useful data on temperature extremes. The study may estimate the probability distribution of severe temperature events and learn more about the possible effects of

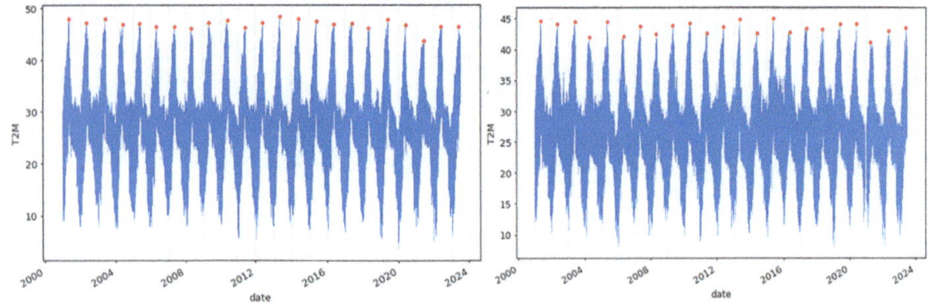

Fig. 5.2 Trend of hourly temperature at 2 m from 2001 to 2021 for (**a**) Nagpur and (**b**) Hyderabad

climate change on temperature extremes by examining temperature data from NASA POWER. Various studies like performance assessment of evapotranspiration, model-derived long-term daily temperature data, and impact on sugarcane yield simulation were performed using NASA/POWER satellite data (Monteiro et al. 2018; Srivastava et al. 2020; White et al. 2008). The research on temperature extremes may be conducted using a variety of methods and datasets, but it's vital to estimate extreme value. These model- and satellite-based products have shown to be precise enough to deliver reliable solar and meteorological resource data over areas with a lack or absence of surface measurements. The extreme value theory was used to determine the frequency of heat waves using the average air (dry bulb) temperature at 2 m (T2M) above the earth's surface. NASA power temperature data at 2 m are used to make heat wave predictions for the cities of Nagpur and Hyderabad from 2001 to 2021(hourly) using the extreme value theory as shown in Fig. 5.2.

5.2.3 EVT- Block Maxima Approach

Extreme deviations from the median of probability distributions are the focus of the statistical field known as EVT (Makatjane 2022). It aims to determine the probability of events that are more severe compared to those that have already been witnessed.

A method used in extreme value theory called "Block Maxima" focuses on modelling the highest values found within non-overlapping pieces of data (Ghorbel and Trabelsi 2008). This approach assumes that extreme occurrences happen in groups or blocks and seeks to determine how the maximum values are distributed throughout these groups (Dombry 2013). The dataset is separated into non-overlapping, fixed-size blocks. While assuming asymptotic independence between blocks, the block size is set to capture extreme events (Muela et al. 2017). The application in question as well as the underlying data properties will determine the block size. The highest value is noted for each block. The severe incident that was observed inside that specific block is represented by this maximum value (Bali 2003). The extreme results from each block are considered a collection of random variables. The simplified formula used to probability function P (x) distribution is given by:

$$P(x) = \left(\frac{1}{\sigma}\right) * t(x)^{\xi+1} * e^{-t(x)} \tag{5.1}$$

$$\left(1 + \xi * \left(\frac{x - \mu}{\sigma}\right)\right)^{-\frac{1}{\xi}} \text{if } \xi \neq 0$$

$$e^{-\left(\frac{x-\mu}{\sigma}\right)} \text{if } \xi = 0 \tag{5.2}$$

Where $t(\mathrm{x}) =$

Where μ is the location parameter, σ is the scale parameter, ξ is the shape parameter of the fitted distribution.

5.2.4 *Methodology*

The temperature data is only comprised of values that are larger than or equal to the selected threshold, which represents the 95th percentile of the total dataset. This is the cutoff point over which excess (or non-inclusive) data are utilized to estimate the distribution function's parameters for both Hyderabad and Nagpur City using IDLE v-3.10. Using the pyextremes library, GEV distribution was applied to the filtered temperature data. Based on the given information, the pyextremes library calculates the parameters of the GEV distribution. The location parameter (mu), scale parameter (σ), and shape parameter (ξ) of the GEV distribution are the most used parameters (Rasmussen 2001). To determine the magnitude of severe occurrences that can be anticipated to occur within the designated return periods, analyse the return levels (Yoon et al. 2011). Lower return levels signify less extreme occurrences that are more likely to happen, whereas higher return levels signify more extreme events that are less likely to happen (Fig. 5.3).

5.3 Results and Discussion

5.3.1 *Trend Analysis: Sequential Mann-Kendall*

The sequential Mann-Kendall test (Seq-MK), a non-parametric statistical test based on a rank system, was used in this chapter to identify trends in the yearly average temperature data series from 1981 to 2021 (Hu et al. 2020). The test compares the evidence of changes between earlier and later time series data points. The sign values will tend to rise or drop continuously if a trend is present. The Mann-Kendall statistic, which assesses the strength and trajectory of the trend, is calculated as

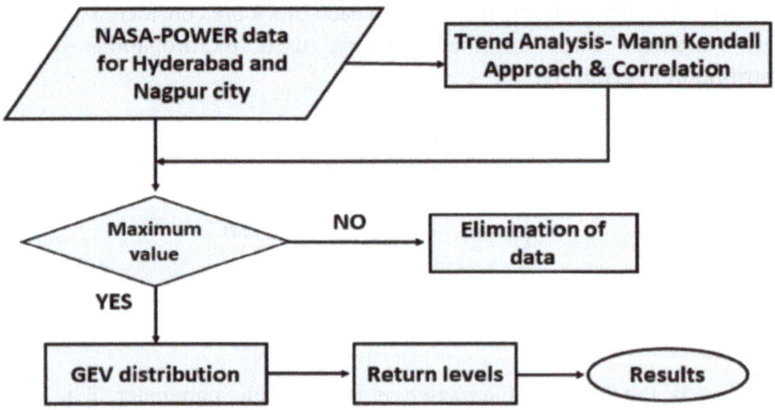

Fig. 5.3 The methodology of heat wave study

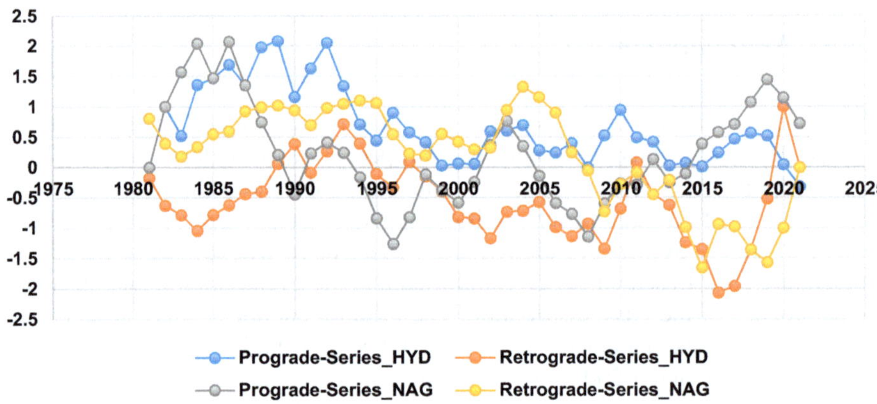

Fig. 5.4 Showing the sequential Mann- Kendall plot of Hyderabad and Nagpur

part of the test (Yanming et al. 2012). While prograde series show an increasing trend, retrograde series show a trend that is decreasing, as shown in Eqs. (5.3) and (5.4) (Reilly et al. 2000).

$$\text{For retrograde series}: S_R = \sum_{i=1}^{n} I \sum_{j=i+1}^{n} \operatorname{sgn}(x_i - x_j) \tag{5.3}$$

$$\text{For prograde series}: \quad S_P = \sum_{i=1}^{n} I \sum_{j=i+1}^{n} \operatorname{sgn}(x_j - x_i) \tag{5.4}$$

Where n denotes the number of data points in the segment, x_i and x_j are the data values at the time points i and j, respectively.

The results shown in Fig. 5.4 illustrate that there may be numerous possible change points in the trend behaviour for the Nagpur series, where the prograde and retrograde series appear to overlap more than five times. This shows that the trend in

Table 5.1 Year-wise detected change points for Nagpur and Hyderabad

Detected change point	Year	Value significant $p \leq 0.05$	
		Nagpur	Hyderabad
1st	1981	0	0
2nd	1990	−0.44721	–
3rd	1994	−0.16423	–
4th	1995	−0.84128	–
5th	1996	−1.26063	–
6th	1997	−0.82385	–
7th	1998	−0.11363	–
8th	1999	−0.38484	0.034986
9th	2000	−0.584	–
10th	2001	−0.24158	–
11th	2005	−0.14013	–
12th	2006	−0.59512	–
13th	2007	−0.77134	–
14th	2008	−1.14588	0
15th	2009	−0.60026	–
16th	2010	−0.26762	–
17th	2011	−0.28894	–
18th	2013	−0.24791	0.030989
19th	2014	−0.10377	–
20th	2015	–	0.014201
21st	2020	–	0.046604
22nd	2021	–	−0.31449

the Nagpur data is not consistent and displays distinct patterns over time. On the other hand, the Hyderabad series only shows an intersecting point from 1981 to 2021, which is close to 2020, showing a more consistent trend behaviour over that particular period. This means that the trend in the data from Hyderabad is rather consistent and has a distinct pattern of change during that period. There are many turning points in Nagpur and Hyderabad time series datasets as shown in Table. 5.1 which are all below the chosen level of significance ($p \leq 0.05$).

5.3.2 Correlation Analysis

The months of March to June and October to December showed a strong correlation when the monthly average temperatures of Hyderabad and Nagpur were compared from 1981 to 2021 using R-v4.2.2 (Wu and Yan 2012). This suggests a possible seasonal pattern or relationship between the temperature data of the two cities during these times.

The significance of finding a high association during these specific times may point to the existence of similar meteorological impacts or elements that have an impact on Nagpur and Hyderabad from 1981 to 2021. It indicates significant correlations (with a value of 0.6 or higher) between the average monthly temperatures at 2 meters for March to July and October to December for the years 1981 to 2021. The significance of this correlation matrix suggests that there is a strong relationship between the average monthly temperatures of Nagpur and Hyderabad during these specific months. The correlation coefficients indicate that the temperature patterns of the two cities tend to move in a similar direction during these periods.

5.3.3 EVT Prediction Results

The following findings were found during the analysis of the dataset of Nagpur using Python's Pyextremes package, as illustrated in Fig. 5.5. For 2, 5, 10, and 25 years, the return values for temperature at 2 m are 46.99 °C, 47.69 °C, 47.96 °C, and 48.17 °C, respectively. These numbers represent the average threshold temperature for a certain return time that is anticipated to be exceeded. The value's relevance is that if a temperature return value of 46.99 °C corresponds to a return time of 2 years,

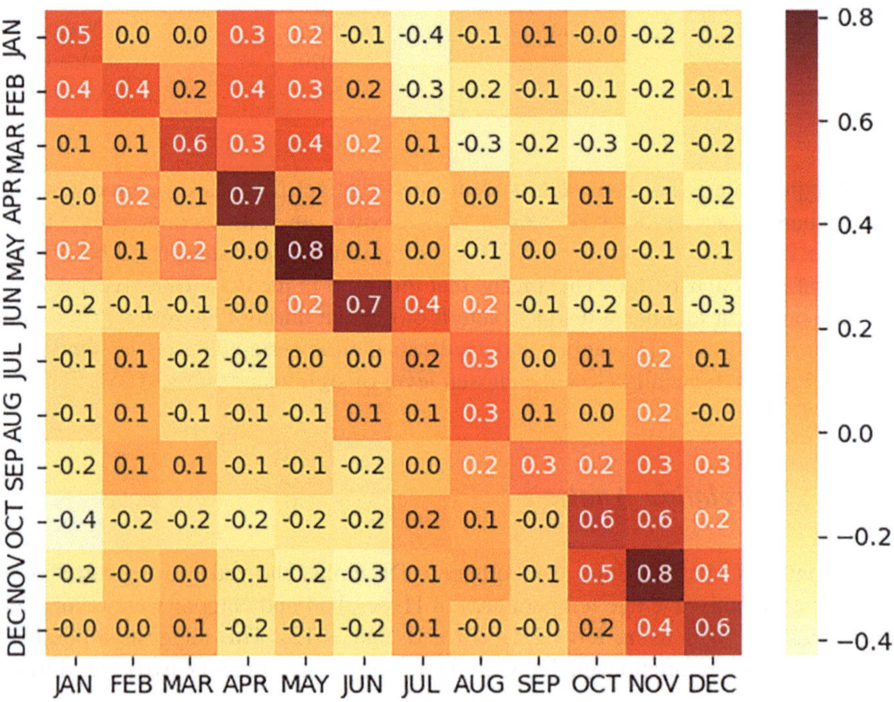

Fig. 5.5 Showing correlation of NASA-POWER data from 1981–2021 (monthly) of Hyderabad and Nagpur city of Central India

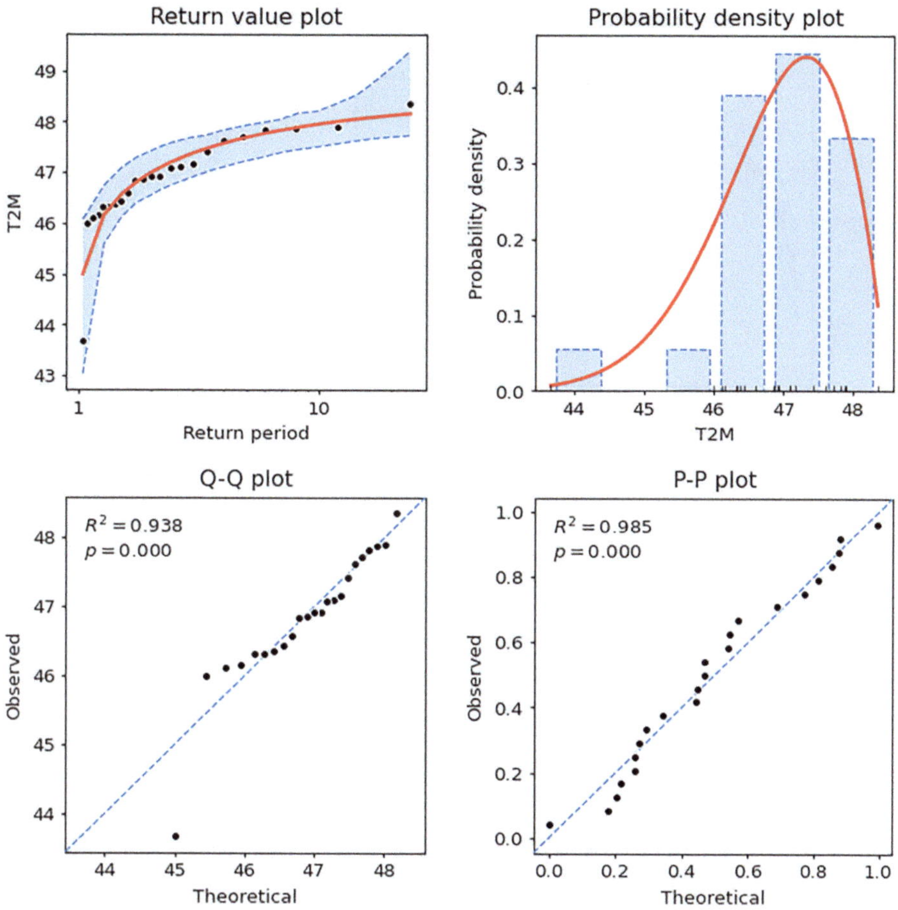

Fig. 5.6 EVT results showing return period with temperature, probability density function, Q-Q plot and P-P plot for Nagpur hourly data

it signifies that a temperature over 46.99 °C is typically anticipated to occur once every 2 years.

In the case of Hyderabad, for 2, 5, 10, and 25 years, the return values for temperature at 2 m are 43.59 °C, 44.32 °C, 44.59 °C, and 44.80 °C, respectively. The higher return value of 44.80 °C for a 25-year return period suggests a higher risk of encountering temperature values exceeding 44.80 °C compared to the other return periods as shown in Figs. 5.6 and 5.7.

Extreme values or extremes are more likely to occur on the peak end of a right-skewed distribution's tail, which is longer than the left tail. This implies that there is a higher probability of observing values that deviate in a positive direction from the mean in both the cases of Nagpur and Hyderabad (Pecoraro et al. 2015; Vrecko et al. 2009).

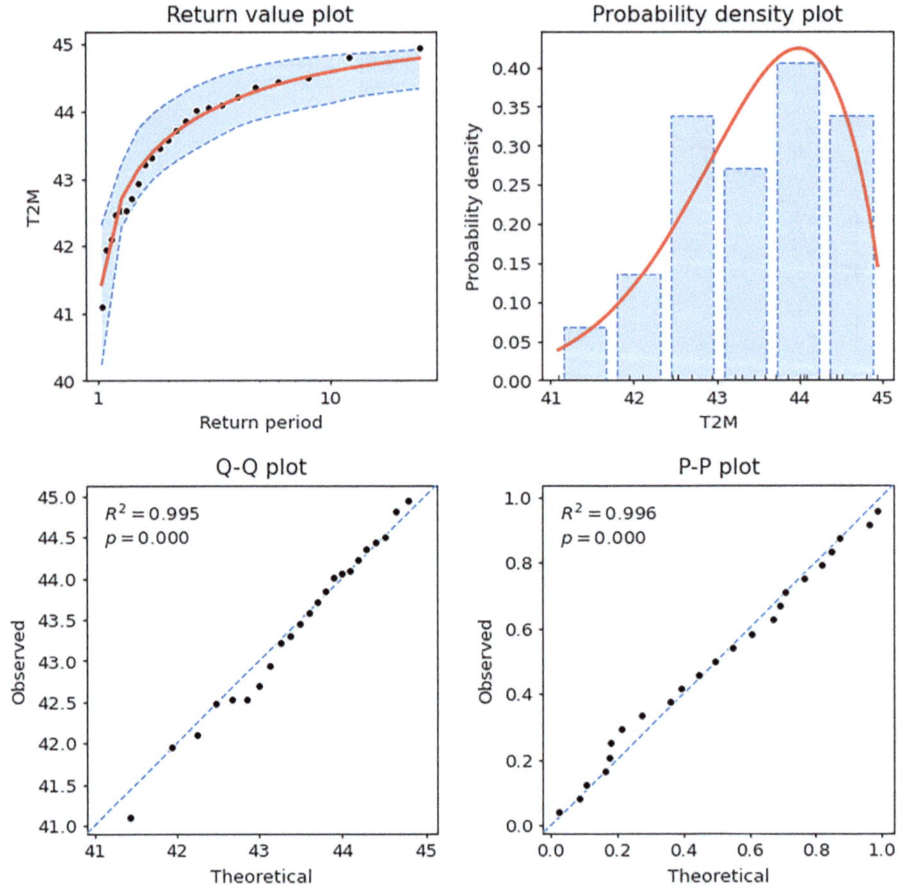

Fig. 5.7 EVT results showing return period with temperature, probability density function, Q-Q plot, and P-P plot for Hyderabad hourly data

The quantiles of a dataset are compared to the quantiles of a theoretical distribution using a Q-Q plot (Quantile-Quantile plot). It assists in determining how closely the observed data are similar to the predicted distribution. The amount of variability in the observed data that can be explained by the assumed distribution is indicated by the R^2 value in the Q-Q plot. The theoretical distribution can account for around 93.8% for Nagpur and 99.5% for Hyderabad of variability in the data, according to an R^2 value of 0.938 and 0.995 for Nagpur and Hyderabad, respectively. It shows that the observed data and the expected distribution match significantly (Oh et al. 2008). Whereas, the cumulative distribution function (CDF) of a dataset is compared to the CDF of a theoretical distribution using a P-P plot (Probability-Probability plot). The assumed distributions can account for about 98.5% for Nagpur and 99.6% Hyderabad of variability in the relevant datasets (Ryu et al. 2011).

5.4 Conclusion

The study provides insight into the characteristics of severe temperature episodes in Hyderabad and Nagpur. The return level of extreme temperature (or heat wave) is 47.96 °C and 44.59 °C for Nagpur and Hyderabad for a return period of 10 years. The frequency and severity of severe temperatures shown by the return values and the agreement between the observed data and thc predicted distributions enable the risk assessment, planning, and decision-making processes. The results help us understand the behaviour and possible effects of severe temperatures in these areas.

Return values are helpful for planning and design in many different sectors. Knowing the return values for severe temperatures can aid with infrastructure planning when constructing buildings, roads, or other structures that can endure or adapt to such extreme conditions. Return values for severe temperatures are useful in agriculture for choosing the right crop kinds or managing irrigation plans. The intensity and frequency of severe temperature occurrences can be assessed by comparing return data at various return dates. Decision-making and setting priorities for actions including risk reduction, adaption strategies, or resource allocation benefit from this knowledge.

Acknowledgement The authors would like to express their gratitude to CSIR-HRDG for providing the fellowship and to CSIR-NEERI, Nagpur, for access to computational facilities, which were instrumental in the completion of this research.

Declaration of Competing Interests The authors affirm that there are no competing financial interests, relationships, or conflicts of interest that might have impacted the outcome or interpretation of this book chapter.

References

Adams CR (n.d.) Impacts of temperature extremes. Cent. Sci. Technol. Policy Res. https://sciencepolicy.colorado.edu/socasp/weather1/adams.html

Arismendi I, Safeeq M, Dunham JB, Johnson SL (2014) Can air temperature be used to project influences of climate change on stream temperature? Environ Res Lett 9:084015. https://doi.org/10.1088/1748-9326/9/8/084015

Bali TG (2003) An extreme value approach to estimating volatility and value at risk*. J Bus 76:83–108. https://doi.org/10.1086/344669

Barriopedro D, García-Herrera R, Ordóñez C, Miralles DG, Salcedo-Sanz S (2023) Heat waves: a growing threat to society and the environment. Eos 104

Benalcazar P, Kamiński J (2019) Short-term heat load forecasting in district heating systems using artificial neural networks. IOP Conf Ser Earth Environ Sci 214:012023. https://doi.org/10.1088/1755-1315/214/1/012023

Benameur S, Benkhaled A, Meraghni D, Chebana F, Necir A (2017) Complete flood frequency analysis in Abiod watershed, Biskra (Algeria). Nat Hazards 86:519–534. https://doi.org/10.1007/s11069-016-2703-4

Burillo D, Chester MV, Ruddell B, Johnson N (2017) Electricity demand planning forecasts should consider climate non-stationarity to maintain reserve margins during heat waves. Appl Energy 206:267–277. https://doi.org/10.1016/j.apenergy.2017.08.141

Buzan JR, Oleson K, Huber M (2015) Implementation and comparison of a suite of heat stress metrics within the community land model version 4.5. Geosci Model Dev 8:151–170. https://doi.org/10.5194/gmd-8-151-2015

Chen H-L, Ramachandra Rao A (2003) Linearity analysis on stationary segments of hydrologic time series. J Hydrol 277:89–99. https://doi.org/10.1016/S0022-1694(03)00086-6

Dhole A, Kadaverugu R, Tomar S (2023) Impact of land cover classes on surface temperature in the vicinity of urban lakes and vegetation patches: A non-parametric regression analysis over decadal data. Earth Sci Inform 16:3947–3961. https://doi.org/10.1007/s12145-023-01140-2

Diffenbaugh NS, Scherer M (2011) Observational and model evidence of global emergence of permanent, unprecedented heat in the 20th and 21st centuries: a letter. Clim Chang 107:615–624. https://doi.org/10.1007/s10584-011-0112-y

Dombry C (2013) Maximum likelihood estimators for the extreme value index based on the block maxima method. 10.48550/ARXIV.1301.5611

Dreesman JM, Tutz G (2001) Non-stationary conditional models for spatial data based on varying coefficients. J R Stat Soc Ser Stat 50:1–15. https://doi.org/10.1111/1467-9884.00256

Furrer E, Katz R, Walter M, Furrer R (2010) Statistical modeling of hot spells and heat waves. Clim Res 43:191–205. https://doi.org/10.3354/cr00924

Ghorbel A, Trabelsi A (2008) Predictive performance of conditional extreme value theory in value-at-risk estimation. Int J Monet Econ Finance 1:121. https://doi.org/10.1504/IJMEF.2008.019218

Hu Z, Liu S, Zhong G, Lin H, Zhou Z (2020) Modified Mann-Kendall trend test for hydrological time series under the scaling hypothesis and its application. Hydrol Sci J 65:2419–2438. https://doi.org/10.1080/02626667.2020.1810253

Huyse L, Chen R, Stamatakos JA (2010) Application of generalized Pareto distribution to constrain uncertainty in peak ground accelerations. Bull Seismol Soc Am 100:87–101. https://doi.org/10.1785/0120080265

Iyengar VS (1999) HOT: heuristics for oblique trees. In: Proceedings 11th international conference on tools with artificial intelligence. Presented at the 11th international conference on tools with artificial intelligence TAI 99. IEEE Comput. Soc, Chicago, pp 91–98. https://doi.org/10.1109/TAI.1999.809771

Kadaverugu R, Gurav C, Rai A, Sharma A, Matli C, Biniwale R (2021) Quantification of heat mitigation by urban green spaces using InVEST model – a scenario analysis of Nagpur City, India. Arab J Geosci 14:82. https://doi.org/10.1007/s12517-020-06380-w

Kadaverugu R, Lama S, Purohit V, Mopati S, Matli C, Biniwale R (2022) Potential of Green Roofs in Heat Stress Mitigation—A Case of Nagpur City. In: Dikshit, A.K., Narasimhan, B., Kumar, B., Patel, A.K. (eds) Innovative Trends in Hydrological and Environmental Systems. Lecture Notes in Civil Engineering, vol 234. Springer, Singapore. https://doi.org/10.1007/978-981-19-0304-5_27

Kim HH (1992) Urban heat Island. Int J Remote Sens 13:2319–2336. https://doi.org/10.1080/01431169208904271

Knowlton K, Rotkin-Ellman M, King G, Margolis HG, Smith D, Solomon G, Trent R, English P (2009) The 2006 California heat wave: impacts on hospitalizations and emergency department visits. Environ Health Perspect 117:61–67. https://doi.org/10.1289/ehp.11594

Lal V, Gupta S, Gupta O, Bhatnagar S (2012) Forecasting incidence of dengue in Rajasthan, using time series analyses. Indian J Public Health 56:281. https://doi.org/10.4103/0019-557X.106415

Lee Y, Yoon S, Murshed MS, Kim M-K, Cho C, Baek H-J, Park J-S (2013) Spatial modeling of the highest daily maximum temperature in Korea via max-stable processes. Adv Atmospheric Sci 30:1608–1620. https://doi.org/10.1007/s00376-013-2216-y

Li K, Amatus G (2020) Spatiotemporal changes of heat waves and extreme temperatures in the main cities of China from 1955 to 2014. Nat Hazards Earth Syst Sci 20:1889–1901. https://doi.org/10.5194/nhess-20-1889-2020

Li X, Li Z, Huang W, Zhou P (2020) Performance of statistical and machine learning ensembles for daily temperature downscaling. Theor Appl Climatol 140:571–588. https://doi.org/10.1007/s00704-020-03098-3

Liao W, Wang D, Liu X, Wang G, Zhang J (2017) Estimated influence of urbanization on surface warming in eastern China using time-varying land use data: INFLUENCE OF URBANIZATION ON SURFACE WARMING IN EASTERN China. Int J Climatol 37:3197–3208. https://doi.org/10.1002/joc.4908

Makatjane K (2022) Forecasting uncertainty intervals for return period of extreme daily electricity consumption. Int J Energy Econ Policy 12:217–225. https://doi.org/10.32479/ijeep.12901

Manisha P (2018) A scientific study of water bodies in Hyderabad: a geoinformatics approach

Monteiro LA, Sentelhas PC, Pedra GU (2018) Assessment of NASA/POWER satellite-based weather system for Brazilian conditions and its impact on sugarcane yield simulation: sugarcane yield simulation with nasa/power satellite-based data. Int J Climatol 38:1571–1581. https://doi.org/10.1002/joc.5282

Muela SB, Martín CL, Sanz RA (2017) An application of extreme value theory in estimating liquidity risk. Eur Res Manag Bus Econ 23:157–164. https://doi.org/10.1016/j.iedeen.2017.05.001

Mulder C, Perrelli R, Rocha MD (2012) External vulnerability, balance sheet effects, and the institutional framework – lessons from the Asian crisis. Int Rev Econ Finance 21:16–28. https://doi.org/10.1016/j.iref.2011.04.002

Murari KK, Ghosh S, Patwardhan A, Daly E, Salvi K (2015) Intensification of future severe heat waves in India and their effect on heat stress and mortality. Reg Environ Chang 15:569–579. https://doi.org/10.1007/s10113-014-0660-6

Nalla S (n.d.) climate change in a mega city a temporal analysis of long term temperature variability in greater hyderabad, India

Neftci SN (2000) Value at risk calculations, extreme events, and tail estimation. J Deriv 7:23–37. https://doi.org/10.3905/jod.2000.319126

Nori-Sarma A, Anderson GB, Rajiva A, ShahAzhar G, Gupta P, Pednekar MS, Son J-Y, Peng RD, Bell ML (2019) The impact of heat waves on mortality in Northwest India. Environ Res 176:108546. https://doi.org/10.1016/j.envres.2019.108546

Oh S, Carriere KC, Park T (2008) Model diagnostic plots for repeated measures data using the generalized estimating equations approach. Comput Stat Data Anal 53:222–232. https://doi.org/10.1016/j.csda.2008.07.022

Pandey MD (2001) Extreme quantile estimation using order statistics with minimum cross-entropy principle. Probabilistic Eng Mech 16:31–42. https://doi.org/10.1016/S0266-8920(00)00004-7

Pecoraro A, Di Felice F, Felli M, Salvatore F, Viviani M (2015) An improved wake description by higher order velocity statistical moments for single screw vessel. Ocean Eng 108:181–190. https://doi.org/10.1016/j.oceaneng.2015.07.038

Rasmussen PF (2001) Generalized probability weighted moments: application to the generalized Pareto distribution. Water Resour Res 37:1745–1751. https://doi.org/10.1029/2001WR900014

Reilly FK, Wright DJ, Chan KC (2000) Bond market volatility compared to stock market volatility. J Portf Manag 27:82–92. https://doi.org/10.3905/jpm.2000.319786

Roberts SJ (2000) Extreme value statistics for novelty detection in biomedical signal processing. In: First international conference on advances in medical signal and information processing. Presented at the first international conference on advances in medical signal and information processing. IEE, Bristol, pp 166–172. https://doi.org/10.1049/cp:20000333

Ryu H-S, Lee J-S, Kang CG (2011) Performance analysis of relay systems in an interference-limited environment. In: 2011 IEEE 73rd vehicular technology conference (VTC spring). Presented at the 2011 IEEE vehicular technology conference (VTC 2011-spring). IEEE, Budapest, pp 1–5. https://doi.org/10.1109/VETECS.2011.5956735

Sandeep A, Prasad VS (2018) Intra-annual variability of heat wave episodes over the east coast of India: intra-annual variability of heat wave episodes. Int J Climatol 38:e617–e628. https://doi.org/10.1002/joc.5395

Sillmann J, Croci-Maspoli M, Kallache M, Katz RW (2011) Extreme cold winter temperatures in Europe under the influence of North Atlantic atmospheric blocking. J Clim 24:5899–5913. https://doi.org/10.1175/2011JCLI4075.1

Srivastava PK, Singh P, Mall RK, Pradhan RK, Bray M, Gupta A (2020) Performance assessment of evapotranspiration estimated from different data sources over agricultural landscape in northern India. Theor Appl Climatol 140:145–156. https://doi.org/10.1007/s00704-019-03076-4

Thibault J, Grandjean BPA (1991) A neural network methodology for heat transfer data analysis. Int J Heat Mass Transf 34:2063–2070. https://doi.org/10.1016/0017-9310(91)90217-3

USDA (n.d.) Extreme weather. https://www.climatehubs.usda.gov/content/extreme-weather

Van Der Wiel K, Selten FM, Bintanja R, Blackport R, Screen JA (2020) Ensemble climate-impact modelling: extreme impacts from moderate meteorological conditions. Environ Res Lett 15: 034050. https://doi.org/10.1088/1748-9326/ab7668

Vrecko D, Klos A, Langer T (2009) Impact of presentation format and self-reported risk aversion on revealed skewness preferences. Decis Anal 6:57–74. https://doi.org/10.1287/deca.1090.0141

Wang J, Yan Z, Quan X-W, Feng J (2017) Urban warming in the 2013 summer heat wave in eastern China. Clim Dyn 48:3015–3033. https://doi.org/10.1007/s00382-016-3248-7

White JW, Hoogenboom G, Stackhouse PW, Hoell JM (2008) Evaluation of NASA satellite- and assimilation model-derived long-term daily temperature data over the continental US. Agric For Meteorol 148:1574–1584. https://doi.org/10.1016/j.agrformet.2008.05.017

Wu SJ, Yan ZF (2012) Indoor thermal environment simulation of Xi'an residential building in summer. Adv Mater Res 512–515:2882–2886. https://doi.org/10.4028/www.scientific.net/AMR.512-515.2882

Yanming Z, Jun W, Xinhua W (2012) Study on the change trend of precipitation and temperature in Kunming City Based on Mann-Kendall analysis. In: Zhang T (ed) Future computer, communication, control and automation, advances in intelligent and soft computing. Springer, Berlin Heidelberg, pp 505–513. https://doi.org/10.1007/978-3-642-25538-0_71

Yoon J-Y, Cho J-B, Jun B-C (2011) An alternative study of the determination of the threshold for the generalized Pareto distribution. Korean J Appl Stat 24:931–939. https://doi.org/10.5351/KJAS.2011.24.5.931

Zhang K, Chen Y-H, Schwartz JD, Rood RB, O'Neill MS (2014a) Using forecast and observed weather data to assess performance of forecast products in identifying heat waves and estimating heat wave effects on mortality. Environ Health Perspect 122:912–918. https://doi.org/10.1289/ehp.1306858

Zhang K, Li Y, Schwartz JD, O'Neill MS (2014b) What weather variables are important in predicting heat-related mortality? A new application of statistical learning methods. Environ Res 132:350–359. https://doi.org/10.1016/j.envres.2014.04.004

Zhang R, Sun C, Zhu J, Zhang R, Li W (2020) Increased European heat waves in recent decades in response to shrinking Arctic Sea ice and Eurasian snow cover. NPJ Clim Atmospheric Sci 3:7. https://doi.org/10.1038/s41612-020-0110-8

Zhou XY, Schmidt F, Jacob B (2012) Extrapolation of traffic data for development of traffic load models: assessment of methods used during background works of Eurocode. In: IABMAS2012, 6th conference on bridge maintenance, safety and management, p 7

Zschenderlein P, Pfahl S, Wernli H, Fink AH (2020) A Lagrangian analysis of upper-tropospheric anticyclones associated with heat waves in Europe. Weather Clim Dyn 1:191–206. https://doi.org/10.5194/wcd-1-191-2020

Chapter 6
Risk Assessment and Vulnerability Mapping of Soil Erosion for Climate-Resilient Disaster Management with Geospatial Techniques

Santosh Chougale

Abstract Soil erosion is a significant issue that poses a threat to the environment. For management and soil conservation, the erosion-prone area must be assessed and mapped. To determine the rate of soil erosion and compute the amount of sediment yield owing to erosion, several models were developed. The most popular analytical model for calculating the average annual soil loss is the Revised Universal Soil Loss Equation. An attempt has been made to predict possible yearly soil loss in Chikodi Taluk. An actual model of five factors was used for the RUSLE equation: Soil Erodibility (K), Runoff-Rainfall Erosivity (R), Land use/land cover (C), Topographic (LS), and Supporting Conservation Practice (P) for the prediction. In the ArcGIS environment, the revised universal soil loss equation was used. The result shows that the study area's annual average of soil erosion is 30 t /ha /year. The map of estimated soil loss will be a useful tool for developing sustainable development strategies.

Keywords RUSLE · Chikodi · GIS · Remote sensing · Soil erosion

6.1 Introduction

Soil erosion is a widespread issue that has become a top concern in many countries. Soil erosion, broadly speaking, is the loss of soil due to a mix of natural and artificial factors (Zachar 1982). One of the processes that happens naturally and affects agriculture and the environment is soil erosion. Although the cycle of soil formation takes several years, it can be countered in a few significant storms by rainfall erosion, which leaves the soil with reduced residues and lower yields. Soil erosion can be caused by geomorphologic processes; however, human interference is the primary reason for accelerated soil erosion. Soil erosion has been caused by overgrazing,

S. Chougale (✉)
Department of Marine Geology, Mangalore University, Mangalagangothri, Karnataka, India

D. Kumar, N. P. Bassill (eds.), *Artificial Intelligence and Machine Learning for Climate Disaster Management*, Disaster Resilience and Green Growth,
https://doi.org/10.1007/978-981-95-1588-2_6

inappropriate land cultivation, deforestation, and rapid population growth. Soil erosion not only decreases water availability but also limits agricultural productivity. Most of the region is occupied by agriculture in the study area, and due to low soil quality and lower crop yields, it can even wash away seeds and small plants. The use of increased quantities of pesticides and artificial fertilizers may lead to waterway contamination through runoff. Soil also loses its ability to hold water because it removes organic matter found on the top layer of soil and leaves the field susceptible to weather conditions like drought. With the loss of fertile soil and land pollution, waterways and aquatic species have been blocked by sedimentation in rivers. Degraded land retains less water and induces flooding. Sustainable land resource use helps to prevent soil loss, erosion, and desertification and reduces negative impacts on livestock and crop yield. Soil quality is one of the biggest problems for the farmer, whose livelihood depends mainly on well-managed crops.

Soil erosion assessments are a requirement for land use planning; in reality, erosion assessment should be a part of a soil resource assessment. This allows the delineation of essential soil loss remediation areas. Preventing soil erosion requires the effective implementation of management practices. In India, the total area impacted by soil erosion and other forms of soil degradation is approximately 53%. According to the Environment and Forest Ministry (2001), the degradation causes terrain deformation in 16.4 mha and the loss of topsoil in 132 mha of land. To improve evidence-based performance and priorities, quantitative analysis of soil erosion and its spatial variability is crucial.

Normally river basin geomorphic characteristics are added to the creation of hydrological models under insufficient data conditions to solve various river basin hydrological problems (Bagyaraj et al. 2011; Dutta and Roy 2012; Altaf et al. 2014). Huge data and computational works would be required in the conventional methods for defining erosion risk zones even for a micro-watershed. However, it is expensive and time-consuming to evaluate soil erosion risk using conventional methods. Using various models based on the suitability of regions worldwide to classify soil erosion zones, many scholars have worked on the subject of soil erosion. Globally, the models used for calculating soil erosion are overly varied. While conventional methods produce point information, GIS incorporates spatial analysis functionality for spatially distributed data, and the Remote Sensing (RS) method enables hydrological parameters to be measured at a spatial scale. Soil erosion mapping, monitoring, and prediction rely heavily on data from remote sensing systems. The geospatial methods provide a cost-effective way to evaluate soil erosion where there are no accessible territories or direct field methods are expensive.

The technique and empirically based model have been used in recent years to assess soil erosion (De Vente and Poesen 2005). One of the empirical models used to prioritize watersheds is the Sediment Yield Index (Rawat et al. 2014). The Sediment Yield Index (SYI) approach is based primarily on weighted parameters of soil overlays, rainfall erosivity, topography, and land use/land cover (Naqvi et al. 2015). For assessing sediment production rate, empirical models based on morphometric parameters have been suggested in the past years by Josh and Das (1982). For the prioritization of sub-watersheds, the sediment production rate was calculated. The concept of Sediment Delivery Ratio is often applicable to the identification of risk areas for soil erosion. Delivery ratio values are affected by various factors such as the physiography of the catchment, transport system, sources of sediment, texture of eroded material, and land cover (Walling 1983, 1988; Richards 1993). The variables such as catchment area, slope, and land cover are also used as parameters in distribution ratio equations (Maner 1958; Roehl 1962; Williams and Berndt 1972; Hadley 1985; Kothyari and Jain 1997). Process-based models, such as the Water Erosion Prediction Project (WEPP) (Nearing et al. 1989) and the European Soil Erosion Model (EUROSEM) (Morgan et al. 1999) have been used to identify areas susceptible to soil erosion. Wischmeier, Smith, and others derived the equation for Universal Soil Loss (USLE). Through worldwide recognition, USLE has been used as the major tool in the United States and many other nations worldwide to measure the soil erosion rate. In addition to additional data and resources available, researchers developed the USLE using the same methodology as RUSLE with the changes in determining factors. These include some new approaches to soil erodibility factor, cover-management factor and derived a new length factor equation to slope and new conservation-practice principles (Renard et al. 1997). During the year 1996, the USDA published the description of RUSLE in a new Agriculture Handbook (No. 703).

Good and sustainable soil, land, and water resource management is best done on a watershed basis. The RUSLE could be incorporated into GIS software to spatially visualize the erosion-prone areas. Several researchers obtained the RUSLE model for the estimation of soil erosion in watersheds (Dabral et al. 2008; Prasannakumar et al. 2011; Balasubramani et al. 2015; Sumantra and Padmini 2015; Ganasri and Ramesh 2016; Gelagay and Minale 2016; Mengie et al. 2022). Recent studies for estimating soil loss at administrative boundaries are carried out (Adediji et al. 2010; Suresh Kumar et al. 2017; Lee et al. 2017; Koirala et al. 2019).

Most of the Chikodi taluk is covered with agricultural land, and sloping land production and uncontrolled deforestation are major causes of soil erosion. Till now no studies have been conducted regarding soil erosion in Chikodi Taluk. The purpose of this study is to estimate the potential soil loss in Chikodi taluk using the RUSLE model in a GIS environment.

6.2 Site of Study

Geographically, the Chikodi taluk is located at 74°20′0″–74°40′0″ E longitude and 16°20′0″– 16°40′0" N latitude in Belagavi district, Karnataka. Chikodi has an area of 1263 square kilometres and an average elevation of 683 metres above sea level. The study area is surrounded by Athani, Raibag, Gokak, Nipani, and Hukkeri talukas and lies on the border between Maharastra and Karnataka state. The study area is covered with basalt. The major river in the study area is the Krishna River with its

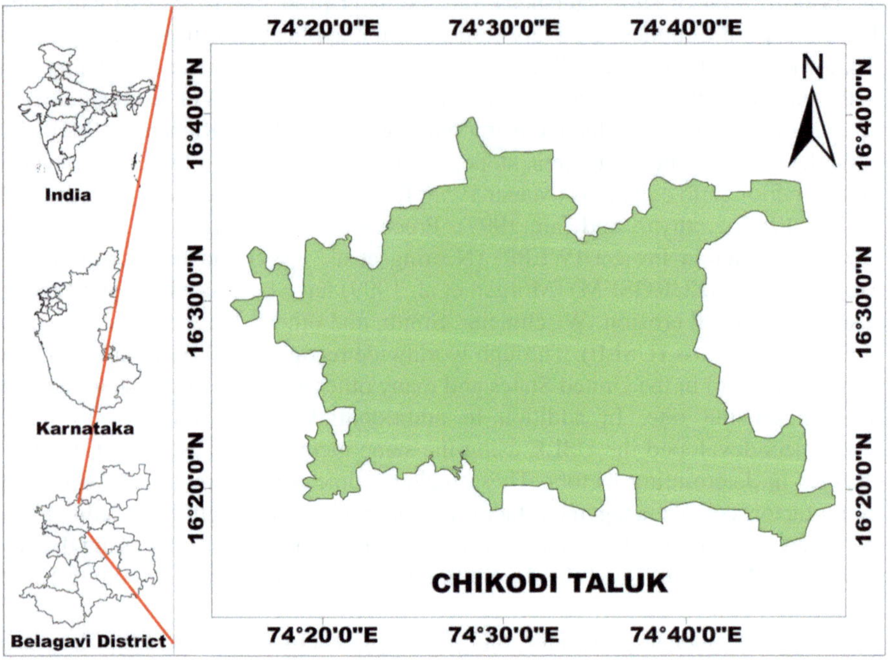

Fig. 6.1 Study area

tributaries: Vedaganga, Doodganga, and Panchganaga. The agro-climatic study area falls under the Northern Transition Zone of Karnataka (Fig. 6.1).

6.3 Datasets Used

The toposheets E43U6, E43U7, E43U10, E43U11, E43U14, and E43U15 were collected from Survey of India, Bengaluru. Sentinel Satellite image (10 m resolution) and Digital Elevation Model (30 m resolution) were downloaded from the USGS Earth Explorer website. National Bureau of Soil Survey and Land Use Planning, Bengaluru, prepared a soil map. From the District Statistical Office, Belagavi collected rainfall data.

6.4 Methodology

Toposheets were geo-referenced and rectified and the base map was prepared on a 1: 50,000 scale of SOI toposheets (E43U6, E43U7, E43U10, E43U11, E43U14, and E43U15). To isolate, the AOI layer is superimposed on different thematic layers.

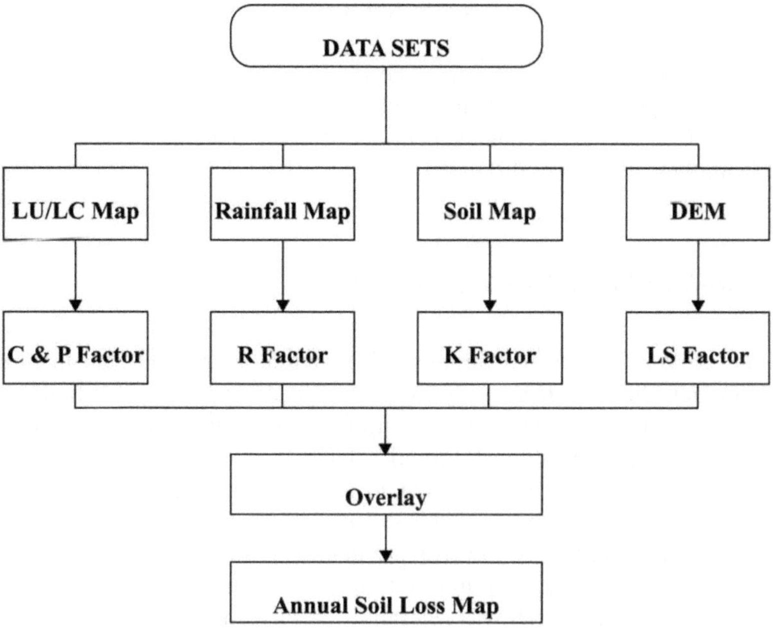

Fig. 6.2 Flow chart

The RUSLE model was adopted using factors such as crop management, conservation practice, rainfall erosivity, topography, and soil erodibility. The majority of the factors were calculated using specific techniques and taken from published research papers that were adapted for Indian circumstances. Slope and flow accumulation is generated from SRTM DEM. Land use land cover classification was done by a supervised classification technique using a Sentinal 2A image. ArcGIS software was used for data processing, storing, analysis, displaying, and viewing the result. For conservation practice and crop management factors land use land cover map was used. All five layers were integrated into ArcGIS software using values/formulae and raster calculator option and an average annual soil map for the study area. The methodology adopted is shown in Fig. 6.2.

6.5 Results

The long-term rates of inter-rill and rill erosion from field or farm-size units subject to different management practices are projected globally using the RUSLE method. The RUSLE model, whose equation is provided below, was extensively used to estimate average yearly soil loss.

$$A = R^{*}K^{*}LS^{*}C^{*}P \tag{6.1}$$

Where A is mean soil loss over a given period R, typically on an annual basis (t/ha/year); R is the rainfall factor (MJ mm ha/h/year); K is the soil erodibility; LS is the topography; C is the cover management and P is the practice supporting. All five factors with their results are discussed below.

6.5.1 Rainfall Erosivity Factor (R)

According to Wischmeier and Smith (1978) continuous rainfall statistics must be used to compute the rainfall erosivity factor, which takes into consideration the effect of rainfall intensity on soil erosion. The primary source of kinetic energy is rainfall intensity, and various analytical formulas have been developed to estimate the erosivity of rainfall for various global locations. When there is a high rate of precipitation and large drop sizes, the soil may disintegrate more quickly than it would with typical rainfall. A powerful storm that produces the majority of sheet or rill erosion is the reason for the high rate of runoff (Biswas and Pani 2015). The annual erosivity of the rainfall was determined in this study with the linear relationship developed by Parveen and Kumar (2012) and Singh et al. (1981). The derived formulae are provided below:

$$R = 79 + 0.363Rn \tag{6.2}$$

Where Rn is the amount of rainfall in mm. R factor was computed using average yearly rainfall data spanning 15 years. In ArcGIS IDW interpolation technique was adopted for the R factor map. The values of the R factor range from 278.81 to 921.225 (MJ mm ha/h/year). R map shows western region exhibits a higher rainfall compared to the eastern part of the area. In Fig. 6.3, a rainfall erosivity map is presented.

6.5.2 Soil Erodibility Factor (K)

The intrinsic vulnerability of soil to erosion is known as the soil erodibility factor determined by the mineralogical, physical, chemical, and morphological characteristics of the soil (Pérez-Rodríguez et al. 2007). In Renard et al.'s (1997) research paper, the K factor illustrates how soil properties and soil profile characteristics affect soil loss. The K factor is based on the organic content, texture, and structure of soil. The K factor map is delineated on a geo-referenced soil map. The study found two different types of soil: clay and loamy, and K values were used following Stone and Hilborn's recommendations (Stone and Hilborn 2000). For clay soil, the K value was 0.49, and for loamy soil, it was 0.67. The soil erosion map is presented in Fig. 6.4.

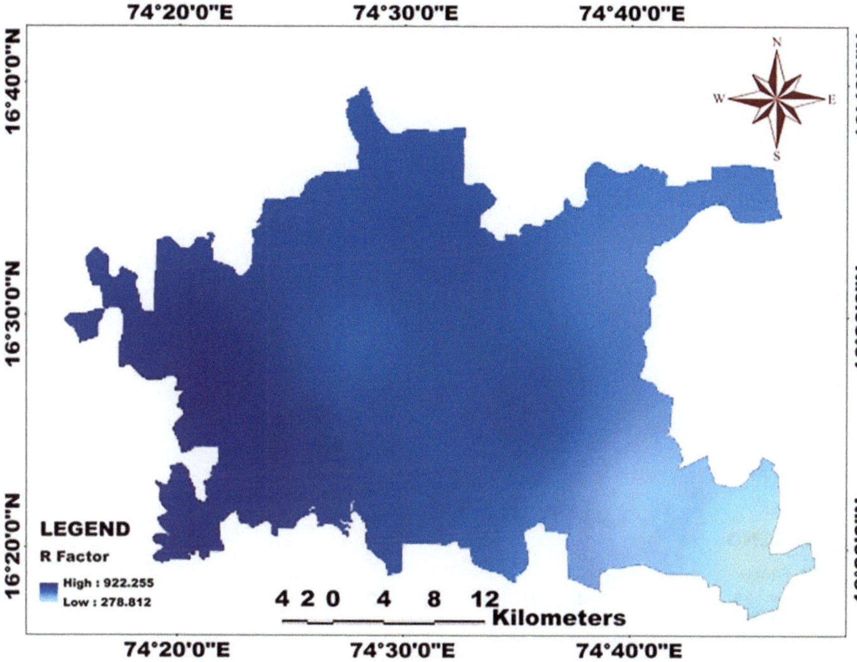

Fig. 6.3 Rainfall erosivity map

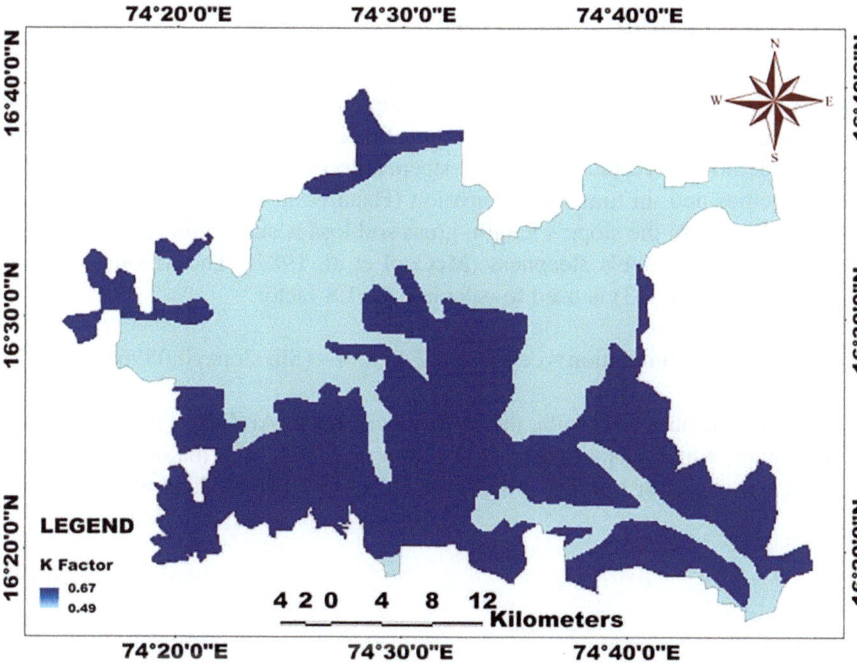

Fig. 6.4 Soil erodibility map

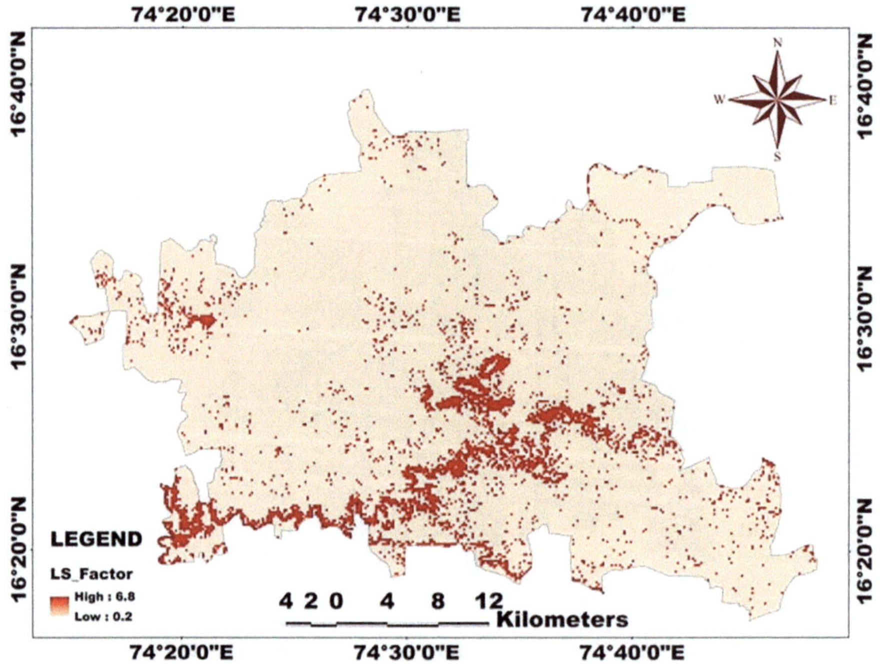

Fig. 6.5 Topographic map

6.5.3 Topographic Factor (LS)

The LS factors in RUSLE represent how topography affects erosion. The slope percentage and river accumulation are inputs that calculate the topographic factor. It has been demonstrated that longer and steeper slopes can result in higher overland flow velocities and, in turn, higher erosion (Haan et al. 1994). Additionally, compared to changes in the slope's length, gross soil loss is substantially more sensitive to changes in the slope's steepness (McCool et al. 1987). The Moore and Burch (1986) method (Eq. 6.3) is used to calculate the LS factor.

$$LS = (\text{flow accumulation} \times \text{cell size}/22.13)0.4 \times (\text{Sin slope}/0.0896)1.3 \quad (6.3)$$

Applying the above formula, the raster calculator in ArcGIS is used to calculate the LS factor, which has a range of 0.2 to 6.8 (Fig. 6.5). The majority of the study area has low slopes or low LS, which makes low to moderate land loss.

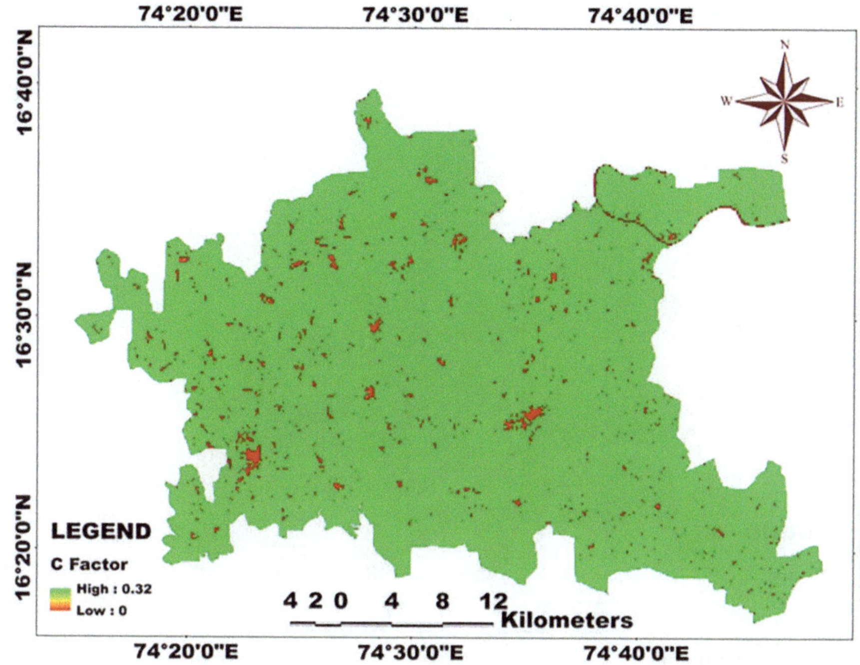

Fig. 6.6 Cover management map

6.5.4 Cover Management Factor (C)

A significant factor that can be used to track soil loss depending on an area's pattern of land use is the cover management factor (C). According to Khanchoul et al. (2020), the coverage management factor illustrates how farming and management techniques affect the rate of farmland erosion in agricultural lands. Seasons and agricultural production systems have an impact on how much vegetation and land cover reduce soil erosion in forested areas (Ganasri and Ramesh 2016). Based on Pandey et al. (2007), the C factor is the ratio of soil loss in continuously tilled fallow land to soil loss in areas with protective cover and management. The study's literature (Devatha et al. 2015; Gelagay and Minale 2016) was used to select the C factor value, which was given a higher value for agriculture 0.32, wasteland 0.1, built-up 0.01, and water bodies 0. The cover management map is presented in Fig. 6.6.

6.5.5 Conservation Practice Factor (P)

The factor has to do with the actions that limit water runoff and lessen soil erosion. It is the ratio of soil loss on croplands caused by a particular support practice to the

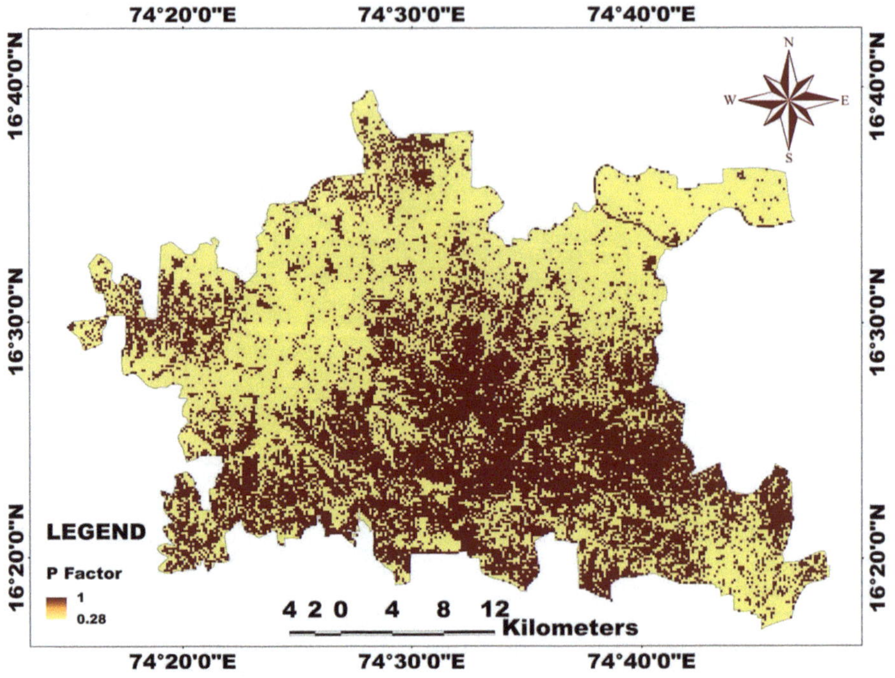

Fig. 6.7 Conservation practice map

corresponding loss caused by slope-parallel tillage. In agricultural fields, use of bunds is the primary conservation strategy, and agricultural land was given a value of 0.28 and other land as 1 (Parveen and Kumar 2012). Areas which are lacking conservation practices receive the maximum value, which ranges from 0 to 1. The values are typically determined by the slope of the terrain and the cultivation system (Shi et al. 2004). The conservation practice map is presented in Fig. 6.7.

6.6 Estimated Annual Soil Loss

The integration of RUSLE and GIS technique is used to estimate annual soil loss using five factors for the study area namely R, K, LS, C, and P. All layers have been reclassified, assigned the values, and integrated into the raster calculator option of ArcGIS software for the calculation of annual soil loss. The estimated soil loss ranges from 0 to 1276.71 t/ha/year, with an annual average of 30 t/ha/year (Fig. 6.8).

Soil erosion is very high with a steep slope in the southern part of the study area, while less erosion rest of the area with gentle slope. A steeper slope will enhance the flow resulting in a bigger power and amount of water to transport the soil. Potential soil erosion risk severity increases from the upper to the lower reaches of the taluk.

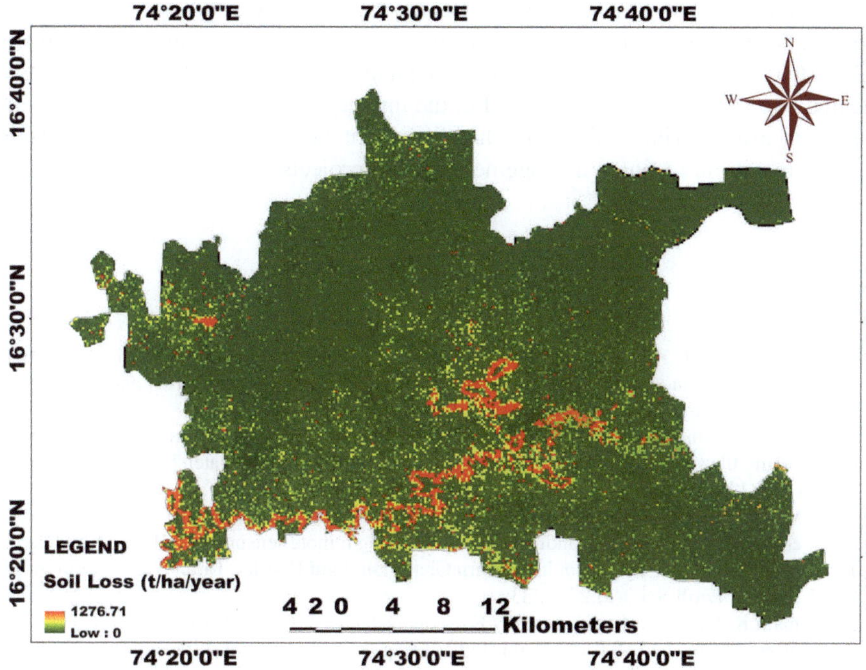

Fig. 6.8 Annual soil loss map

The higher soil erosion rates are mainly attributed to the relatively high values of the K coefficient (loamy soil), combined with the high values of the LS (high relief) and C (low vegetation cover) coefficients. The most common causes of soil erosion are high altitudes, steep slopes, vegetation removal and inadequate conservation activities. The results help to understand soil erosion risk, and its spatial pattern, and identify erosion-prone areas. Therefore, comprehensive management of land and water resources is essential in the high erosion zones for the sustainability of agriculture as well as the protection of soil resources. The results provide valuable information for sustainable management strategies in the study area.

6.7 Conclusion

RUSLE and GIS-based methods offer a potentially helpful method for locating soil erosion zones. The long-term average annual soil erosion rates on slopes can be predicted by the empirically based RUSLE model, which takes into account topography, soil type, crop system, rainfall patterns, and management techniques. All five factors are integrated into ArcGIS software using a raster calculator. The estimated soil loss ranges from 0 to 1276.71 t/ha/year with 30 t/ha/year annual mean. The final soil erosion map shows that high-risk areas are located in the southern portion study area with steep zones. To prevent degradation, new technologies must be developed and put to use, along with efficient management techniques. In areas that are severely affected, the research findings will aid in the implementation of appropriate erosion control measures. The study's conclusions might be helpful to different parties engaged in natural resource management-related projects.

References

Adediji A, Tukur AM, Adepoju KA (2010) Assessment of revised universal soil loss equation (RUSLE) in Katsina area, Katsina state of Nigeria using remote sensing (RS) and geographic information system (GIS). Iranica J Energy Environ 1(3):255–264

Altaf S, Meraj G, Romshoo SA (2014) Morphometry and land cover based multi-criteria analysis for assessing the soil erosion susceptibility of the western Himalayan watershed. Environ Monit Assess 186(12):8391–8412

Bagyaraj M, Gurugnanam B, Nagar A (2011) Significance of morphometry studies, soil characteristics, erosion phenomena and landform processes using remote sensing and GIS for Kodaikanal Hills, a global biodiversity hotpot in Western Ghats, Dindigul District, Tamil Nadu, South India. Res J Environ Earth Sci 3(3):221–233

Balasubramani K, Veena M, Kumaraswamy K, Saravanabavan V (2015) Estimation of soil erosion in a semi-arid watershed of Tamil Nadu (India) using revised universal soil loss equation (rusle) model through GIS. Model Earth Syst Environ 1(3):10

Biswas SS, Pani P (2015) Estimation of soil erosion using RUSLE and GIS techniques: a case study of Barakar River basin, Jharkhand, India. Model Earth Syst Environ 1(4):42. https://doi.org/10.1007/s40808-015-0040-3

Dabral PP, Baithuri N, Pandey A (2008) Soil erosion assessment in a hilly catchment of north eastern India using USLE, GIS and remote sensing. Water Resour Manag 22(12):1783–1798

De Vente J, Poesen J (2005) Predicting soil erosion and sediment yield at the basin scale: scale issues and semi-quantitative models. Earth Sci Rev 71(1–2):95–125

Devatha CP, Deshpande V, Renukaprasad MS (2015) Estimation of soil loss using USLE model for Kulhan watershed, Chattisgarh-a case study. Aquatic Proc 4:1429–1436

Dutta S, Roy S (2012) Determination of erosion surfaces and stages of evolution of Sangra drainage basin in Giridih district, Jharkhand, India. Int J Geomat Geosci 3(1):63

Ganasri BP, Ramesh H (2016) Assessment of soil erosion by RUSLE model using remote sensing and GIS-A case study of Nethravathi Basin. Geosci Front 7(6):953–961

Gelagay HS, Minale AS (2016) Soil loss estimation using GIS and remote sensing techniques: a case of Koga watershed, northwestern Ethiopia. Int Soil Water Conserv Res 4(2):126–136

Haan CT, Barfield BJ, Hayes JC (1994) Design hydrology and sedimentology for small catchments. Elsevier

Hadley RF (1985) Recent developments in erosion and sediment yield studies. Unesco

Jose CS, Das DC (1982) Geomorphic prediction models for sediment production rate and intensive priorities of watersheds in Mayurakshi catchment. In: Proceedings of international symposium on hydrological aspects of mountainous watersheds, school of hydrology, vol 1. UOR, Roorkee, pp 15–23

Khanchoul K, Balla F, Othmani O (2020) Assessment of soil erosion by Rusle model using Gis: a case study of Chemorah Basin, Algeria. Malaysian J Geosci 4(2):70–78

Koirala P, Thakuri S, Joshi S, Chauhan R (2019) Estimation of soil erosion in Nepal using a RUSLE modeling and geospatial tool. Geosciences 9(4):147

Kothyari UC, Jain SK (1997) Sediment yield estimation using GIS. Hydrol Sci J 42(6):833–843

Lee E, Ahn S, Im S (2017) Estimation of soil erosion rate in The Democratic People's Republic of Korea using the RUSLE model. For Sci Technol 13(3):100–108

Maner SB (1958) Factors affecting sediment delivery rates in the Red Hills physiographic area. EOS Trans Am Geophys Union 39(4):669–675

McCool DK, Brown LC, Foster GR, Mutchler CK, Meyer LD (1987) Revised slope steepness factor for the universal soil loss equation. Trans ASAE 30(5):1387–1396

Mengie MA, Hagos YG, Malede DA, Andualem TG (2022) Assessment of soil loss rate using GIS–RUSLE interface in Tashat watershed. Northwest Ethiopia J Sediment Environ 7(3):617–631

Ministry for Environment and Forests (MoEF) (2001) State of environment report: land degradation. In: Environmental information system (ENVIS), Ministry of Environment and Forests. Government of India, New Delhi

Moore ID, Burch GJ (1986) Physical basis of the length-slope factor in the universal soil loss equation 1. Soil Sci Soc Am J 50(5):1294–1298

Morgan RPC, Quinton JN, Smith RE, Govers G, Poesen JWA, Auerswald K, Chisci G, Torri D, Styczen ME (1999) Reply to discussion on 'the European soil erosion model (EUROSEM): a dynamic approach for predicting sediment transport from fields and small catchments'. Earth Surface Proce Landforms: J Br Geomorphol Res Group 24(6):567–568

Naqvi HR, Athick AMA, Ganaie HA, Siddiqui MA (2015) Soil erosion planning using sediment yield index method in the nun Nadi watershed, India. Int So Water Conserv Res 3(2):86–96

Nearing MA, Foster GR, Lane LJ, Finkner SC (1989) A process-based soil erosion model for USDA-water erosion prediction project technology. Trans ASAE 32(5):1587–1593

Pandey A, Chowdary VM, Mal BC (2007) Identification of critical erosion prone areas in the small agricultural watershed using USLE, GIS and remote sensing. Water Resour Manag 21(4): 729–746

Parveen R, Kumar U (2012) Integrated approach of universal soil loss equation (USLE) and geographical information system (GIS) for soil loss risk assessment in upper south Koel Basin, Jharkhand. J Geogr Inf Syst 4(06):588

Pérez-Rodríguez R, Marques MJ, Bienes R (2007) Spatial variability of the soil erodibility parameters and their relation with the soil map at subgroup level. Sci Total Environ 378(1–2): 166–173

Prasannakumar V, Shiny R, Geetha N, Vijith H (2011) Spatial prediction of soil erosion risk by remote sensing, GIS and RUSLE approach: a case study of Siruvani river watershed in Attapady valley, Kerala, India. Environ Earth Sci 64:965–972. https://doi.org/10.1007/s12665-011-0913-3

Rawat KS, Tripathi VK, Mishra AK (2014) Sediment yield index mapping and prioritization of Madia subwatershed, Sagar District of Madhya Pradesh (India). Arab J Geosci 7(8):3131–3145

Renard K, Foster G, Weesies G, McCool D, Yoder D (1997) predicting soil erosion by water: a guide to conservation planning with the revised universal soil loss equation (RUSLE). US Department of Agriculture, Agriculture Handbook No.703USDA, USDA, Washington DC

Richards K (1993) Sediment delivery and the drainage network. In: Beven K, Kirkby MJ (eds) Channel network hydrology, pp 221–254

Roehl JE (1962) Sediment source areas, and delivery ratios influencing morphological factors. Int Assoc Hydro Sci 59:202–213

Shi ZH, Cai CF, Ding SW, Wang TW, Chow TL (2004) Soil conservation planning at the small watershed level using RUSLE with GIS: a case study in the three gorge area of China. Catena 55(1):33–48

Singh G, Chandra S, Babu R (1981) Soil loss and prediction research in India, Bulletin No. T-12/D9, Central Soil and Water Conservation Research Training Institute, Dehradun

Stone RP, Hilborn D (2000) Universal soil loss equation-factsheet. http://www.omafra.gov.on.ca/english/engineer/facts/00-001.htm

Sumantra B, Padmini P (2015) Estimation of soil erosion using RUSLE and GIS techniques: a case study of Barakar River Basin, Jharkhand, India. Modeling Earth Systems and Environment, 1

Suresh Kumar BV, Sunil Kumar RK, Kaliraj S (2017) Environmental impact assessment (EIA) and assessment of soil loss in Sandur taluk due to mining activity in and around Bellary District, South India. Int J 6(1):587–595

Walling DE (1983) The sediment delivery problem. J Hydrol 65(1–3):209–237

Walling DE (1988) Erosion and sediment yield research – some recent perspectives. J Hydrol 100(1–3):113–141

Williams JR, Berndt HD (1972) Sediment yield computed with universal equation. J Hydraul Div 98(12):2087

Zachar D (1982) Soil erosion. Developments in soil science, vol 10. Elsevier Scientific, Amsterdam

Chapter 7
A Geospatial Approach for DEM-Based Surface Hydrological Modelling of Sub-Watersheds Towards Climate Disaster Management

Ashish Mani, Ruchi Badola, Syed Ainul Hussain, and Maya Kumari

Abstract The management, development, and prioritization of watersheds are essential for preserving not only the water resources but also the surrounding soil and vegetation. This depends on the hydrological evaluation and modelling of a particular watershed. The focus of the study for the watershed management of Bandal River sub-watersheds is based on morphometric analysis method, a Digital Elevation Model (DEM)-based surface hydrological modelling. Bandal River is in the northeast of the Doon Valley and flows around 16 km to meet Song River at Maldevta, Dehradun. This river frequently results in flash floods, cloud bursts, and landslides, especially in the monsoon season due to the valley region and sloping terrain. The combined area of all four sub-watersheds of Bandal River is 82.35 km^2. The morphometric analysis of the sub-watersheds was carried out using Shuttle Radar Topography Mission (SRTM) DEM data at a 30 m resolution. Several other significant hydrological and topographical components of sub-watersheds namely, slope, aspect, elevation, and drainage density have been analysed. This study concludes that the Bandal River sub-watersheds have dendritic and parallel drainage patterns with moderate to high relief. The watershed's dendritic and parallel drainage

A. Mani (✉)
Ganga Aqualife Conservation Monitoring Centre, Wildlife Institute of India, Dehradun, Uttarakhand, India

Amity School of Natural Resources & Sustainable Development (ASNRSD), Amity University Uttar Pradesh (AUUP), Noida, Uttar Pradesh, India
e-mail: ashish.mani@s.amity.edu

R. Badola · S. A. Hussain
Ganga Aqualife Conservation Monitoring Centre, Wildlife Institute of India, Dehradun, Uttarakhand, India
e-mail: ruchi@wii.gov.in

M. Kumari
Amity School of Natural Resources & Sustainable Development (ASNRSD), Amity University Uttar Pradesh (AUUP), Noida, Uttar Pradesh, India
e-mail: mkumar10@amity.edu

D. Kumar, N. P. Bassill (eds.), *Artificial Intelligence and Machine Learning for Climate Disaster Management*, Disaster Resilience and Green Growth,
https://doi.org/10.1007/978-981-95-1588-2_7

patterns facilitate understanding of a variety of topographical aspects, such as runoff and infiltration rate. The sub-watersheds 2 and 3 are elongated, sub-watershed 4 is less elongated, and sub-watershed 1 is oval. Additionally, sub-watersheds with an elongated shape are more susceptible to soil erosion and landslides with higher runoff potential. In comparison to other methods, the DEM-based method at the sub-watershed scale is more accurate and efficient for surface hydrological assessment. Various managers and decision-makers in sustainable watershed management would benefit from this study.

Keywords Geospatial techniques · Morphometric analysis · Digital elevation model (DEM) · Surface hydrology · Watershed management

7.1 Introduction

Due to rising water demand, climate change, and urbanization, the globe is heading toward a water scarcity crisis (Boretti and Rosa 2019; Grill et al. 2019; Mani et al. 2021; Mani et al. 2023a, b, c). Watershed characterization and water resource management become essential if we need to protect the world from water crises (Sivakumar 2011; Srinivasan et al. 2012; Dobriyal et al. 2017). Especially in India, where the demand and supply of water resources are increasing day by day, proper hydrological evaluation and modelling of water resources are essential (Mall et al. 2006). To ensure an adequate supply of water resources, the management of watersheds and sub-watersheds is of paramount significance (Gebre et al. 2015).

Digital Elevation Models (DEM) have become integral to hydrological modelling due to their ability to provide detailed information about the terrain, including elevation, slope, aspect, and flow patterns (Moore et al. 1991; Mani et al. 2023a, b, c). Morphometric analysis: DEM-based surface hydrological modelling is a precise and efficient method for watershed characterization (Arabameri et al. 2020). It helps in determining the morphological and topographical characteristics of watershed and sub-watersheds and their effects on the dynamics of surface water (Shekar and Mathew 2022).

The conventional approaches to planning watersheds included laborious data collection, predictions, and assessments based on hard copies of aerial photographs. This was a time-consuming and error-prone task. Also, the error rates were very high (Mani and Kumar 2020). A geospatial approach, on the other hand, begins with a digital representation of the DEM and multispectral satellite images. They can be found and accessed with ease in both public and private domains (Guertin et al. 2000). In the geospatial approach, assumptions may be confirmed and various investigations can be done by the desktop evaluation itself rather than sending a survey crew to the site and there is also improved image display. Geospatial-based models have made suitable techniques for morphometric analysis and watershed management available because to their accessibility and potential (Chowdary et al. 2009).

Many studies have used this DEM-based surface hydrological modelling method to examine watersheds and sub-watersheds in different landscapes. These studies have shown that this method is an extremely effective scientific tool for defining and developing the characteristics of watersheds and sub-watersheds. Singh et al. (2014) examine the use of remote sensing DEM data for analysing the hydrological characteristics of Orr watersheds. This chapter emphasizes that this technique has better precision than conventional methods. According to the research done by Mani et al. (2022), the use of DEM-based morphometric analysis gives vital information into the areal, linear, and relief characteristics of the Suswa River basin.

This research focuses on the DEM-derived surface hydrological modelling based on morphometric analysis of the Bandal River sub-watersheds for watershed management. We aim to enhance watershed management of the Bandal sub-watersheds through this evaluation. These scientific databases can also be useful sources for upcoming hydrological research. Also, the findings of this study will help decision-makers in other fields by improving their comprehension of natural resources.

7.2 Methodology

7.2.1 Study Area

The study area lies on the northeast of the Doon Valley of Uttarakhand, India. Its coordinates fall between Latitude (from 30°20′20″ N to 30°26′50″ N) and Longitude (from 78°7′30″ E to 78°15′30″ E). The Bandal River, a small tributary of the Song River, originates from the Dhanaulti mountain ranges of the Lesser Himalayas, flows in the southwest direction, and merges with the Song River at Maldevta, Dehradun, which further merges with the Ganga River. The total length of the Bandal River is around 16 km. The overall area of all four sub-watersheds of Bandal River is 82.35 km^2 with an elevation range from 719 m to 2581 m. The annual average rainfall in the area is around 2000 mm. The annual temperature ranges from −1 °C in winters to 42 °C in summers. The region is surrounded by hills, boulders, and deciduous forest cover. The Main Boundary Thrust (MBT) passes close to the sub-watersheds, and the presence of the Krol and Tal geological formations in the area makes the sub-watersheds geology prone to landslides (Dudeja et al. 2011). Additionally, in recent years, due to its scenic landscape, Bandal River Valley has faced rapid urban development, which is a major cause of landslides in the region. High biotic pressure is also affecting the river and is associated with stream water quality and biodiversity. Figure 7.1 provides a map of the study area.

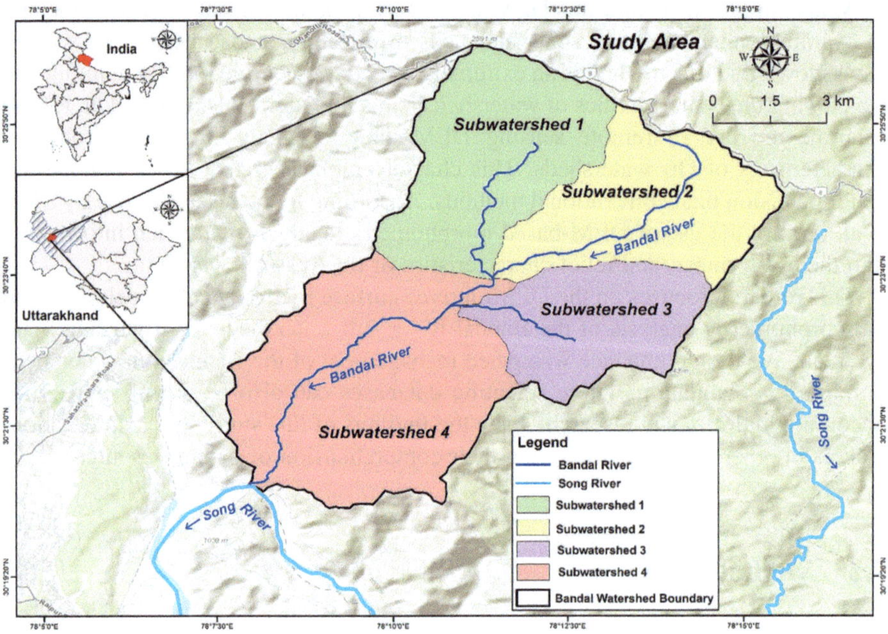

Fig. 7.1 The study area map

Table 7.1 Data type and data source

S. no.	Type of data	Detail of data	Source
1	SRTM DEM	30 m resolution of date 23/09/2014	United States geological survey (USGS) earth explorer https://earthexplorer.usgs.gov/
2	Survey of India Toposheet	Toposheet no.: 53 J3 and 53 J7 at 1: 50,000 scale	Survey of India (SOI), Dehradun, India

7.2.2 Data and Methods

This study employs an integrated approach using DEM data and Survey of India toposheets for geodatabase creation and subsequent hydrological analyses (Table 7.1). For delineating the sub-watershed boundary and drainage network, the Spatial Analysis Tools (SATs) in ArcGIS desktop software have been used. Additionally, the Sub-watersheds map, Slope map, Aspect map, Elevation map, and Drainage Density map have all been created using ArcGIS desktop software. The formulas for performing morphometric analysis have been shown in Table 7.2.

Table 7.2 Morphometric analysis formulas

S. no.	Parameter	Formula	References
1	Stream order (w)	*Hierarchical rank*	Strahler (1964)
2	Stream length (L_u)	*Length of the stream*	Horton (1945)
3	Mean stream length (L_{sm})	$L_{sm} = L_u/N_u$	Strahler (1964)
4	Stream length ratio (R_L)	$R_L = L_u / (L_u - 1)$	Horton (1945)
5	Bifurcation ration (R_b)	(R_b) = nu/nu +1	Schumm (1956)
6	Mean bifurcation ratio (R_{bm})	R_{bm} = *average of bifurcation ratios of all orders*	Strahler (1957)
7	Drainage density (D_d)	$D_d = L_u/A$	Horton (1945)
8	Drainage texture (T_d)	$T_d = N_u/P$	Horton (1945)
9	Stream frequency (F_s)	$F_s = N_u/A$	Horton (1945)
10	Elongation ratio (R_e)	$Re = 2\surd(A/\pi)/L_b$	Schumm (1956)
11	Circularity ratio (R_c)	$R_c = 4\ \pi\ A/P^2$	Miller (1953)
12	Form factor (F_f)	$F_f = A/L^2$	Horton (1945)
13	Basin relief (R)	$R = H — h$	Strahler (1952)
14	Relief ratio (Rr)	$R_r = R/L$	Schumm (1956)
15	Compactness coefficient (cc)	$Cc = 0.2821*P/(A)0.5$	Horton (1945)

7.3 Results and Discussion

7.3.1 Morphometric Analysis

The morphometric analysis explains the linear aspects, areal aspects, and relief aspects of the drainage basin, which is useful to understand its hydrological characteristics (Mani et al. 2023a, b, c).

7.3.1.1 Linear Aspect

The one-dimensional linear aspect defines the linear characteristics of the sub-watersheds such as Stream Number, Stream Length, Stream Order, Length ratio, and Bifurcation ratio. The total number of streams in the Bandal sub-watersheds is 1763 (Table 7.3). The highest number of streams is in sub-watershed 4. All four sub-watersheds follow Horton's first law: the stream number decreases with increasing stream order, as represented by the graph in Fig. 7.2. The Bandal sub-watershed's overall stream order ranges from 1 to 6 as shown in Fig. 7.3. The mean bifurcation ratio ranges from 3.97 to 4.92. Sub-watershed 3 has the lowest value of mean bifurcation ratio 3.97 and sub-watershed 4 has the highest value 4.92. The bifurcation ratio explains the drainage pattern types in the sub-watersheds. For this study, the patterns are dendritic and parallel in shape.

Table 7.3 Linear aspects table

Sub-watershed 1	Stream order (w)	No. of streams (nu)	Length of streams (km) (Lu)	Mean length of streams (km) (Lsm)	Cumulative mean stream length (km)	Length ratio (RL)	Bifurcation ratio (RbF)	Mean bifurcation ratio (Rbm)
	1	321	57.37	0.18	0.18			
	2	67	20.27	0.30	0.48	0.35	4.79	
	3	14	13.83	0.99	1.47	0.68	4.79	4.27
	4	4	4.88	1.22	2.69	0.35	3.50	
	5	1	5.27	5.27	7.96	1.08	4.00	
	Total	**407**	**101.62**					
Sub-watershed 2	Stream order (w)	No. of streams (nu)	Length of streams (km) (Lu)	Mean length of streams (km) (Lsm)	Cumulative mean stream length (km)	Length ratio (RL)	Bifurcation ratio (RbF)	Mean bifurcation ratio (Rbm)
	1	293	56.72	0.19	0.19			
	2	58	15.13	0.26	0.45	0.27	5.05	
	3	13	9.96	0.77	1.22	0.66	4.46	4.19
	4	4	4.14	1.03	2.25	0.42	3.25	
	5	1	5.19	5.19	7.44	1.25	4.00	
	Total	**369**	**91.14**					
Sub-watershed 3	Stream order (w)	No. of streams (nu)	Length of streams (km) (Lu)	Mean length of streams (km) (Lsm)	Cumulative mean stream length (km)	Length ratio (RL)	Bifurcation ratio (RbF)	Mean bifurcation ratio (Rbm)
	1	233	41.03	0.18	0.18			
	2	46	10.78	0.23	0.41	0.26	5.07	
	3	12	6.23	0.52	0.93	0.58	3.83	3.97
	4	3	6.39	2.13	3.06	1.03	4.00	
	5	1	3.35	3.35	6.41	0.52	3.00	
	Total	**295**	**67.77**					

Sub-watershed 4	Stream order (w)	No. of streams (nu)	Length of streams (km) (Lu)	Mean length of streams (km) (Lsm)	Cumulative mean stream length (km)	Length ratio (RL)	Bifurcation ratio (RbF)	Mean bifurcation ratio (Rbm)
	1	554	96.15	0.17	0.17			
	2	108	33.28	0.31	0.48	0.35	5.13	
	3	25	15.24	0.61	1.09	0.46	4.32	4.92
	4	4	8.28	2.07	3.16	0.54	6.25	
	6	1	9.59	9.59	12.75	1.16	4.00	
	Total	**692**	**162.54**					

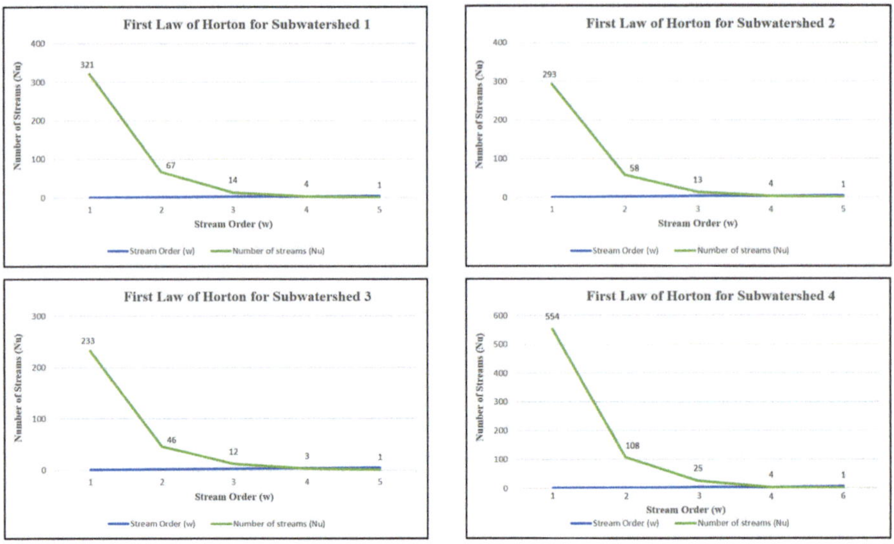

Fig. 7.2 First law of Horton for the Bandal River sub-watersheds

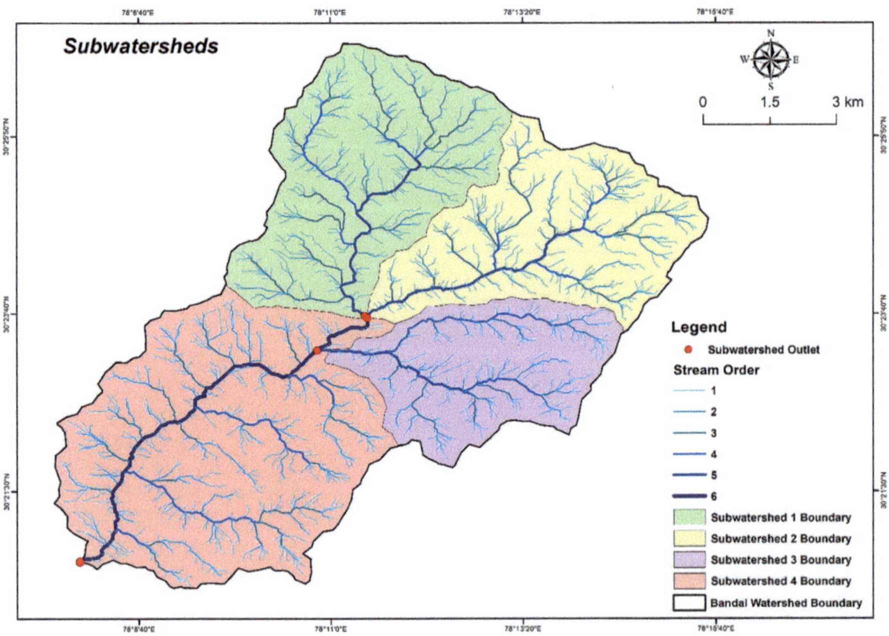

Fig. 7.3 Map of the Bandal River sub-watersheds

Table 7.4 Areal aspects table

Sub-watersheds	Area (km^2)	Perimeter (km)	Length (km)	Form factor (Ff)	Elongation ratio (Re)	Circularity ratio (Rc)	Drainage density (Dd) (Km/Km^2)	Stream Frequency (Fs)	Drainage Texture (Dt)	Compactness Coefficient (Cc)
Sub-watershed 1	19.99	19.96	6.15	0.53	0.82	0.63	5.08	20.36	20.39	1.26
Sub-watershed 2	17.46	20.51	7	0.36	0.67	0.52	5.22	21.13	17.99	1.38
Sub-watershed 3	13.35	17.05	6	0.37	0.69	0.58	5.07	22.10	17.30	1.32
Sub-watershed 4	31.55	26.88	8.29	0.46	0.76	0.55	5.15	21.93	25.74	1.35

7.3.1.2 Areal Aspect

The areal aspect is two-dimensional. It includes the areal characteristics of the sub-watersheds such as area, drainage density, drainage texture, perimeter, elongation ratio, circularity ratio, stream frequency, form factor, and compactness coefficient. The area of four sub-watersheds of Bandal River are 19.99 km^2 of sub-watershed 1, 17.46km^2 of sub-watershed 2, 13.35km^2 of sub-watershed 3, and 31.55km^2 of sub-watershed 4 (Table 7.4). The elongation ratio has been classified into four classes: circular (> 0.9), oval (0.9–0.8), less elongated (0.8–0.7), and elongated (< 0.7) as per Schumm (1956). The shapes of the sub-watersheds 2 and 3 are elongated, with values of 0.67 and 0.69, sub-watershed 4 is less elongated, with a value of 0.76; and sub-watershed 1 is oval, with a value of 0.82. The sub-watersheds' circular shape is determined by the circulatory ratio. It has a direct relation with the elongation ratio. Sub-watersheds 1 and 2 have the highest and lowest circulatory ratio values 0.63 and 0.52, respectively. The flow intensity of the basin is determined by the form factor. The form factor values for sub-watersheds 1 and 2 are 0.53 and 0.36, respectively. As per Singh et al. (1945), the stream frequency is directly related to drainage density. For sub-watersheds 3 and 1, the greatest and lowest values of stream frequency are 22.10 and 20.36, respectively. For sub-watersheds 2 and 3, the highest and lowest values of drainage density are 5.22 km/km^2 and 5.07 km/km^2, respectively. The drainage texture value of Bandal sub-watersheds falls between 17 and 26 which implies that Bandal sub-watersheds have very fine drainage texture (Smith 1950). The Overall compactness of the sub-watersheds is determined by the compactness coefficient. Sub-watershed 1 has the lowest value, 1.26 whereas sub-watershed 2 has the highest value, 1.38. As per Sinha and Eldho (2021) Greater than 1 value of compactness coefficient is associated with a higher risk of landslide and erosion therefore, Bandal sub-watersheds are vulnerable to landslide and erosion. Less infiltration, less permeability, and higher runoff potential are indicated by lower values of the elongation ratio, form factor, and circulation ratio, whereas more infiltration, high permeability, and lesser runoff potential are indicated by higher values of these variables. (Mundetia et al. 2018; Das et al. 2021) (Table 7.5).

Table 7.5 Relief aspects table

Sub-watersheds	Height of sub-watershed mouth (z) m	Maximum height of the sub-watershed (Z) m	Basin relief (R) m	Relief ratio (Rr)
Sub-watershed 1	1101	2581	1480	240.65
Sub-watershed 2	1100	2561	1461	208.71
Sub-watershed 3	1008	2224	1216	202.67
Sub-watershed 4	719	1993	1274	153.68

7.3.1.3 Relief Aspect

The third-dimensional aspect is relief. For the Bandal River sub-watersheds, the basin relief value ranges from 1216 m to 1480 m. The sub-watershed 1 has having maximum basin relief value of 1480 m and sub-watershed 3 has a minimum basin relief value of 1216 m, which indicates the Bandal River sub-watersheds have moderate to high relief. The relief ratio values are ranging from 153.38 for sub-watershed 4 to 240.65 for sub-watershed 1. The high value of the relief ratio implies that the maximum area of the Bandal River sub-watersheds is having a moderate to steep slope, which suggests that the area is prone to landslides (Shit et al. 2016).

7.3.2 Slope

Understanding the movement of water across the surface of the Earth requires an understanding of slope, a fundamental term in hydrology. Usually, it refers to the steepness or gradient of the terrain. As indicated in Fig. 7.4, the Bandal sub-watersheds slope is categorized into five classes: (< = 5°) is very gentle,

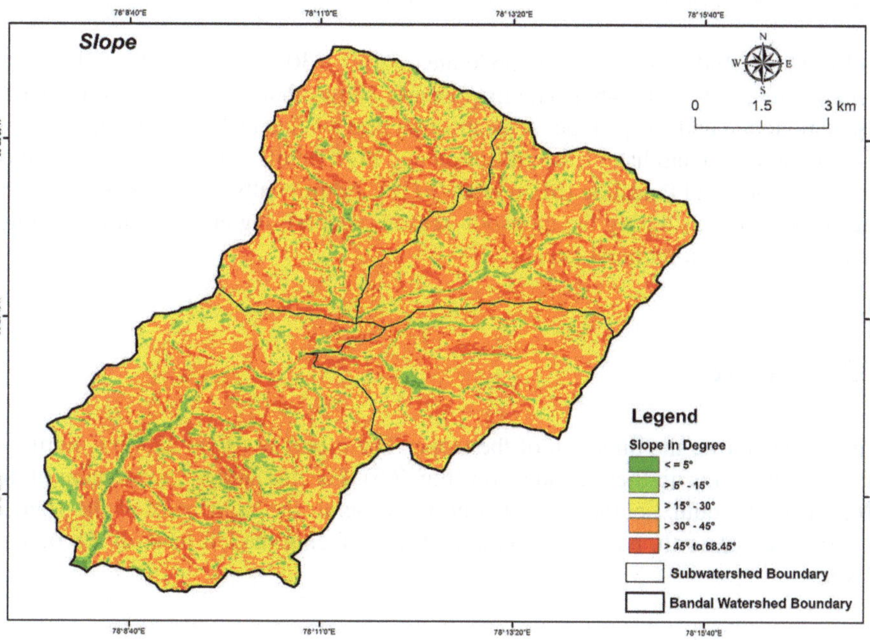

Fig. 7.4 Slope map of the Bandal River sub-watersheds

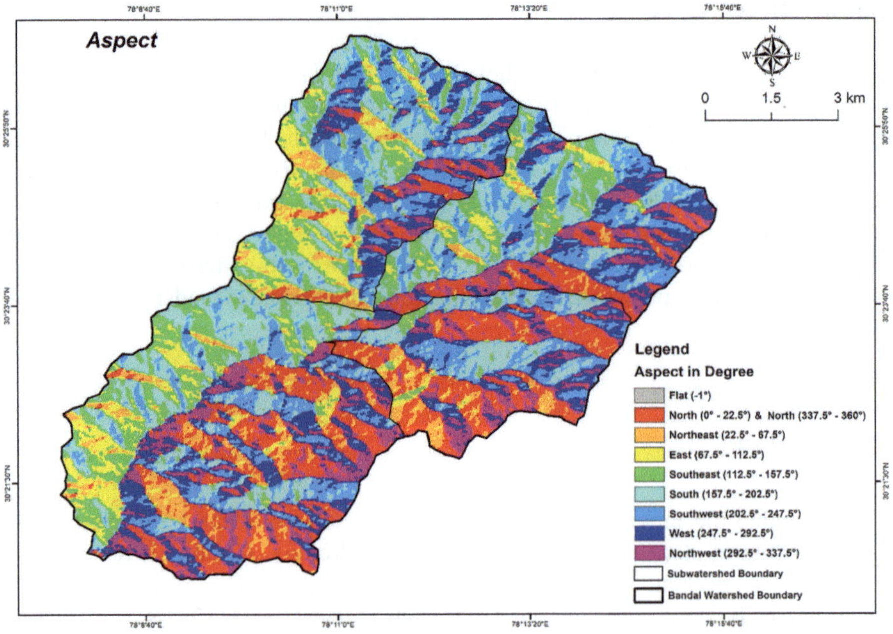

Fig. 7.5 Aspect map of the Bandal River sub-watersheds

followed by gentle ($> 5^\circ - 15^\circ$), moderate ($> 15^\circ - 30^\circ$), steep ($> 30^\circ - 45^\circ$), and very steep ($> 45^\circ - 68.45^\circ$). The map clearly shows that the majority of the basin has a moderate and steep slope. A steep slope implies that the sub-watersheds are in the valley region and have less infiltration and high runoff with greater possibilities of soil erosion and landslide (Mani et al. 2022). Additionally, a steep slope contributes to flash floods during the monsoon season, which bring about calamities in the area because of cloud burst events.

7.3.3 *Aspect*

The aspect shows the direction of the slope. The aspect at (0° to 22.5°) is north, at (22.5° to 67.5°) is northeast, and so on (Fig. 7.5). For this study, the direction of the slope is east, south, southeast, and southwest facing the right bank of the Bandal River, whereas north, northwest, west, and south facing the left bank of the Bandal River.

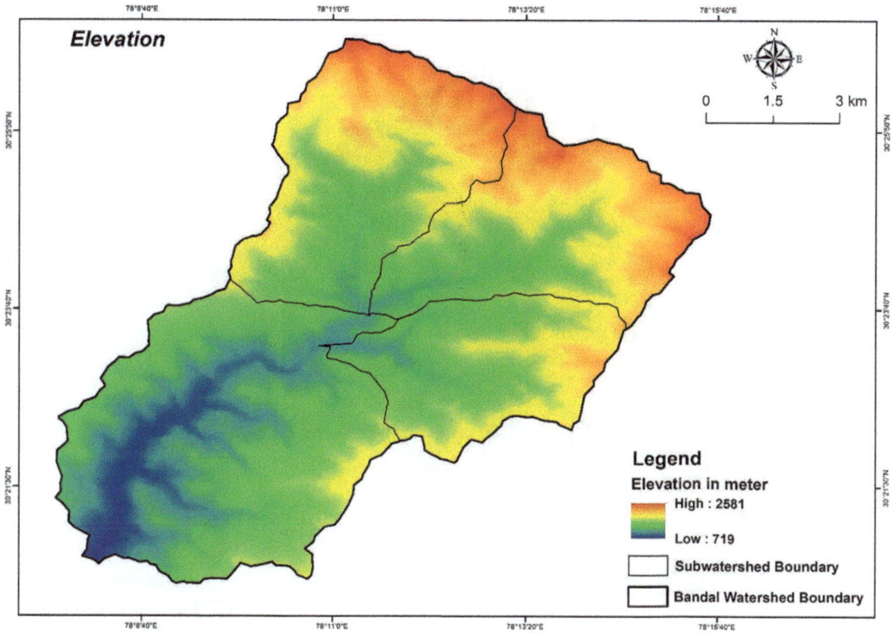

Fig. 7.6 Elevation map of the Bandal River sub-watersheds

7.3.4 *Elevation*

Elevation is another topographical factor significant for the study of sub-watersheds. It is an essential component of hydrological study and is crucial for managing water resources and determining the impacts of environmental changes on hydrological processes. The elevation of Bandal River sub-watersheds ranges from 719 m to 2581 m (Fig. 7.6). Most of the area has higher elevation, which means that Bandal River flows through the valley region in the southwest direction. For describing a region's topography and understanding its impact on the movement, storage, and availability of water, elevation data analysis is essential (Yeh et al. 2009).

7.3.5 *Drainage Density*

In hydrology research, drainage density is a crucial parameter that is used to describe the shape and drainage flow of stream networks in a particular area. It is defined as the total length of stream channels (or rivers) within a watershed or sub-watershed divided by the total area of that watershed. As indicated in Fig. 7.7, the drainage density of Bandal River sub-watersheds is categorized into five classes: (< = 5 km/

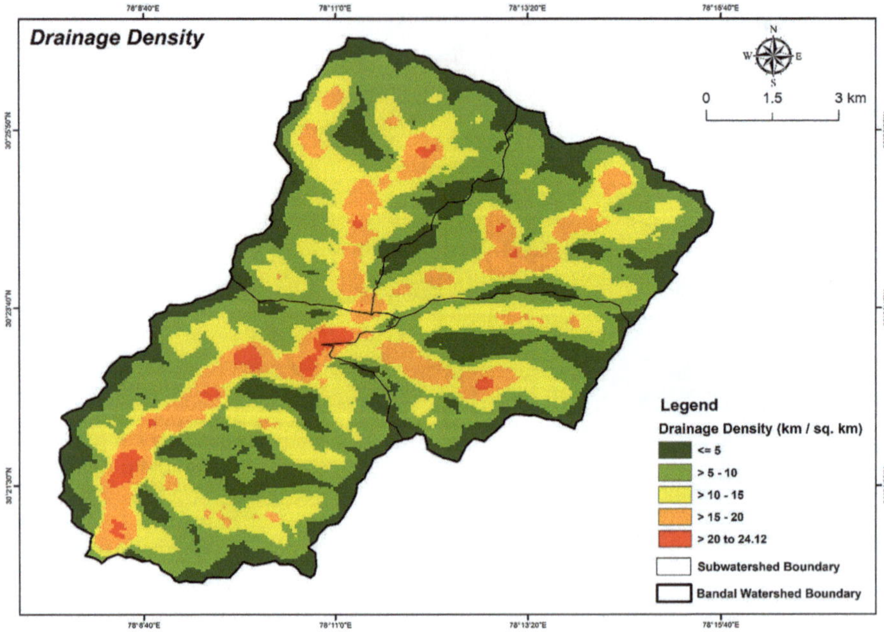

Fig. 7.7 Drainage density map of the Bandal River sub-watersheds

km^2) is very low, followed by low (> 5 km/km^2–10 km/km^2), medium (> 10 km/ km^2–15 km/km^2), high (> 15 km/km^2–20 km/km^2), and very high (> 20 km/km^2–24.12 km/km^2). Most of the areas in the sub-watersheds have low drainage density, which implies higher slope, lesser infiltration, and higher runoff potential due to the valley region. As an important parameter for hydrologists and water resource managers, drainage density provides insights into the spatial distribution of stream networks, which influences hydrological behaviour and is essential for evaluating and managing water resources, especially in the context of watershed management and environmental conservation (Singh et al. 2013).

7.4 Conclusion

In conclusion, it is critical to manage watersheds and sub-watersheds sustainably to protect not only our limited water supplies but also the ecological well-being of the nearby soil and vegetation. The creation of precise watershed modelling and a thorough hydrological assessment is essential for achieving these goals. The findings of the study focus on various hydrological characteristics of the Bandal River sub-watersheds. These sub-watersheds are notable for their combination of dendritic and parallel drainage patterns with moderate to high relief. Understanding numerous topographical aspects, such as gradient, surface runoff, and infiltration rate, can be facilitated by the watershed's dendritic and parallel drainage pattern. Also, this study holds significant implications for the vulnerability of these sub-watersheds to soil erosion, landslides, cloud burst, flash flood, and their runoff potential. Particularly elongated forms are linked to increased sensitivity to these hydrological challenges. This DEM-based surface hydrological modelling approach provides an efficient framework for well-informed decision-making, especially for stakeholders, managers, and policymakers involved in the management of watersheds and natural resources. The knowledge gained from this research is positioned to influence and guide future actions as we continue to face environmental issues and the urgency to conserve our priceless water resources. We can take preventive steps to reduce hazards, improve ecological resilience, and guarantee the long-term health and sustainability of the Bandal River sub-watersheds by incorporating these findings into watershed management policies. Ultimately, this study provides a way towards more efficient and resilient watershed management practices, serving as a vital resource for individuals committed to the responsible care of our natural environment and water resources.

Acknowledgement The authors express gratitude to Amity School of Natural Resources and Sustainable Development, Amity University, Noida, and acknowledge the support of the NMCG-WII Project (Phase—II), Ganga Aqualife Conservation Monitoring Centre (GACMC), and Wildlife Institute of India, Dehradun.

References

Arabameri A, Tiefenbacher JP, Blaschke T, Pradhan B, Tien Bui D (2020) Morphometric analysis for soil erosion susceptibility mapping using novel GIS-based ensemble model. Remote Sens 12(5):874

Boretti A, Rosa L (2019) Reassessing the projections of the world water development report. NPJ Clean Water 2:15

Chowdary VM, Ramakrishnan D, Srivastava YK, Chandran V, Jeyaram A (2009) Integrated water resource development plan for sustainable management of Mayurakshi watershed, India using remote sensing and GIS. Water Resour Manag 23(8):1581–1602

Das B, Singh S, Jain SK, Thakur PK (2021) Prioritization of sub-basins of Gomti river for soil and water conservation through morphometric and LULC analysis using remote sensing and GIS. J Indian Soc Remote Sens 49:2503–2522

Dobriyal P, Badola R, Tuboi C, Hussain SA (2017) A review of methods for monitoring streamflow for sustainable water resource management. Appl Water Sci 7(6):2617–2628

Dudeja D, Bartarya SK, Biyani AK (2011) Hydrochemical and water quality assessment of groundwater in Doon Valley of outer Himalaya, Uttarakhand, India. Environ Monit Assess 181:183–204. https://doi.org/10.1007/s10661-010-1823-7

Gebre T, Kibru T, Tesfaye S, Taye G (2015) Analysis of watershed attributes for water resources management using GIS: the case of Chelekot micro-watershed, Tigray, Ethiopia. J Geogr Inf Syst 7(02):177

Grill G et al (2019) Mapping the world's free-flowing rivers. Nature 569(7755):215–221

Guertin DP, Miller SN, Goodrich DC (2000) Emerging tools and technologies in watershed management. In: USDA Forest Service proceedings RMRS-P-13: proceedings of the conference on land stewardship in the 21st century: the contributions of watershed management, Tucson, pp 194–204

Horton RE (1945) Erosional development of streams and their drainage basins: Hydrophysical approach to quantitative morphology. Geol Soc Am Bull 56:275–370

Mall RK, Gupta A, Singh R, Singh RS, Rathore LS (2006) Water resources and climate change: an Indian perspective. Curr Sci:1610–1626

Mani A, Kumar D (2020) Morphometric analysis of Manali watershed of Beas River basin for watershed management. VayuMandal 46:21–29

Mani A, Kumar D, Kumar D (2021) Evaluating decadal change in green cover of Dehradun City. In: Sharma P (ed) Geospatial technology and smart cities. The urban book series. Springer, Cham. https://doi.org/10.1007/978-3-030-71945-6_26

Mani A, Kumari M, Badola R (2022) Morphometric analysis of Suswa River basin using geospatial techniques. Eng Proc 27(1):65

Mani A, Bansal D, Kumari M, Kumar D (2023a) Land use land cover changes and climate change impact on the water resources: a study of Uttarakhand state. In: River conservation and water resource management. Springer Nature Singapore, Singapore, pp 1–16

Mani A, Kumari M, Badola R (2023b) A GIS-based assessment of Asan River basin for watershed management. In: Proceedings of 42nd INCA international congress on digital cartography to harness blue economy. Dehradun, INCA

Mani A, Kumari M, Badola R (2023c) Evaluating urban topography and land use changes for Urban River management using geospatial techniques. Eng Proc 58(1):12

Miller VC (1953) A quantitative geomorphic study of drainage basin characteristics in the Clinch Mountain area Virginia and Tennessee. New York (NY): Department of Geology, ONR, Columbia University, Virginia and Tennessee (Proj. NR 389–402, Technical Report 3)

Moore ID, Grayson RB, Ladson AR (1991) Digital terrain modelling: a review of hydrological, geomorphological, and biological applications. Hydrol Process 5(1):3–30

Mundetia N, Sharma D, Dubey SK (2018) Morphometric assessment and sub-watershed prioritization of Khari River basin in semi-arid region of Rajasthan, India. Arab J Geosci 11:1–18

Schumm SA (1956) Evolution of drainage systems and slopes in Badlands at Perth Amboy, New Jersey. Geol Soc Am Bull 67:597–646

Shekar PR, Mathew A (2022) Morphometric analysis for prioritizing sub-watersheds of Murredu River basin, Telangana state, India, using a geographical information system. J Eng Appl Sci 69(1):1–30

Shit PK, Bhunia GS, Maiti R (2016) Potential landslide susceptibility mapping using weighted overlay model (WOM). Model Earth Syst Environ 2:1–10

Singh P, Thakur JK, Singh UC (2013) Morphometric analysis of Morar River basin, Madhya Pradesh, India, using remote sensing and GIS techniques. Environ Earth Sci 68(7):1967–1977

Singh P, Gupta A, Singh M (2014) Hydrological inferences from watershed analysis for water resource management using remote sensing and GIS techniques. Egypt J Remote Sens Space Sci 17(2):111–121

Sinha RK, Eldho TI (2021) Assessment of soil erosion susceptibility based on morphometric and landcover analysis: a case study of Netravati River basin, India. J Indian Soc Remote Sens 49: 1709–1725

Sivakumar B (2011) Water crisis: from conflict to cooperation – an overview. Hydrol Sci J 56(4): 531–552

Smith KG (1950) Standards for grading texture of erosional topography. Am J Sci 248:655–668

Srinivasan V, Lambin EF, Gorelick SM, Thompson BH, Rozelle S (2012) The nature and causes of the global water crisis: syndromes from a meta-analysis of coupled human-water studies. Water Resour Res 48(10)

Strahler AN (1952) Hypsometric (area-altitude) analysis of erosional topography. Geol Soc Am Bull 63:1117–1142

Strahler AN (1957) Quantitative analysis of watershed geomorphology. *Trans Am Geophys Union* 38:913–920

Strahler AN (1964) Quantative geomorphology of drainage basins and channel networks. In: Te Chow V (ed) Hand book of applied hydrology, 1st edn. New York, McGraw Hill Book Company, pp 439–476

Yeh HF, Lee CH, Hsu KC, Chang PH (2009) GIS for the assessment of the groundwater recharge potential zone. Environ Geol 58:185–195

Chapter 8
Urban Heat Island Mitigation Through Green Infrastructure for Sustainable Development

Prashant Shukla, Samanpreet Kaur, Abhishek Pandey, Pramod Kumar Mishra, and Mukesh Siag

Abstract Urban heat island (UHI) presents a significant concern to the well-being of urban residents and the sustainability of urban environments due to urbanisation and industrialisation worldwide, and it has drawn considerable attention recently. UHI is caused by several factors, including the consequences of global climate change and the widespread conversion of natural green spaces into paved surfaces as a result of rapid urban growth. The UHI effect lowers air quality, increases water demand, increases energy consumption owing to cooling requirements, and increases morbidity and death from heat stress. Green infrastructure or adding natural and semi-natural features to urban areas is one viable option. This work addresses the approach for UHI source and mitigation with an evaluation of how green infrastructure helps to lessen the negative effects. This addresses the benefits of green infrastructure comprising social, economic, and environmental aspects like increased community well-being, better air and water quality, and increased urban biodiversity. Understanding the relationship between green infrastructure and UHI reduction helps to formulate well-informed plans. This approach can be replicated in other cities to create more resilient, sustainable, and pleasant urban environments.

Keywords Urban heat Island · Green infrastructure · Geographical information system · Climate change · Urbanisation · Industrialisation

---

P. Shukla (✉) · A. Pandey
Division of Agricultural Engineering, ICAR-Indian Agricultural Research Institute, New Delhi, India

S. Kaur · M. Siag
Department of Soil and Water Engineering, Punjab Agricultural University, Ludhiana, Punjab, India
e-mail: samanpreet@pau.edu

P. K. Mishra
Acharya Narendra Deva University of Agriculture and Technology, Kumarganj, Ayodhya, Uttar Pradesh, India
e-mail: pramod.btag@gmaol.com

D. Kumar, N. P. Bassill (eds.), *Artificial Intelligence and Machine Learning for Climate Disaster Management*, Disaster Resilience and Green Growth,
https://doi.org/10.1007/978-981-95-1588-2_8

8.1 Introduction

The functionality and livability of cities and urban regions are compromised by the global phenomenon known as the Urban Heat Island (UHI). Significantly higher temperatures in cities than in the surrounding countryside are the hallmark of this heat island effect (Nakayama and Fujita 2010; Synnefa et al. 2011). Urban landscape changes, such as a reduction in vegetation and evapotranspiration, a rise in the frequency of dark surfaces with poor albedo, and an increase in anthropogenic heat output, are blamed for the creation of the heat island effect (Stone Jr et al. 2014). Because of the extensive use of heat-absorbing materials and urbanisation, UHIs pose a significant environmental threat and result in unfavourable effects like higher temperatures, deteriorated air quality, and greater health risks. Consequently, the specific surface conditions of an urban area will directly influence the selection of strategies for mitigating UHI effects (Fig. 8.1).

A temperature curve that shows the differences in temperature between rural and urban areas and back again can be used to illustrate the Urban Heat Island (UHI) phenomena. The graph's central peak, which resembles urban structures, forms an "island", surrounded by cliffs that signify the barrier between urban and rural areas (Oke 1981). The substantial temperature difference that cities endure is clearly shown by this graphical depiction. According to Yang et al. (2016), the effect is becoming more and more of a concern, which emphasises the need for ways to adapt to and lessen the negative environmental effects of UHIs.

Understanding UHI's effects on energy use, the environment, and human health requires study. Urban planning, design methods, and policies that mitigate heat-related health concerns, reduce energy consumption, and promote social fairness are informed by this understanding. To create habitable, climate-resilient cities and solve the issues posed by climate change and rapid urban expansion, UHI research supports public awareness, community involvement, and well-informed decision-making.

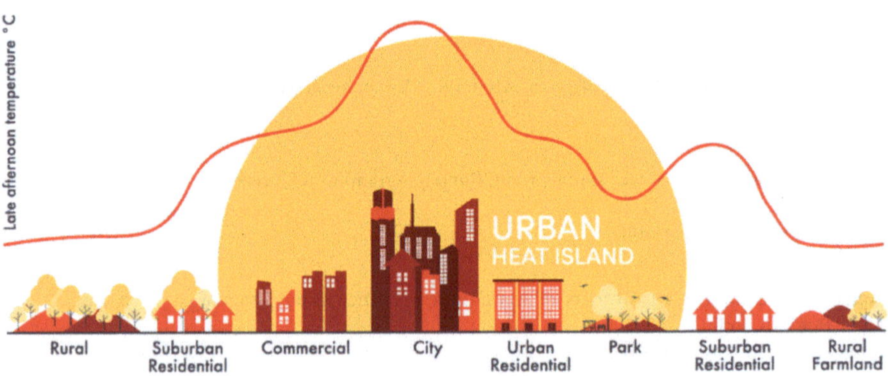

Fig. 8.1 Impact of different land uses on the air and surface temperature

Urban regions, which are defined by the widespread use of materials like asphalt and concrete, absorb and hold heat, releasing it at night to create higher temperatures. Buildings and infrastructure construction can create canyons that trap heat, and tall buildings can obstruct natural cooling mechanisms like wind flow. Natural vegetation, such as grass and trees, are frequently removed during urbanisation, which raises surface temperatures and reduces cooling through evapotranspiration and shade. Urban regions warm even more as a result of human activity producing waste heat, such as transportation and industrial processes. Lower reflectivity, or albedo, is a characteristic of urban surfaces, particularly pavements and buildings, which allows them to absorb more sunlight and raise temperatures. Furthermore, urban areas may have changed microclimates and changing wind patterns, which would reduce effective heat dissipation and raise temperatures.

Mitigation strategies are currently of great concern because of the possibly detrimental effects of UHI on human lives. Adding natural and semi-natural elements to urban areas is known as "green infrastructure", and it is one viable approach. A network of naturally occurring and semi-natural components that are deliberately incorporated into urban areas to provide ecosystem services, improve the quality of the environment, and lessen the effects of urbanisation is known as "green infrastructure". To promote a healthy cohabitation between urban development and nature, reduce the impact of the Urban Heat Island effect, and restore ecological balance, it incorporates elements such as parks, green roofs, urban forests, and permeable pavements. This work presents a methodology to look at UHI's sources and effects while evaluating how green infrastructure can help lessen its negative effects.

8.1.1 Causes and Consequences of Urban Heat Island

Heat is absorbed and retained by urban areas, which are defined by impervious surfaces like concrete and asphalt, leading to noticeably higher temperatures than on natural surfaces. Building and infrastructure materials have a large thermal mass, which allows them to collect and release heat. Tall buildings also impede natural airflow and retain heat. Reduced greenery, such as parks and trees, is a common result of urbanisation, which raises temperatures by lessening the cooling effects of flora. Human activities, including industrial processes and transportation, generate substantial waste heat, adding to the overall temperature increase. Urban surfaces' lower albedo, or reflectivity, compared to rural areas, leads to increased heating as darker surfaces absorb more sunlight. Poor urban planning, including the arrangement of buildings and roads, can limit natural ventilation, creating localised pockets of higher temperatures. Increased human density in urban areas results in heightened heat generation through daily activities like transportation and energy consumption. Elevated levels of air pollution in urban areas impact sunlight absorption and reflection, contributing to atmospheric warming. Certain weather conditions, like calm and clear nights, can exacerbate the UHI effect by allowing heat to accumulate without natural dissipation processes (Fig. 8.2).

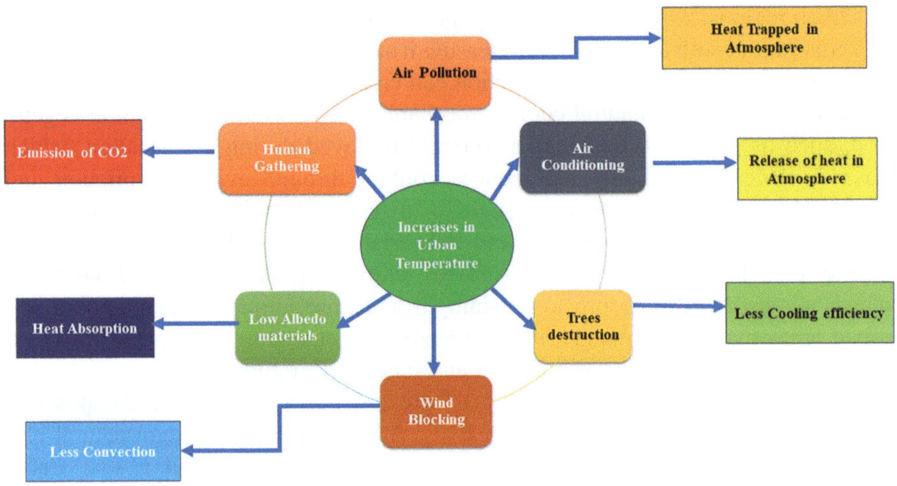

Fig. 8.2 Factors contributing to increased urban temperature

8.1.2 Satellite-Based Monitoring of Urban Heat Island

Monitoring Urban Heat Island (UHI) through satellite-based methods proves to be a valuable and effective approach for evaluating and comprehending the spatial and temporal dynamics of temperature fluctuations in urban regions. Earth-observing satellites equipped with thermal infrared sensors, can gauge land surface temperature (LST) and pinpoint regions influenced by the UHI phenomenon. These satellites capture thermal radiation emitted from the Earth's surface, yielding comprehensive data for extensive geographical areas. Through remote sensing technologies, thermal maps are generated to illustrate temperature variations across urban landscapes, aiding in the identification of heat islands and the measurement of temperature differences between urban and non-urban zones. Satellite-based observations offer a comprehensive viewpoint, assisting urban planners and policymakers in formulating effective strategies to address heat-related challenges.

8.2 Study Area

The study has been carried out in parts of central Punjab, India. Ludhiana in the semiarid region (Fig. 8.3).

Ludhiana, an industrial city in Punjab, experiences subtropical weather with summertime showers coming from the southeast. Westerly atmospheric depressions bring on winter rains. Located in the middle of the Punjab, Ludhiana is characterised by a notably flat topography and a lack of notable topographic features. The city of Ludhiana has a larger population density. Paddy (June–October) and wheat

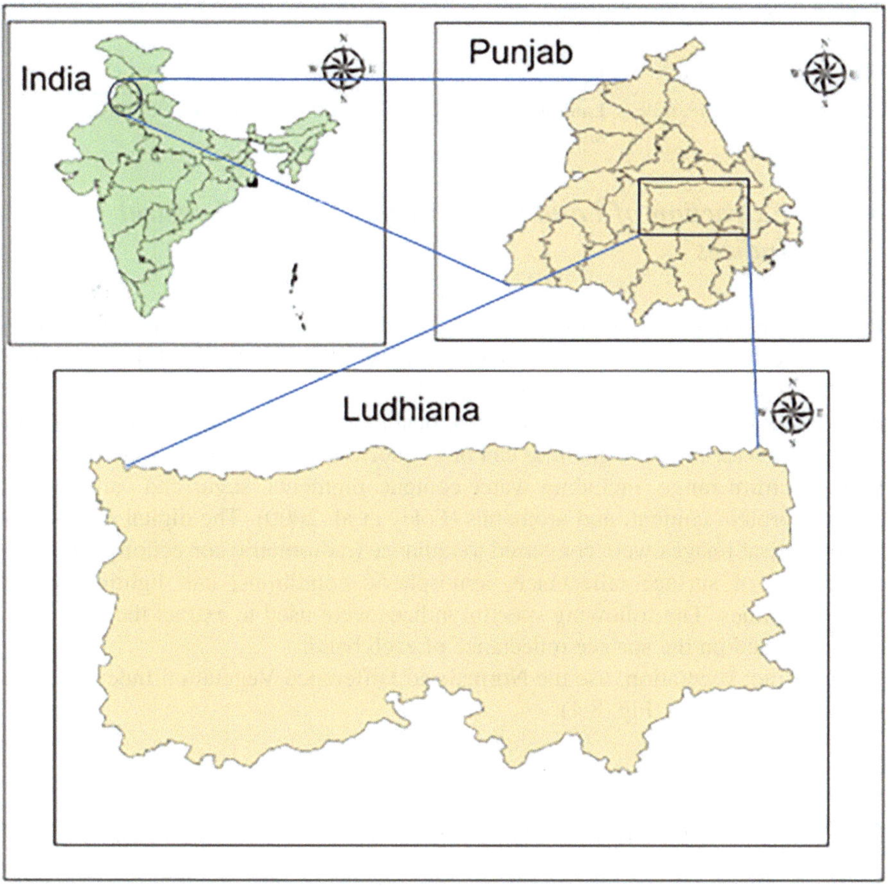

Fig. 8.3 Location of Study Area (i.e. Ludhiana district of Punjab, India)

(November–April) are the principal cropping systems in the city's rural districts. The Ludhiana district's latitude and longitude are N 30° 54′ 43.344" and E 75° 51' 13.644".

8.2.1 Satellite Data

The United States Geological Survey (USGS) offers geographic maps in GeoTIFF format, available in the public domain. These maps were downloaded for the study to acquire images of the specified region, ensuring they were devoid of clouds. The two thermal bands (Band 10 and Band 11) of the independent Thermal Infrared Sensor (TIRS) onboard Landsat 8/9 are recorded at a spatial resolution of 100 m (resampled to 30 m). Earth Explorer provided the Landsat satellite data for April and May of 2023 (USGS) as shown in Table 8.1.

Table 8.1 Summary of satellite data

Year	Month	Path, Row	Satellite	Band combination	Tier
2023	April	148, 039	Landsat 8/9	B1, B2, B3, B4, B5, B6, B7, B8, B9, B10, B11	1

8.2.2 Extraction of Land Cover Features Using Spectral Indices

Remote sensing of vegetation is mostly used to measure electromagnetic wave reflectance from canopies using passive sensors. The reflectance of the light spectrum from plants is known to be influenced by the type of plant, the amount of water in the tissues, and other intrinsic features. In addition to the development and vigour of plants, a wide range of qualities can be connected to indicators collected from this light spectrum range, including water content, pigments, sugar and carbohydrate content, protein content, and aromatics (Foley et al. 2000). The digital numbers of these Landsat images were converted to radiance (radiometric correction) to exclude the effects of surface reflectance, atmospheric conditions, and lighting on the radiance values. The following spectral indices were used to extract the land cover features based on the surface reflectance of each band:

To extract vegetation, use the Normalised Difference Vegetation Index (NDVI): (NIR-R)/ (NIR + R) (Fig. 8.4).

8.2.3 Retrieval of LST from Landsat Data

LST is defined as the temperature of several millimetres (mm) of the Earth's surface (Kumar et al. 2017). As a result, it plays a significant role in surface energy balances and affects numerous biophysical processes, including evapotranspiration, photosynthesis, and plant respiration. Remote sensing, in contrast to point data, enables the extraction of LST for broad regions with adequate temporal and geographical resolution (Li et al. 2013). Many techniques have been developed to calculate LST using satellite data, including the radiative transfer equation (RTE), split-window (SW), and single-channel (SC) techniques. Over the past few decades, several studies have examined thermal analysis using data from satellites like MODIS, ASTER, Landsat TM, and ETM (Barsi et al. 2003; Cristóbal et al. 2009; Jimenez-Munoz et al. 2008; Jiménez-Muñoz and Sobrino 2009; Oguz 2013, 2015; Li et al. 2013; Dhaloiya et al. 2023).

In this study, LST is extracted from Landsat 8 satellite imagery using the ArcGIS Model Builder's RTE approach. After providing a few settings, users need to input band4, band5, and band10; the tool will then automatically get the final LST imagery (Fig. 8.5).

The LST calculation was done by following these steps:

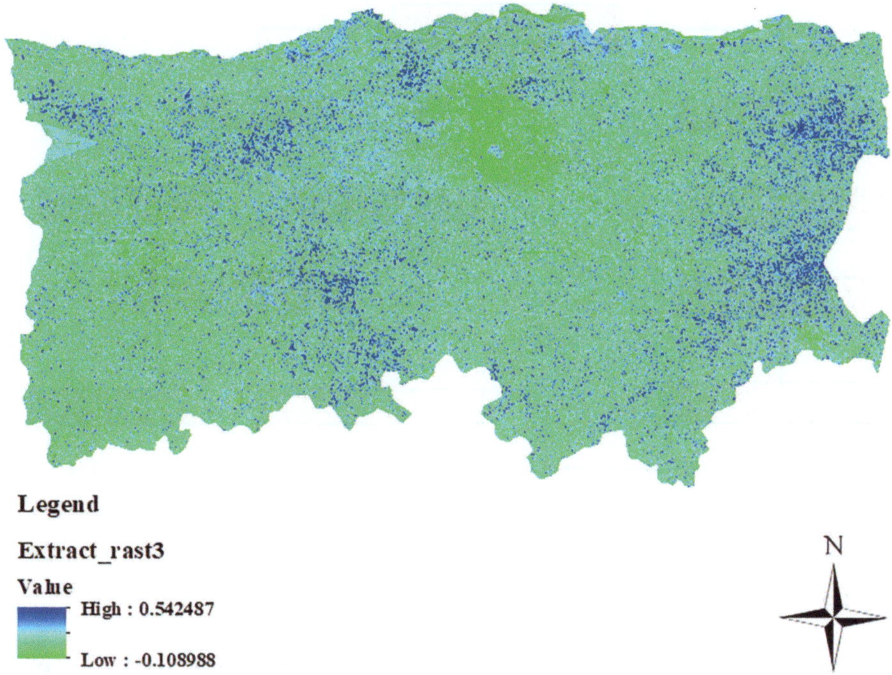

Fig. 8.4 NDVI map of Ludhiana district of Punjab

1. To calculate the Top of Atmosphere (TOA) spectral radiance, use the formula TOA (L) = ML * Q_{cal} + AL, in which Q_{cal} is band 10, and AL is the band-specific additive rescaling factor from the metadata. ML is the band-specific multiplicative rescaling factor that was retrieved from the metadata.
2. The formula BT = (K_2 / (ln (K_1 / L) + 1)) − 273.15 is used to convert TOA to Brightness Temperature (BT), where K_1 and K_2 are band-specific thermal conversion constants that are derived from the metadata and L stands for TOA.
3. The Normalised Difference Vegetation Index (NDVI) computation is essential for determining vegetation proportion (P_v) and emissivity (ε) in the future. The formula for calculating NDVI is NDVI = (Band 5—Band 4) / (Band 5 + Band 4). It emphasises that to get reliable results, NDVI values must be represented as floating-point numbers.
4. Determination of the vegetation proportion (P_v) involves the formula P_v = Square ((NDVI—$NDVI_{min)}$ / ($NDVI_{max}$—$NDVI_{min}$)), with $NDVI_{min}$ and $NDVI_{max}$ being the minimum and maximum values of the NDVI image, respectively. These values can be directly observed in software tools like ArcGIS, QGIS, ENVI, or Erdas Imagine, or accessed through the raster properties.
5. Emissivity (ε) is calculated using the formula ε = 0.004 * P_v + 0.986, and this calculation is implemented in the raster calculator.

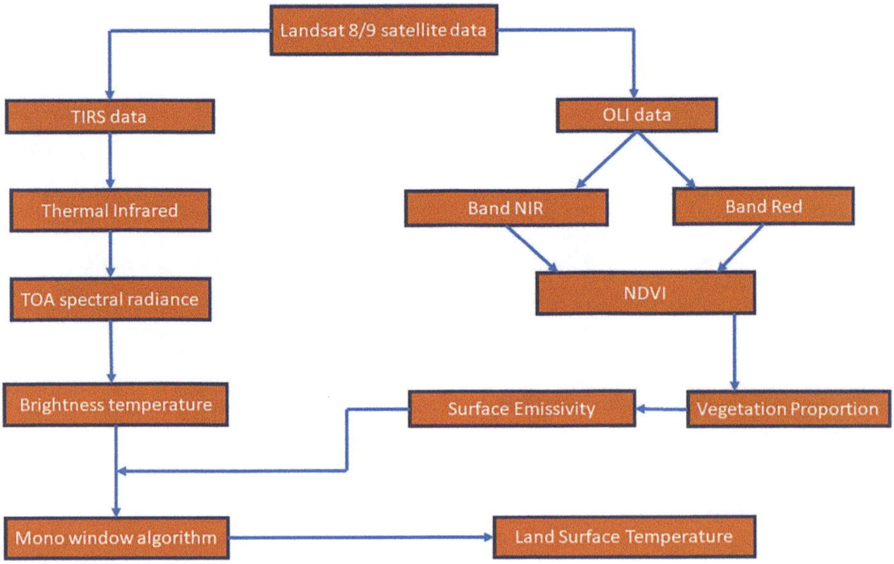

Fig. 8.5 Flow Chart for LST calculation

6. The LST is obtained through the formula LST = (BT / (1 + (0.00115 * BT / 1.4388) * Ln(ε))). Applying this equation yields the final map of Land Surface Temperature, with the note that it does not equate to air temperature.

To summarise, the method presented entails a methodical series of actions to retrieve important information from Landsat data, primarily concentrating on determining Land Surface Temperature (LST). The Normalised Difference Vegetation Index (NDVI) is computed by starting with the computation of Top of Atmosphere (TOA) spectral radiation and transforming it into Brightness Temperature (BT) afterwards. When calculating the proportions of vegetation (Pv) and emissivity (ε), which are critical components in the LST calculation, the NDVI values play a critical role (Fig. 8.6).

This procedure makes use of sophisticated mathematical transformations and remote sensing techniques to enable the production of an extensive map of land surface temperature. It is important to emphasise that surface temperatures, not air temperatures, are the focus of the resulting LST map. The Ludhiana district's April 2023 temperature variation, which ranges from 23 to 44 °C, shows a significant difference in temperature over the month. Furthermore, the finding that temperatures in metropolitan centres are noticeably higher than in surrounding areas raises the possibility of an urban heat island (UHI) effect. The natural environment's alterations and human activity are the main causes of this phenomenon.

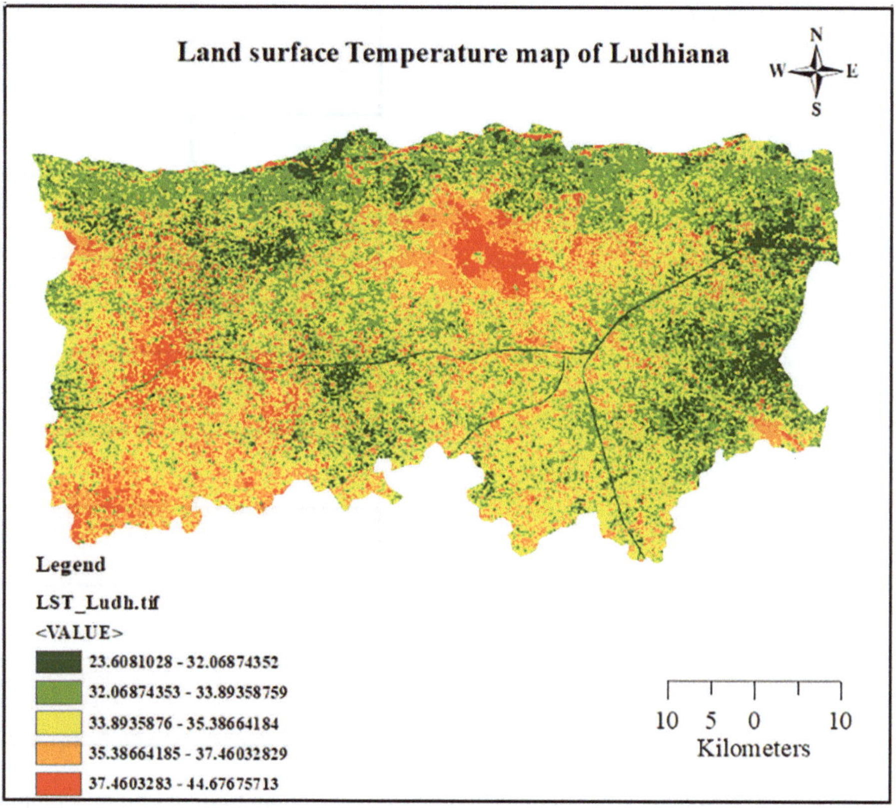

Fig. 8.6 Land surface temperature map of Ludhiana district of Punjab

8.2.4 Retrieval of LULC from Landsat Data

Using Landsat 8/9 imagery to create a LULC map in ArcGIS requires a methodical process. The research region was originally drawn inside the ArcGIS framework. After that, pre-processing techniques such as geometric alignment and radiometric and atmospheric corrections were applied to the acquired Landsat 8/9 satellite pictures. The crucial stage is classifying land cover using supervised techniques. Representative samples are used to train the classifier. Refinement and smoothing are two post-classification improvements that are used to improve accuracy and aesthetic appeal (Fig. 8.7).

Accuracy assessment becomes imperative, using tools such as the Confusion Matrix for validation against ground truth data. The final LULC map was generated in ArcGIS, integrating the classification outcomes into the study area and employing mapping tools for a refined presentation. Validation through field surveys or additional ground truth data ensures precision, and the entire process is meticulously

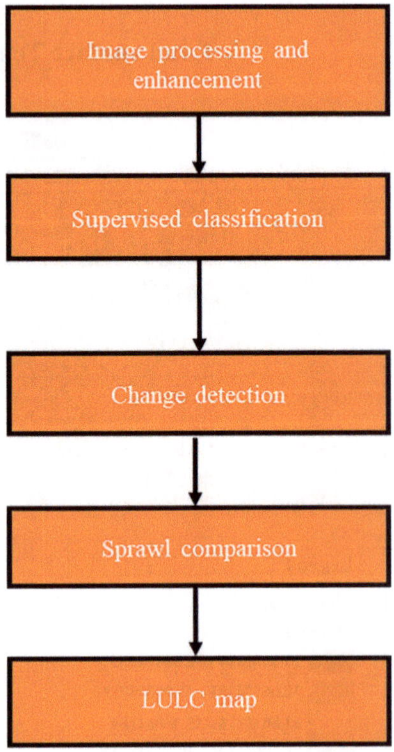

Fig. 8.7 Flow chart for making a LULC map

documented, providing a comprehensive report on methodology, findings, and any encountered constraints. The various land covers of the study area were classified into four categories: urban, vegetation, water, and barren land.

In April 2023, urban regions extended across 66,845.7 hectares, vegetation covered 81,546.57 hectares, waterbodies spanned over 1211.58 hectares, and bare land, the largest at 204,760.35 hectares. During this month, agricultural land exhibited a radiation pattern akin to bare land due to the absence of vegetation in fields for April month (Fig. 8.8).

8.2.5 Relation of Land Surface Temperature with LULC

The correlation between LST and LULC in Ludhiana district, Punjab, is influenced by various factors. Changes in land use and cover are pivotal in altering surface properties, thereby affecting temperature patterns. Noteworthy relationships include the Urban Heat Island (UHI) Effect, where urban areas with extensive impervious surfaces exhibit higher LST compared to rural areas. Additionally, the presence of green spaces, agricultural land, and water bodies plays a role in temperature regulation. Variations in agricultural practices and shifts in land use over time contribute

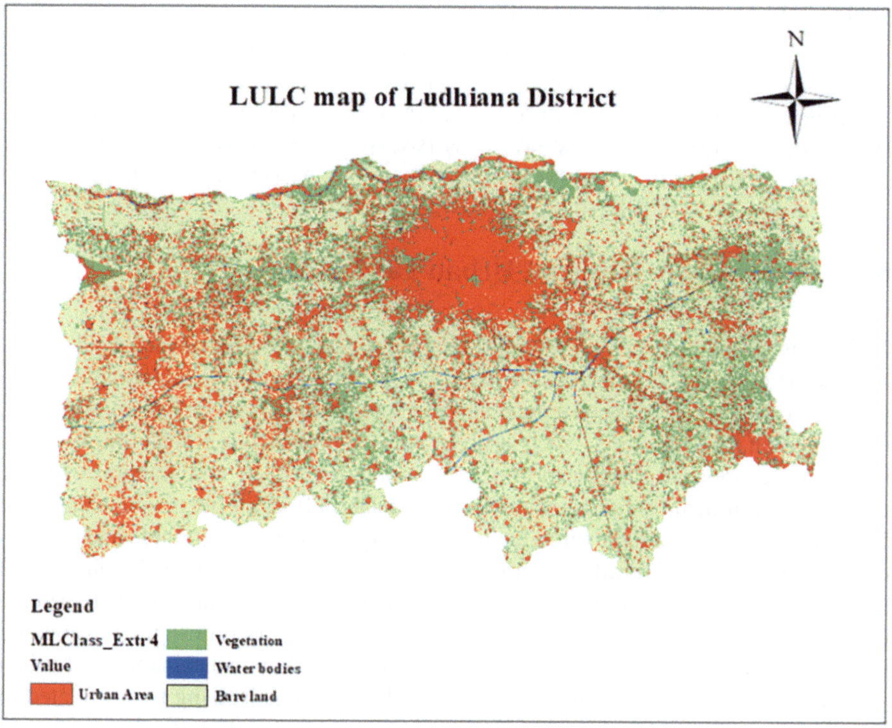

Fig. 8.8 Land use and land cover map of Ludhiana district of Punjab

to LST fluctuations. To comprehensively investigate this relationship, a detailed remote sensing and GIS analysis utilising satellite imagery and geographic information systems is recommended for quantifying land cover, assessing temperature patterns, and identifying correlations or trends. Table 8.2 shows the LULC and LST of the Ludhiana District.

The table summarises information on various land use types. Urban areas spanning 66,845.7 hectares have a mean Land Surface Temperature (LST) of 35.91 °C, with a range from 23.60 °C to 44.48 °C and a standard deviation of 1.88. The vegetation that covers 81,546.57 hectares has a mean LST of 33.10 °C, with a standard deviation of 1.78, and a range of 26.56 °C to 43.22 °C. The mean LST of waterbodies larger than 1211.58 hectares is 31.56 °C, with a standard deviation of 1.58 and a range of 27.39 °C to 39.19 °C. The largest area of bare land, covering 204,760.35 hectares, has a mean LST of 34.45 °C, with a standard deviation of 1.30 and a range of 28.58 °C to 44.76 °C. This information offers a succinct summary of the temperature fluctuations, area coverage, and thermal properties among various land use classes.

Table 8.2 LULC and LST of Ludhiana District

Land use	Area (ha)	LST (MIN)	LST (MAX)	Mean LST	Standard deviation
Urban area	66845.7	23.601003	44.476891	35.914848	1.884249
Vegetation	81546.57	26.562876	43.218407	33.09607	1.781489
Waterbody	1211.58	27.387547	39.186096	31.563299	1.581449
Bare land	204760.35	28.583429	44.759705	34.453224	1.300012

8.3 Strategies to Mitigate Urban Heat Islands

Numerous strategies have been proposed to mitigate the urban heat island effect. Seifeddine et al. (2023) emphasises how cool materials, such as asphalt and reflective concrete, can effectively reduce surface temperatures. According to Wai et al. (2022), green infrastructure—such as green walls and roofs—is crucial for providing shade and evapotranspiration to cool metropolitan areas. A combination of these tactics can significantly reduce the urban heat island effect's effects. Understanding UHI dynamics is essential to architecture planning to create comfortable and energy-efficient spaces. By carefully selecting materials, ventilation, and insulation, as well as by adding vegetation, reflective surfaces, and green roofs, architects can lessen the effects of urban heat islands. Urban planning and open space design decisions are informed by UHI research, which supports the resilience and sustainability of urban growth. Furthermore, this information aids architects in addressing issues related to health and well-being in hot urban contexts. to heat-related stress. By designing buildings with heat-resistant materials, natural ventilation, and shade, architects help to create healthier urban settings. To incorporate UHI mitigation measures into city planning and design standards and ensure visually appealing, environmentally sustainable, and resilient urban areas, collaboration between urban planners and architects is important. In summary, the study of UHI is essential to architectural planning because it informs choices about the layout of buildings and sites as well as the design of urban areas that put the needs of people and the environment first.

8.4 Conclusions

The findings suggest that vegetation increment can significantly lower land surface temperature (LST). The study indicates a worrying decrease in water supplies, which highlights the critical necessity for conservation given their importance. The LULC and LST map of the research area highlights the correlation between decreasing LST and an increase in green and water-covered land. This reduction aids the global effort to battle climate change in LST as it helps to mitigate the Urban Heat Island (UHI) impact. A thorough analysis of the many green infrastructure techniques—from permeable pavements and community gardens to urban forests and green roofs—reveals their many benefits. In addition to helping to cool metropolitan areas, these initiatives improve biodiversity, improve air quality, and improve community well-being in general. The chapter emphasises the necessity of an integrated and comprehensive approach to urban planning that places a high priority on integrating green infrastructure components into the built environment. Establishing sustainable, resilient, and livable cities requires cooperation between legislators, urban planners, landscape architects, and the community. The chapter's use of case studies and actual data shows the real-world benefits of green infrastructure in reducing urban heat islands and calls for its broad implementation.

In the future, it will be crucial to view green infrastructure as a crucial component of urban development that tackles a wider range of environmental and social issues, rather than just a means of reducing the effects of urban heat islands. The ideas presented in this chapter highlight the critical role that green infrastructure will play in determining the future urban landscape as we work towards more sustainable and climate-resilient cities. There is a chance to create cities that are not only more lively, healthy, and peaceful for present and future generations, but also intentionally integrated via cooperative efforts.

References

Barsi JA, Barker JL, Schott JR (2003) An atmospheric correction parameter calculator for a single thermal band earth-sensing instrument. In: IGARSS 2003. 2003 IEEE international geoscience and remote sensing symposium. Proceedings (IEEE Cat. No. 03CH37477), vol 5. IEEE, pp 3014–3016

Cristóbal J, Jiménez-Muñoz JC, Sobrino JA, Ninyerola M, Pons X (2009) Improvements in land surface temperature retrieval from the Landsat series thermal band using water vapor and air temperature. J Geophys Res Atmos 114(D8)

Dhaloiya A, Denis DM, Duhan D, Kumar R, Singh MC, Malik A (2023) Monitoring vegetation health, water stress, and temperature variation through various indices using landsat 8 data. Indian J Ecol 50(3):802–810

Foley JA, Levis S, Costa MH, Cramer W, Pollard D (2000) Incorporating dynamic vegetation cover within global climate models. Ecol Appl 10(6):1620–1632

Jimenez-Munoz JC, Cristobal J, Sobrino JA, Sòria G, Ninyerola M, Pons X (2008) Revision of the single-channel algorithm for land surface temperature retrieval from Landsat thermal-infrared data. IEEE Trans Geosci Remote Sens 47(1):339–349

Jiménez-Muñoz JC, Sobrino JA (2009) A single-channel algorithm for land-surface temperature retrieval from ASTER data. IEEE Geosci Remote Sens Lett 7(1):176–179

Kumar R, Mishra V, Buzan J, Kumar R, Shindell D, Huber M (2017) Dominant control of agriculture and irrigation on urban heat island in India. Scientific reports 7(1):14054

Li ZL, Tang BH, Wu H, Ren H, Yan G, Wan Z et al (2013) Satellite-derived land surface temperature: current status and perspectives. Remote Sens Environ 131:14–37

Oke TR (1981) Canyon geometry and the nocturnal urban heat Island: comparison of scale model and field observations. J Climatol 1(3):237–254

Oguz H (2013) LST calculator: a program for retrieving land surface temperature from Landsat TM/ETM+ imagery. Environ Eng Manag J 12(3):549–555

Oğuz H (2015) A software tool for retrieving land surface temperature from ASTER imagery ASTER Uydu Görüntüsünden Yer Yüzey Sıcaklığını Hesaplayan Bir Yazılım Aracı Geliştirilmesi. J Agric Sci Tarım Bilimleri Dergisi 21(4):471

Seifeddine K, Amziane S, Toussaint E, Ouldboukhitine SE (2023) Review on thermal behavior of cool pavements. Urban Clim 51:101667

Stone B Jr, Vargo J, Liu P, Habeeb D, DeLucia A, Trail M et al (2014) Avoided heat-related mortality through climate adaptation strategies in three US cities. PLoS One 9(6):e100852. https://doi.org/10.1371/journal.pone.0100852

Synnefa A, Karlessi T, Gaitani N, Santamouris M, Assimakopoulos DN, Papakatsikas C (2011) Experimental testing of cool colored thin layer asphalt and estimation of its potential to improve the urban microclimate. Building and Environment 46(1):38–44

Nakayama T, Fujita T (2010) Cooling effect of water-holding pavements made of new materials on water and heat budgets in urban areas. Landsc Urban Plan 96(2):57–67, ISSN 0169-2046. https://doi.org/10.1016/j.landurbplan.2010.02.003

Wai CY, Tariq MAUR, Muttil N (2022) A systematic review on the existing research, practices, and prospects regarding urban green infrastructure for thermal comfort in a high-density urban context. Water 14(16):2496

Yang L, Qian F, Song DX, Zheng KJ (2016) Research on urban heat-Island effect. Procedia Eng 169:11–18

Chapter 9
Identification of Soil Erosion-Susceptible Areas with Analytical Hierarchy Process (AHP) Modelling for Climate Disaster Management

Gurav Chandrakant, Dhende Atish, and Kulkarni Dhaval

Abstract Previous literature provides evidence related to the slump movement arising due to heavy rainfall in the Jotiba-Panhala hill range in a part of western ghat. The trend of this hill range is east-west, and it divides two easterly flowing rivers, Panchaganga and Warna. The present work is undertaken for the delineation of land degradation vulnerability sites using 11 thematic layers, viz., drainage density, slope, land cover, Normalized Difference Vegetation Index (NDVI), soil texture and depth, rainfall, land surface temperature (LST), multi-resolution index of valley bottom flatness (MRVBF), Terrain Ruggedness Index (TRI), and soil organic carbon (SOC). The land degradation risk zones have been analysed using the Analytic Hierarchy Process (AHP) and Geographical Information Systems (GIS), and the obtained results have been validated with the help of fieldwork observations and Google Earth image evidence. Based on normalised weight in AHP, the risk zones have been classified into five classes, viz., very low risk, low risk, moderate risk, high risk, and very high-risk zones of land degradation. It was finally observed that the very high-risk to high-risk zone of the present study area shows most of the sites where creep movement took place in the last few years. This site has been verified by using Google Earth imagery.

Keywords Creep · Land degradation · Slump · LULC · NDVI

G. Chandrakant (✉)
Department of Geology, Rajaram College, Kolhapur, Maharashtra, India

D. Atish · K. Dhaval
Department of Geology and Geoinformatics, School of Earth Sciences, Punyashlok Ahilyadevi Holkar Solapur University, Solapur, Maharashtra, India

D. Kumar, N. P. Bassill (eds.), *Artificial Intelligence and Machine Learning for Climate Disaster Management*, Disaster Resilience and Green Growth,
https://doi.org/10.1007/978-981-95-1588-2_9

Abbreviation

AHP	Analytic Hierarchy Process
DD	Drainage Density
DSM	Digital Soil Maps
DVBP	Deccan Volcanic Basaltic Plateau
GIS	Geographical Information Techniques
HDP	Dissected Plateau
LDP	Low Dissected Plateau
LDVI	Land Degradation Vulnerability Index
LST	Land Surface Temperature
LUS	Land Use Planning
MDDLP	Moderately Dissected Denudational Lower Plateau
MDP	Moderately Dissected Plateau
MRVBF	Multi-Resolution Index of Valley Bottom Flatness
NBSS	National Bureau of Soil Survey
NDVI	Normalized Difference Vegetation Index
PPC	Pediment Pediplain Complex
SOC	Soil Organic Carbon
TRI	Terrain Ruggedness Index
UNCCD	United Nations Convention to Combat Desertification

9.1 Introduction

Land degradation can be defined as "A short-term or stable decline of land fertility through the decline of deep-rooted biological, chemical, and physical characteristics" (Sandeep et al. 2021; Tagore et al. 2012). It is a complex and challenging area for environmental scientists (Higginbottom and Symeonakis 2014). The study of land degradation and environmental protection has been undertaken by various central and state agencies since 1976, but these agencies have used different definitions, methodologies, data sources, and scales (Maji et al. 2010). The United Nations Convention to Combat Desertification (UNCCD), involving nearly 195 countries in the world, stated that land degradation and desertification are critical environmental issues in the twentieth century (UNCCD 1994, 2002). For this, the United Nations also recommended "Zero net land degradation" policies to decrease the rate of land degradation. Along with political frameworks, and whilst admirable, the scientific study is the key point for such environmental issues (Grainger 2015). It is very important to find the degradation process, and it is also necessary to use early warning technology for the continuing deterioration of land conditions. For this, measurements must adhere to the principles of repetitiveness, objectivity, and consistency (Hill et al. 2008). Ten million hectares of cropland are lost because of land degradation every year.

The most pressing ecological problem of the modern era is soil erosion. Soil erosion is a key operator of soil degradation (Wei et al. 2007). It is a temporary or permanent loss of ecological as well as biological productivity of the land (Das et al. 2023). It depends upon the geological setting and meteorological conditions of the region (Wei et al. 2007). In India, 70% of the population lives in rural areas, and their main occupation is agriculture and related work. The quality of agricultural land is deteriorating due to farmers being small and marginal. In India, agriculture based on rainfall occupies nearly 58% of the cultivated area (Venkateswarlu, 2012). In terms of rainfed agricultural land, Jharkhand has the highest percentage of cultivated area at 91%, followed by Maharashtra at 81%, Chhattisgarh at 74%, Karnataka at 71%, and Andhra Pradesh at 53% (Kumar et al., 2011). The quality of agricultural land is deteriorating due to farmers being small and marginal. This marginal agricultural land decreases further due to population growth. Similarly, other factors like soil erosion, adverse impacts of climate change, heightened nutrient mining, increasing water scarcity, and accumulation of toxic elements in soil and water are also responsible (Maji et al. 2010). New studies show that emerging countries like India cover more than 36.7% or 120.72 million hectares and are under different groups of land degradation like salinity, soil acidity, waterlogging, erosion, and alkalinity (Maji et al. 2010). As a result of cutting down trees, these communities emerge: low-quality soil, type of vegetation cover, agricultural land, and climatic dryness (Salvati and Zitti 2005).

Mathematical modelling and evaluation of land degradation vulnerability play critical roles in the sustainable development and management of natural resources using recently developed approaches (Sandeep et al. 2021). Researchers have used the Analytic Hierarchy Process (AHP) in the last few years to advance remote sensing systems (AHP) and Geographic Information System (GIS) for land degradation (Das et al. 2023; Sandeep et al. 2021). As a multidisciplinary modelling approach, AHP's applicability is subject specific and so open to all fields of study. The demarcation of land can also benefit from GIS technology and datasets derived from remote sensing due to its portability, speed, and ability to examine vast areas, even those in remote locations (de Jong et al. 2011; Higginbottom and Symeonakis 2014).The present study area is a part of the Western Ghats in India. The Western Ghats, located on the west coast of India, are one of the world's eight biodiversity hotspots. Land degradation is a major concern, particularly along the Indian west coast, due to heavy rainfall and steep slopes. The primary factors contributing to this increasing degradation are soil erosion, seawater intrusion, salinization, and rampant urbanization (Venkataraman, 2012; Verma et al., 2018; Mahajan et al., 2021). The land degradation study of the Western Ghats has been carried out by Das et al. (2023) in the west coast river Mandovi, and in the upper Savitri river by Durugwar et al. (2025) with the help of AHP analysis. In the present study area, landslide zonation mapping is carried out with the help of geomorphological parameters, slope, and creep movement field evidence by Gurav et al. (2017). The Creep movements occur after heavy rainfall in a short time in this region, and they give preliminary indications of a landslide-prone area. The groundwater potential zones have also been studied by Gurav and Babar (2018) in Jotiba and the adjoining part. The objective of the present work is to find land degradation sites through the integration of various thematic layers by using AHP, RS, and GIS in the Jotiba-Panhala hill range in Kolhapur district, Maharashtra, India. Our specific objectives are to provide:

(a) A brief understanding of the terrain characteristics of the Jotiba and Panhala hill range using various thematic layers
(b) An evaluation of land degradation susceptibility sites using AHP modelling and GIS techniques
(c) Validation of the results obtained through remote sensing, GIS techniques, and AHP modelling for land degradation for the present case

9.2 Study Area

The Jotiba-Panhala hill range lies between 16° 42′ 15.677″ to 16° 55′ 11.395″ north latitudes and 73° 54′ 7.725″ to 74° 18′ 23.839″ east longitudes; the trend is east to west. For this study, a geographical area of 624.12 km^2 (62,412.1 ha) is being studied. It plays an important role as a water divider between the Panchaganga and Warna rivers. The hill has a relief of 466 m and an altitude range of 542–1008 m above mean sea level (msl). Along the riverbed, you will find black cotton soil, while up on the plateau and in a few isolated spots, you will find lateritic and reddish mud. The rivers Kasari, Panchaganga, Kadvi, and Warna flow in this region, and they developed distinct geomorphic units in the Deccan Volcanic Basaltic Plateau (DVBP).

The Jotiba-Panhala hill is characterised by tropical conditions, and it has three distinct periods, viz., winter starts from October to February, summer from March to May, and the rainy season starts from June to September. The average annual rainfall of Kolhapur district is 1800 mm/year. The main crop in the region is sugarcane; along with this, in the rainy season, rice is grown, and in the summer season, crops like groundnut are cultivated in the region. A total of 124 village boundaries cover the present study area in Kolhapur district, Maharashtra, India. As per the Census (1991, 2001, 2011), total population in these 124 villages is 261,057, 319,338, and 348,168 in the years 1991, 2001, and 2011, respectively. The average population growth rate from 1991 to 2001 is 18.3%, while from 2001 to 2011, it is 8.28%. Out of 124 villages, negative population from 2001 to 1991 was found in 5 villages, and from 2011 to 2001, the population slightly increased in 15 villages.

9.3 Geology of the Area

The Jotiba-Panhala area is a part of the Deccan Trap, which comes lithostratigraphically from the Mahabaleshwar formation of Late Cretaceous to Early Eocene age (Godbole et al. 1996; Gurav et al. 2017; Gurav and Babar 2018). The geology of the study area has been successively studied by Athavale and Natu (2005; Gurav et al. 2017) and Gurav and Babar (2018). Vesicular/amygdaloidal basalt and compact basalt have occurred in the area. At 880 m above sea level in this region, columnar joints are developed in basalt. Seven lava flows are observed in the study area, which are separated by the red bole bed (Gurav et al. 2017). Both the hilltops are covered by lateritic rock (Fig. 9.1).

9.4 Datasets and Methodology

The objective taken for the present study is a very complex natural environmental hazard. It is a challenging task. For solving this problem, 11 factors were taken into account, such as drainage density, slope, land use land cover (LULC), Normalized Vegetation Difference Index (NDVI), soil texture and depth, rainfall, land surface temperature (LST), multi-resolution index of valley bottom flatness (MRVBF),

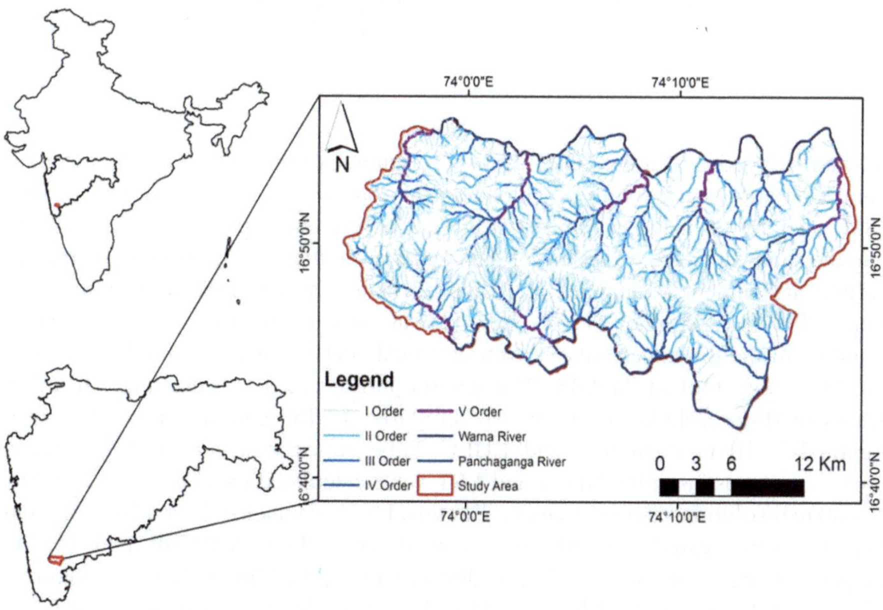

Fig. 9.1 The map shows Jotiba-Panhala hill range drainages in Kolhapur district, Maharashtra

Table 9.1 List of data, sites used for data generation, with their spatial resolution for the present study

Sr. no	Parameters and symbols	Data sources and spatial resolutions	Dataset
1	Villages, drainages	Survey of India toposheets 1: 50000 47H/13, 47H/14, 47L/1, 47L/2, 47L/5 and 47L/6	SOI Nakshe
2	Drainage density	Survey of India layer and ArcGIS software tool 30 m	SoI toposheets
3	Geomorphology	Reference sheet	Bhuvan
4	Soil	Maharashtra soil map in 1996, 1:500000	
5	Slope	SRTM DEM has a 30 m resolution	USGS
6	Multi-Resolution Index of Valley Bottom Flatness (MRVBF)		
7	Terrain Ruggedness Index (TRI)		
8	Rainfall	1 km resolution	World time
9	Normalized Difference Vegetation Index (NDVI)	Sentinel 2A USGS 10 m resolution	USGS
10	Land surface temperature (LST)	Landsat 8 USGS band 10	USGS
11	Land use land cover (LULC)	Sentinel 2A Esri data 2017 to 2021	Esri
12	Soil organic carbon (SOC)	Soil grid data having a 1 km × 1 km resolution	SoilGrids

Terrain Ruggedness Index (TRI), and soil organic carbon (SOC). The first stage of data generation started with downloading the toposheets from the Survey of India (SoI) on 1:50000 listed in Table 9.1. It was georeferenced using ArcGIS software, and the same software was used for data generation, thematic layer preparation, and weight assignment. Drainages are digitised using the onscreen digitisation technique. For this, Strahler (1964) method was adopted because of its simplicity. Finally, all the drainages have been used to calculate drainage (line) density using the hydrology tool in ArcGIS. The geomorphological data were taken from the Bhokosh (https://bhukosh.gsi.gov.in). Similarly, LULC data are also taken from Sentinel-2, 10 m resolution and LULC Time series downloader (https://www.arcgis.com/apps/instant/). NDVI and LST were calculated using Landsat 8, which has 30 m resolution satellite imagery. For this, Landsat bands 4, 5, and 10 were used, and it was processed using the tool map algebra. Slope, MRVBF, and TRI are prepared using open-source software QGIS by using the Shuttle Radar Topography Mission (SRTM) Digital Elevation Models (DEM) having 30 m resolution. Soil texture and soil depth maps were prepared by the Maharashtra soil map in 1996.

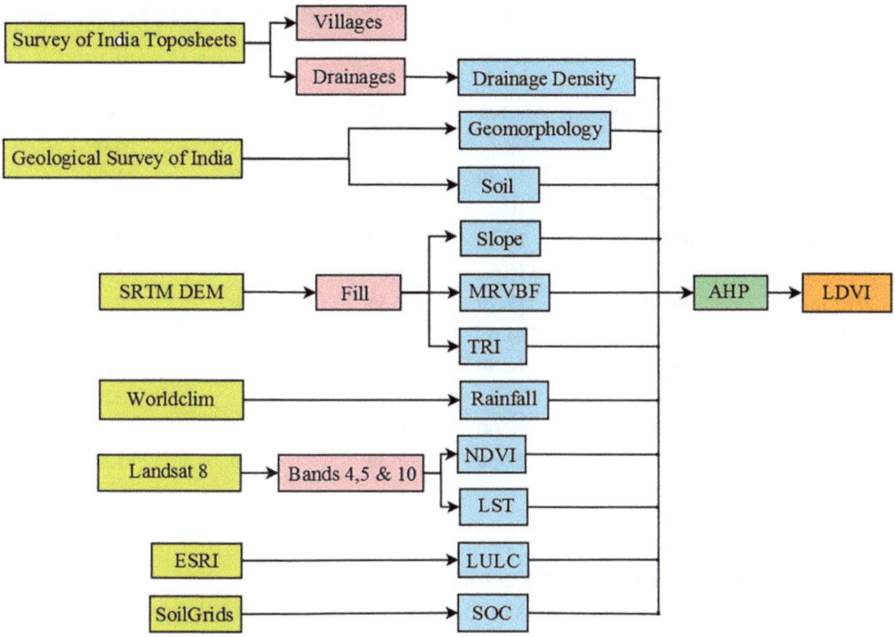

Fig. 9.2 Flowchart of the present methodology adapted for studying land degradation

It was first georeferenced using ArcGIS tools, and the soil zones were extracted and then converted to raster data. For the study, the soil organic carbon (SOC) soil grid data were used with 1 km × 1 km resolution. Rainfall data have been extracted from World Climate Data using village point data. Finally, all the thematic layers are analysed using an 11 × 11 matrix for computing the weight and Consistency Index (CI) in AHP analysis. For calculating the consistency ratio, the Random Index (RI) value taken from the Saaty (1980) index is 1.51. For this, all the layers were converted to 30 m resolution. The adapted methodology for solving the problem of the present work is simply shown in the flowchart (Fig. 9.2). The sites observed during the fieldwork locations were marked using the handheld Global Positioning System (GPS), and the observed sites of slump and landslides were digitised using the polygon tool in Google Earth.

As stated earlier, to study the zones or sites of land degradation, 11 effective thematic layers were adopted, which have been successively implemented for solving similar types of problems in different parts of the globe (Das et al. 2023; Saha et al. 2019; Sandeep et al. 2021). The TRI, slope, and MRVBF layers have been used to study the terrain, which is prepared using the SRTM DEM having 30 m resolution. It is one of the most important topographical variables which show the differential variation of erosion rate.

Table 9.2 Land use land cover data from January 2017 to 2021, ESRI free LULC 10 m spatial resolution

	Classes				
Years and %	Agricultural	Barren	Build-up	Vegetation	Water bodies
2017	424.7	108.3	41.4	43.8	6.00
2017 in %	**68.00**	**17.30**	**6.60**	**7.00**	**1.00**
2018	427.8	110.4	43.6	36.5	5.80
2018 in %	**68.60**	**17.70**	**7.00**	**5.90**	**0.90**
2019	438.0	110.6	45.6	24.4	5.50
2019 in %	**70.20**	**17.70**	**7.30**	**3.90**	**0.90**
2020	432.6	104.6	47.8	32.9	6.30
2020 in %	**69.30**	**16.80**	**7.70**	**5.30**	**1.00**
2021	422.3	112.4	50.2	32.9	6.30
2021 in %	67.70	18.00	8.00	5.30	1.00
Avg. %	**68.70**	**17.50**	**7.30**	**5.50**	**1.00**

Land Degradation Assessment Factors

To assess the land degradation sites in the present study area, 11 thematic layers have been used, which were previously applied by various researchers for studying the land degradation sites in different topographic terrains. These thematic layers are drainage density, slope, land cover, NDVI, soil texture and depth, rainfall, LST, MRVBF, TRI, and SOC, which are shown in Tables 9.1 and 9.2. First, the topographic maps taken from the SoI online maps portal on a 1:50000 scale were downloaded and projected using the ArcGIS software. The drainages have been manually digitised for the generation of a drainage density map. The satellite imagery and DEM have been processed for projection using the same software and preparation of various thematic layers. The thematic layer generated using proper methodology has been processed for resampling, and this has been classified based on its relation to the land degradation process. The projected DEM was then processed to fill the sink spaces available in the DEM that had been removed using the ArcGIS hydrology toolbox. Then the projected fill DEM was used for the preparation of terrain analysis thematic layers such as slope, TRI, and MRVBF. The present land cover study was undertaken using the Esri Sentinel 2A data from 2017 to 2021, based on the average land cover study carried out, and the LULC map was generated in January 2021. The pedological characteristics that are the most influential factors of land degradation, like soil depth and texture, were collected from the SoI soil map of Maharashtra, which was published in the year 1996. The SOC data in raster format for the shallow depth soil layer have been downloaded from Soilgrids at 1 km resolution (Poggio et al. 2021). The mean monthly rainfall was studied in the year 2018 and is taken from the World Climate data in raster format.

AHP Analysis

The analytical hierarchy process (AHP) method was adapted to investigate the land degradation sites in the Jotiba-Panhala hill range. By using this method, the weight of each factor is synthesised. In 1980, Saaty introduced AHP, a multi-criteria decision analysis (MCDA) technique, which is now widely used in many fields by Earth sciences researchers. Some of them are the groundwater potential zones study (Gurav et al. 2021, 2022), land degradation study (Das et al. 2023), site selection for land suitability study (Chandio et al. 2011; Kumar and Anbalagan 2016), and marine construction study (Gumusay et al. 2016). The AHP method, which depends on the subject expert, plays an important role in pairwise comparisons of various analyses (Saaty 2008). It also plays three main functions: structuring complexities, measurements, and synthesis. The AHP consists of a rating factor which is between 1 and 9 depending on the relationship with each criterion. 1 is the lowest rating factor which is used for inherent equal contribution; the second contribution is 2 to 3, which shows one factor is moderately favoured over another for medium; the third range is between 4 and 5, which shows that judgement strongly favoured another; the fourth range is for one factor very strongly favoured over another, showing the range is 5 to 7; and the last range is 7 to 9, which is adapted for one factor over another in the highest degree (Das et al. 2023; Saaty 2004).

The length of the pairwise comparative matrix will depend on the expert and the number of criteria used in decision-making for land degradation. For the study, the Land Degradation Vulnerability Index (LDVI) 11 × 11 criteria matrices have been adapted. In the initial stage of the AHP, a scale is assigned for pairwise comparison of the thematic layer matrix. After the scaling has been done, every layer weight was calculated (C_{wi}) and sub-classes of the thematic layers (SC_{Wij}), the LDVI was calculated using the following formula:

$$\mathrm{LDVI} = \sum\nolimits_{i=1}^{n} \sum\nolimits_{i=1}^{n} C_{wi} \times \mathrm{SC}_{wil}$$

where

n = Total number of classes
m = Number of sub-classes of the ith criterion

Consistency Analysis

Consistency of the pairwise comparison of classes and sub-classes has been studied using the consistency ratio (CR). The CR value can be calculated using the following equation:

$$\mathrm{CR} = \frac{\mathrm{CI}}{\mathrm{RI}}$$

where

RI = random index
CI = consistency index

The CI of the present study was computed using the following formula: where λmax is the largest matrix eigenvalue shown in Table 9.5 and n is the number of thematic layers, i.e. 11.

The random index values are previously stated by Saaty (1980). Up to the eleventh layer, the values are as follows: first two layer values are 0.00, third layer is 0.58, fourth layer is 0.90, fifth layer is 1.12, sixth layer is 1.24, seventh layer is 1.32, eighth layer is 1.41, ninth layer is 1.45, tenth layer is 1.49, and eleventh layer is 1.51.

$$\mathrm{CI} = \frac{\lambda \max - n}{n - 1}$$

If the CR value has come to less than 0.10, it is an indicator of acceptable consistency of pairwise comparison and weight calculation, but if it comes to more than this, it shows the pairwise comparison has to be modified until it is reduced below 0.10.

9.5 Results and Discussion

9.5.1 Climate Study

As far as land degradation is concerned, natural factors depend upon the intensity of the rainfall. The difference in day and night temperature affects the rock's physical properties, and similarly, rainfall also plays a major role in soil development and migration. The soil is filled with water for rainy reasons, and because of the clay contained in it, it loses its strength and is heavy with water that has migrated due to gravity. Jotiba and Panhala hill range rainfall has precipitated from June to September due to the southwest monsoon. The rainfall in the year 2018 was classified into nine major categories: 587–730 mm, 730–872 mm, 872–1014 mm, 1014–1157 mm, 1157–1299 mm, 1299–1442 mm, 1442–1584 mm, 1584–1726 mm, and 1726–1869 mm. The areas under each category are 75.7 km^2 (12.13%), 116.7 km^2 (18.69%), 121.9 km^2 (19.53%), 48.7 km^2 (7.81%), 72.9 km^2 (11.68%), 59.5 km^2 (9.54%), 56.5 km^2 (9.06%), 42.8 km^2 (6.86%), and 29.4 km^2 (4.72%), respectively.

The temperature is not the same all around the earth; its intensity and distribution are indirectly related to the vegetation pattern. The LST thematic layer is classified into four classes, namely 19.8–23.5 °C, 23.5–25.4 °C, 25.4–26.9 °C, and 26.9–31.5 °C. Those areas which receive relatively high temperatures throughout the whole year are highly prone to land degradation. The LST in the study area ranged from 19.8 °C to 31.5 °C. Higher weights were considered for higher temperature values, which showed 26.9–31.5 °C temperature, and the covered area is 133.3 km^2 (21.4%). The area covered under the other three classes was 58.8 km^2, 179.1 km^2, and 252.8 km^2, contributing to 9.43%, 28.7%, and 40.7% of the total area, respectively.

9.5.2 Vegetation Study

The land use land cover extracted using the Esri thematic layer in the year 2017 to 2021, on January 01. L and use is classified into agricultural land, barren land, built-up, vegetation, and water bodies. The average agricultural land area covers nearly 68.7%, barren land is 17.5%, built-up area slightly increases from 2017 to 2021, is 0.3% to 0.4% per year, and the average is 7.3% (Table 9.2). The maximum vegetation percentage in the area occurs in 2018, and the minimum in 2020 is 3.9%. The area covered has been constant in the last 5 years.

The NDVI is one of the important indicators of vegetative condition using satellite imagery. Remote sensing and GIS technique analysis based on band stacking red (band 4) and NIR (band 8) in Landsat 8 using raster calculation plays a very crucial role in studying the NDVI. It is a broadly used satellite earth observation-based index, which provides a pattern of vegetation cover and its greenness (Forkel et al. 2013). NDVI provides a −1 to +1 vegetation index range. The NDVI values of Jotiba and Panhala Hill range from −0.17 to 0.42. The obtained results of NDVI have been categorised into four classes, viz. <0, 0–0.03, 0.03–0.30, and 0.30–0.42. The values of NDVI ranging from 0.03 to 0.30 show unhealthy plants and cover 97.6% of the area.

9.5.3 Terrain Study

The topography of present study area is classified into six classes: Highly Dissected Plateau (HDP), Moderately Dissected Plateau (MDP), Low Dissected Plateau (LDP), Moderately Dissected Denudational Lower Plateau (MDDLP), Pediment Pediplain Complex (PPC), and water bodies with areas covered as 23.4 km^2, 179.7 km^2, 0.05 km^2, 4.0 km^2, 411.56 km^2, and 5.51 km^2, respectively. Based on this, it is found that most of the area covered by the Pediment Pediplain Complex is 65.9% and the lowest area covered by the Moderately Dissected Denudational Lower Plateau is only 0.008%. The elevation of the study area varies from 542 to 1008 m above msl.

Drainage density is defined as the "ratio of the total length of all streams to the total area of the basin" (Horton 1932; Patil et al. 2020). The average Dd of the region is 2.56 km/km^2 (Gurav et al. 2017). The morphometric analysis of Jotiba Hill was

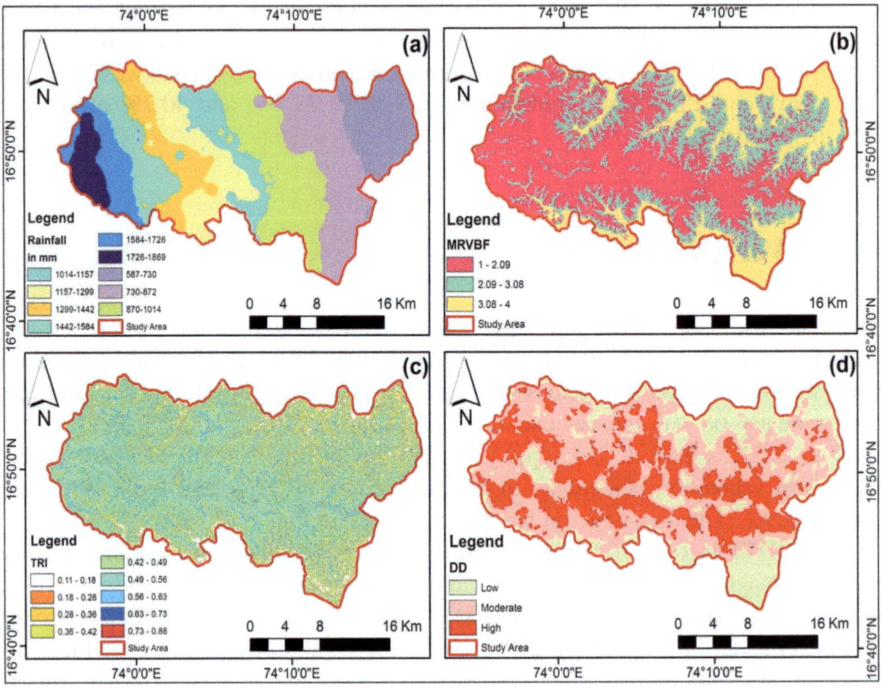

Fig. 9.3 Thematic layers. (**a**) Rainfall, (**b**) MRVBF, (**c**) TRI, and (**d**) drainage density (DD)

undertaken by Gurav and Babar (2018), and they observed that the streams are of first to fifth order, having drainage densities varying from 1.06 to 3.90 km/km^2, clear indications of impermeable lithology and high surface runoff from the hill. For the present study, Dd is classified into three categories, viz., high, moderate, and low, shown in Fig. 9.3, and the area of each category is tabulated for AHP in Table 9.3. High drainage density covered area is 188.7 km^2 and is mostly prone to soil erosion, moderate for the medium it covers 288.7 km^2 and low for the very less covered area is 146.6 km^2.

In general, steep slopes are highly susceptible to high rates of erosion during heavy downpour rather than gentle slopes of the landscape. In the present study area, a steep slope is observed on the south side of the hill as compared to the north side (Gurav et al. 2017). It is also classified into ten classes, as shown in Table 9.3 and Fig. 9.5b. The maximum sloping surface in the study area (218.2 km^2 in 35%) comes under less than 2.14° slope, and the steeper slope, which is shown at 30.7° to 54.4°, is 1.7 km^2 (0.3%).

The MRVBF values show the roughness of the terrain. The lower value shows more ruggedness, while the higher value indicates rugged terrain. The values of MRVBF in the river basin range from 1.0 to 4.0. MRVBF and is categorised into three classes, viz. 1.0–2.09, 2.09–3.08, and 3.08–4.00. MRVBF under the 1.0–2.09

Table 9.3 List of thematic layers, their CR value, weight, and covered area in the Jotiba-Panhala hill range

Thematic layer	Class	Area in km^2	In %	Weight
Rainfall Weight = 14 CR = 0.069	587–730 mm	75.7	12.1	0.32
	730–872 mm	116.7	18.7	0.21
	872–1014 mm	121.9	19.5	0.15
	1014–1157 mm	48.7	7.80	0.11
	1157–1299 mm	72.9	11.7	0.08
	1299–1442 mm	59.5	9.53	0.05
	1442–1584 mm	56.5	9.05	0.03
	1584–1726 mm	42.8	6.86	0.03
	1726–1869 mm	29.4	4.71	0.02
Soil depth and soil Weight = 17 CR = 0.065	Very shallow soil	193.4	31.0	0.50
	Shallow soil	4.3	0.70	0.25
	Moderately deep soil	298.9	47.9	0.16
	Slightly deep soil	127.5	20.4	0.09
Slope Weight = 31 CR = 0.076	0.00–2.14	218.2	35.0	0.30
	2.14–4.72	164.1	26.3	0.20
	4.72–7.72	86.2	13.8	0.15
	7.72–10.9	55.2	8.90	0.11
	10.9–14.1	38.4	6.20	0.08
	14.1–17.4	26.4	4.20	0.05
	17.4–20.8	18.0	2.90	0.04
	20.8–24.9	10.5	1.70	0.03
	24.9–30.7	4.7	0.80	0.02
	30.7–54.4	1.7	0.30	0.02
MRVBF Weight = 03 CR = 0.069	Low ruggedness	140.9	22.6	0.59
	Moderate ruggedness	150.7	24.1	0.25
	High ruggedness	332.7	53.3	0.16
TRI Weight = 04 CR = 0.07	0.11–0.18	17.3	2.80	0.32
	0.18–0.28	37.2	6.00	0.21
	0.28–0.36	54.5	8.70	0.15
	0.36–0.42	79.7	12.8	0.11
	0.42–0.50	152.2	24.4	0.08
	0.50–0.56	162.4	26.0	0.05
	0.56–0.63	75.4	12.1	0.03
	0.63–0.73	37.8	6.10	0.03
	0.73–0.90	8.1	1.30	0.02
LULC Weight = 04 CR = 0.059	Water bodies	6.0	1.00	0.44
	Vegetation	34.1	5.50	0.24
	Agricultural land	429.1	68.7	0.16
	Barren land	109.3	17.5	0.10
	Build-up	45.7	7.30	0.06
	Water bodies	1.7	0.30	0.50
	Moderate healthy	12.2	1.90	0.25

(continued)

Table 9.3 (continued)

Thematic layer	Class	Area in km^2	In %	Weight
NDVI Weight = 02 CR = 0.065	Unhealthy plant	609.4	97.6	0.16
	Barren land	1	0.20	0.09
LST Weight = 02 CR = 0.065	19.8–23.5	58.8	9.40	0.50
	23.5–25.4	179.1	28.7	0.25
	25.4–26.9	252.8	40.5	0.16
	26.9–31.5	133.3	21.4	0.09
SOC g/kg Weight = 06 CR = 0.059	23.2–26.9	98.8	15.8	0.44
	26.9–29.1	180.6	28.9	0.24
	29.1–31.1	164	26.3	0.16
	31.1–33.3	128	20.5	0.10
	33.3–38.1	52.8	8.50	0.06
Drainage density weight = 11 CR = 0.081	Low	146.6	23.5	0.59
	Moderate	288.7	46.3	0.25
	High	188.7	30.2	0.16
Geomorphology weight = 07 CR = 0.068	Water bodies	5.50	0.90	0.40
	PPC	411.60	65.9	0.23
	MDDLP	4.00	0.60	0.16
	LDP	0.00	0.00	0.10
	MDP	179.70	28.8	0.07
	HDP	23.40	3.70	0.04

category covered the maximum area, i.e. 332.7 km^2, which is about 53.3% of the total area, whereas the other classes covered 24.1% and 22.6% of the area, respectively. The higher weights were considered for lower values of MRVBF and vice versa.

The Terrain Ruggedness Index (TRI) of the Jotiba and Panhala hill range was derived from the SRTM DEM of 30 m resolution using QGIS software. It is useful for studying the difference in elevation between two adjacent cells in a DEM. As far as TRI is concerned, the higher the value of TRI, the greater the land degradation that takes place. TRI in the river basin ranges from 0.11 to more than 0.90, with the maximum area under class 0.50–0.56 (162.4 km^2), which is about 26% of the total area. Higher weights were assigned to higher TRI values.

9.5.4 Pedological Characteristics Study

The plateau tops of the study area are covered by lateritic soil cover, while the river and sloping surface areas of the hill range are covered by black cotton soil. For the preparation of digital soil maps (DSM), ICAR-National Bureau of Soil Survey and

Land Use Planning (NBSS and LUP) have classified it into six classes, viz., 0–5 cm, 5–15 cm, 15–30 cm, 30–60 cm, 60–100 cm, and 100–200 cm (ICAR-NBSS and LUP 2021). For the study of land degradation, the present study area is classified into four zones, namely moderately deep soil, shallow soil, slightly deep soil, and very shallow soil. The covered area of moderately deep soil is 28.9 km^2, shallow soil is 4.3 km^2, slightly deep soil is 127.5 km^2, and very shallow soil is 193.4 km^2.

Soil organic carbon (SOC) is the primary indicator of how much better the soil quality is there. The high value of SOC indicates good soil, while the lower value indicates low susceptibility to land degradation. The SOC in the hill range ranges from 23.2 to 38.1 g/kg. Higher weights were assigned to the lower SOC, as they are vulnerable to degradation and vice versa. The SOC was divided into five classes, i.e. 23.2–26.9, 26.9–29.1, 29.1–31.1, 31.1–33.3, and 33.3–38.1 g/kg. The maximum study area is under the class 26.9–29.1 g/kg (180.6 km^2), which constitutes 28.9% of the total area followed by 29.1–31.1 g/kg (164.0 km^2) covering 26.3%, 31.1–33.3 g/kg (128.0 km^2) covering 20.5%, 23.2–26.9 g/kg (98.8 km^2) covering 15.8%, 33.3–38.1 g/kg (52.8 km^2) covering 8.5% (Figs. 9.4 and 9.5).

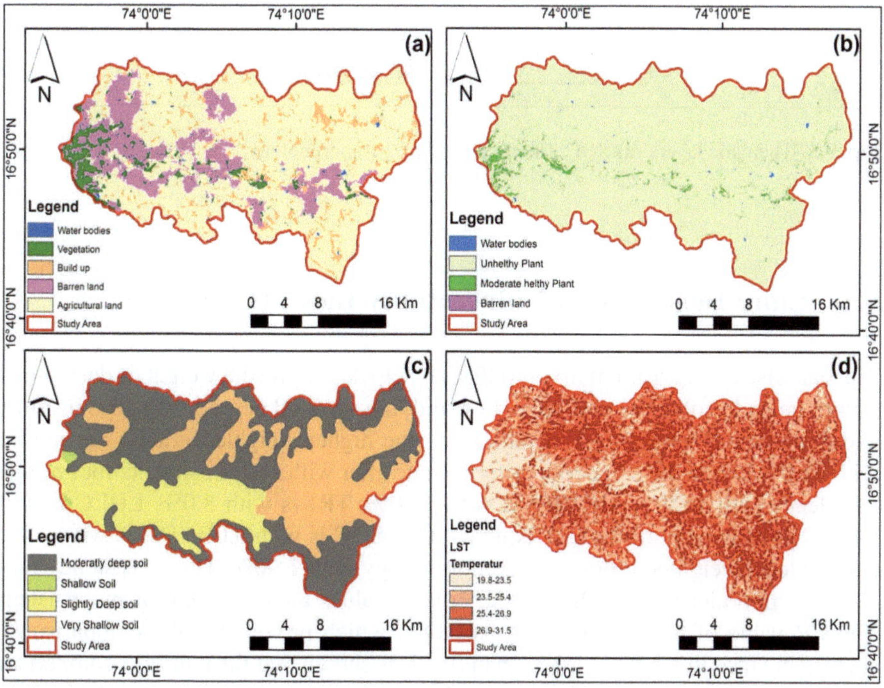

Fig. 9.4 Thematic layers (**a**) LULC-2021, (**b**) NDVI, (**c**) soil, and (**d**) LST

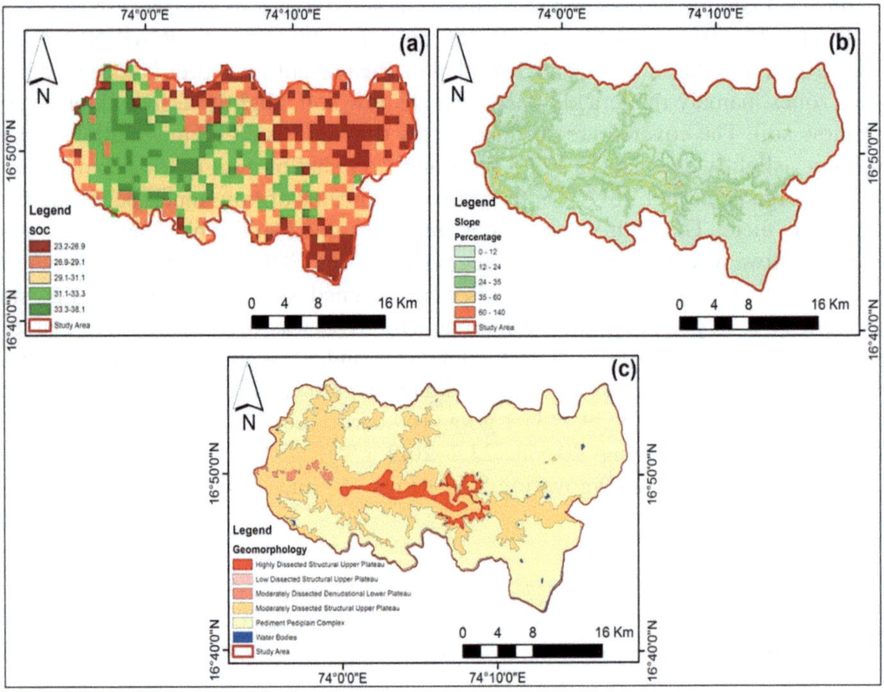

Fig. 9.5 Thematic layers (**a**) SOC, (**b**) slope, and (**c**) geomorphology

9.6 Land Degradation Vulnerability Index (LDVI)

The pairwise comparison matrix (AHP) can help determine the weight of data, which in turn helps to solve the problems. For studying the land degradation sites in Jotiba and Panhala hill range, AHP results show the higher weighted criteria are rainfall with 31.0%, while the soil depth criteria come after with 19.0%, slope comes next to soil depth with 15.0%, MRVBF is with 11.0%, TRI is with 8.0%, LULC is with 5.0%, NDVI is with 4.0%, LST is with 4.0%, SOC is with 3.0%, DD is with 2.0%, and the least weighted criteria is geomorphology with 1.00%. The consistency ratio (CR) is important to test whether the criteria values are consistent or inconsistent. The CR value of the present study is 0.087, which proves that the comparison is consistent since this ratio is lower than 0.1. It is shown that the pairwise comparison matrix has been formed consistently by the decision-maker. The AHP analysis of the present work has come to less than 0.1, indicating that there is no calculation error or an inconsistent pairwise comparison by the decision-maker (Saaty 2008). The final AHP weighting for all criteria can be viewed in Tables 9.3 and 9.5.

Table 9.4 Pairwise comparative matrix

TL	T1	T2	T3	T4	T5	T6	T7	T8	T9	T10	T11
T1	1.00	3.00	3.00	4.00	5.00	6.00	7.00	7.00	8.00	8.00	9.00
T2	0.33	1.00	2.00	3.00	4.00	5.00	6.00	6.00	7.00	7.00	9.00
T3	0.33	0.50	1.00	2.00	3.00	4.00	5.00	5.00	6.00	6.00	8.00
T4	0.25	0.33	0.50	1.00	2.00	3.00	4.00	4.00	5.00	5.00	8.00
T5	0.20	0.25	0.33	0.50	1.00	3.00	3.00	3.00	4.00	4.00	6.00
T6	0.17	0.20	0.25	0.33	0.33	1.00	2.00	2.00	3.00	3.00	5.00
T7	0.14	0.17	0.20	0.25	0.33	0.50	1.00	1.00	2.00	3.00	3.00
T8	0.14	0.17	0.20	0.25	0.33	0.50	1.00	1.00	2.00	2.00	4.00
T9	0.13	0.14	0.17	0.20	0.25	0.33	0.50	0.50	1.00	2.00	3.00
T10	0.13	0.14	0.17	0.20	0.25	0.33	0.33	0.50	0.50	1.00	2.00
T11	0.11	0.11	0.13	0.13	0.17	0.20	0.33	0.25	0.33	0.50	1.00
Sum	2.9	6.0	7.9	11.9	16.7	23.9	30.2	30.3	38.8	41.5	58.0

TL thematic layers, *T1* rainfall, *T2* soil depth and soil, *T3* slope, *T4* MRVBF, *T5* TRI, *T6* LULC, *T7* NDVI, *T8* LST, *T9* SOC g/kg, *T10* drainage density, and *T11* geomorphology

Land degradation analysis of the Jotiba and Panhala hill range area was accessed using the previously stated methodology of 11 thematic layers and AHP analysis. The LDVI was developed as very high, high, moderate, low, and very low vulnerable (Fig. 9.3). The very high vulnerable zone covered area is 13.1 km^2 (2.10%), high covers 50.0 km^2 (8.02%), moderate covers 93.3 km^2 (15.0%), low covers 165.2 km^2 (26.5%), and very low covers 302.2 km^2 (48.5%) (Tables 9.4 and 9.5 and Fig. 9.6).

9.7 Conclusions

In the present study area, land degradation occurs due to anthropogenic activities like the development of roads and land cultivation for the development of agricultural land. Vegetation in the study area has been decreasing in the last few years. The higher temperature received area is the southern and northern sloping surface of the hill range. The steep slope area mostly shows the creep movements, which are the primary evidence of land degradation. The AHP analysis of 11 thematic layers using the normalised weight of the AHP analysis plays an important role in the present study. This method shows 63.1 km^2 of area comes under high to very high risk of land degradation, which is 10.1% of the total area; the remaining portion shows very little land degradation zone.

Table 9.5 Pairwise comparative matrix of 11 criteria for calculating the ratio (CR)

TL	T1	T2	T3	T4	T5	T6	T7	T8	T9	T10	T11	W	λ
T1	0.34	0.50	0.38	0.34	0.30	0.25	0.23	0.23	0.21	0.19	0.16	0.28	12.2
T2	0.11	0.17	0.25	0.25	0.24	0.21	0.20	0.20	0.18	0.17	0.16	0.19	12.3
T3	0.11	0.08	0.13	0.17	0.18	0.17	0.17	0.17	0.15	0.14	0.14	0.15	12.1
T4	0.09	0.06	0.06	0.08	0.12	0.13	0.13	0.13	0.13	0.12	0.14	0.11	11.9
T5	0.07	0.04	0.04	0.04	0.06	0.13	0.10	0.10	0.10	0.10	0.10	0.08	11.8
T6	0.06	0.03	0.03	0.03	0.02	0.04	0.07	0.07	0.08	0.07	0.09	0.05	11.4
T7	0.05	0.03	0.03	0.02	0.02	0.02	0.03	0.03	0.05	0.07	0.05	0.04	11.3
T8	0.05	0.03	0.03	0.02	0.02	0.02	0.03	0.03	0.05	0.05	0.07	0.04	11.3
T9	0.04	0.02	0.02	0.02	0.02	0.01	0.02	0.02	0.03	0.05	0.05	0.03	11.1
T10	0.04	0.02	0.02	0.02	0.02	0.01	0.01	0.02	0.01	0.02	0.03	0.02	11.4
T11	0.04	0.02	0.02	0.01	0.01	0.01	0.01	0.01	0.01	0.01	0.02	0.01	11.4
CI													**0.131**
RI													**1.510**
CR													**0.087**

W Weight, λ lambda, *CI* Consistency Index, *RI* Random Index, and *CR* Consistency Ratio

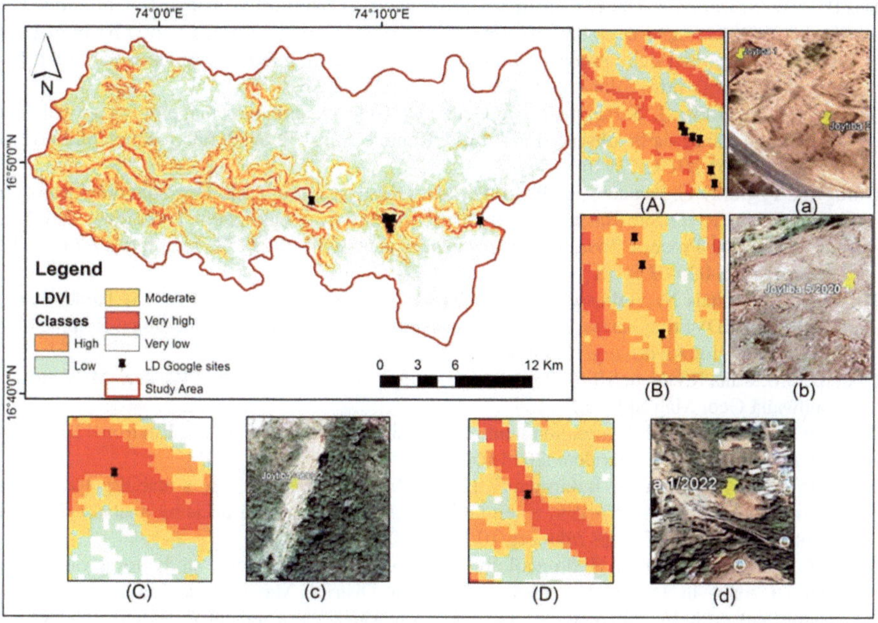

Fig. 9.6 Land degradation vulnerability map of Jotiba and Panhala hill range. Maps (**A**–**D**) show high to very high risk zones with geo-tagging sites, and maps (**a**–**d**) are Google Earth geo-tagging images that show land degradation sites

Acknowledgement The authors thank the Vice Chancellor and Director of the School of Earth Sciences, Punyashlok Ahilyadevi Holkar Solapur University, Solapur, and Head of the Geology department, Rajaram College, Kolhapur, for their help and support.

Author Contribution Conceptualisation: G.C.A.; Methodology: K.D.D.; writing original draft preparation: G.C.A., D.A.K., and K.D.D.; writing, review, and editing: G.C.A., and K.D.D.; Supervision: G.C.A., and K.D.D.

References

Athavale H, Natu SR (2005) Report on transvers mapping in toposheet 47L/1, 2 and 3 in Sangali and Kolhapur District, Maharashtra (No. 050/MAP/CR/MH/2004/2009)
Census Book (1991) District Census Handbook Kolhapur Maharashtra. Census of India. Directorate of Census Operations, Maharashtra, Mumbai
Census Book (2001), District Census handbook, Kolhapur, Maharashtra, Census of India. Directorate of Census Operations, Maharashtra, Mumbai
Census Book (2011), District Census handbook, Kolhapur, Maharashtra. Census of India. Directorate of Census Operations, Maharashtra, Mumbai
Chandio IA, Matori A-N, Lawal DU, Sabri S (2011) GIS-based land suitability analysis using AHP for public parks planning in Larkana city. MAS 5:177. https://doi.org/10.5539/mas.v5n4p177

Das B, Desai S, Daripa A, Gurav C, Kumar U, Khalkho D, Thangavel V, Kumar N, Obi Reddy GP, Kumar P (2023) Land degradation vulnerability mapping in a west coast river basin of India using analytical hierarchy process combined machine learning models. Environ Sci Pollut Res 30:83975–83990. https://doi.org/10.1007/s11356-023-28276-4

de Jong R, de Bruin S, Schaepman M, Dent D (2011) Quantitative mapping of global land degradation using earth observations. Int J Remote Sens 32:6823–6853. https://doi.org/10.1080/01431161.2010.512946

Durugwar Y, Gurav C, Pekam S, Kulkarni D (2025) Assessment of Land Degradation Hotspots in Upper Savitri River Sub-Basin, Raigad District, India Using Geospatial Techniques and Analytic Hierarchy Process, Journal of Geosciences Research 10(2):151–162. https://doi.org/10.56153/g19088-024-0234-78

Forkel M, Carvalhais N, Verbesselt J, Mahecha M, Neigh C, Reichstein M (2013) Trend change detection in NDVI time series: effects of inter-annual variability and methodology. Remote Sens 5:2113–2144. https://doi.org/10.3390/rs5052113

Godbole SM, Rana RS, Natu SR (1996) Lava stratigraphy of Feccan basalt of Western Maharashtra. Gondwana Geol Mag Spl 2:125–134

Grainger A (2015) Is land degradation neutrality feasible in dry areas? J Arid Environ 112:14–24. https://doi.org/10.1016/j.jaridenv.2014.05.014

Gumusay MU, Koseoglu G, Bakirman T (2016) An assessment of site suitability for marina construction in Istanbul, Turkey, using GIS and AHP multicriteria decision analysis. Environ Monit Assess 188:677. https://doi.org/10.1007/s10661-016-5677-5

Gurav C, Babar Md (2018) Hydrogeology and drainage morphomeric study for groundwater potential zones in Joytiba Hill area in Kolhapur District, Maharashtra, India-using remote sensing and GIS. In: 5th national conference on water, environment & society (NCWES – 2018). BS Publications, JNTUH, Hyderabad, pp 205–210

Gurav C, Babar Md, Patil Y, Patil A, Patode HS (2017) Application of remote sensing, geology and geomorphological studies for mass wasting zone analysis in Jotiba-Panhala Hill range area, Kolhapur District, Maharashtra, India. J Appl Geol Geophys (IOSR-JAGG) 5:29–37. https://doi.org/10.9790/0990-0502012937

Gurav C, Babar Md, Gurav N (2021) Remote sensing, geographical information system (GIS) and analytic hierarchy process (AHP) based delineation of groundwater potential zones – a case study of Rena river basin in Latur district, Maharashtra, India. In: Fourth Indian National Groundwater Conference (INGWC-2021). Groundwater Management in Arid and Semi-Arid Regions of Hard Rock Terrains. BS Publications, JNTUH, Hyderabad, pp 69-82.

Gurav C, Babar Md, Pishte JB, Gurav N (2022) Groundwater potential zones (GPZs) delineation in Dhamani River Basin in Kolhapur District, Maharashtra, India, using remote sensing (RS), geographical information system (GIS) and multi-criteria decision analysis (MCDA) techniques. In: Panneerselvam B, Pande CB, Muniraj K, Balasubramanian A, Ravichandran N (eds) Climate change impact on groundwater resources. Springer, Cham, pp 345–371. https://doi.org/10.1007/978-3-031-04707-7_19

Higginbottom T, Symeonakis E (2014) Assessing land degradation and desertification using vegetation index data: current frameworks and future directions. Remote Sens 6:9552–9575. https://doi.org/10.3390/rs6109552

Hill J, Stellmes M, Udelhoven T, Röder A, Sommer S (2008) Mediterranean desertification and land degradation. Glob Planet Chang 64:146–157. https://doi.org/10.1016/j.gloplacha.2008.10.005

Horton RE (1932) Drainage basin characteristics. Trans Am Geophys Union 13:350–361

ICAR-NBSS & LUP (2021) ICAR-NBSS & LUP annual report 2021. ICAR-National Bureau of Soil Survey and Land Use Planning, Nagpur

Kumar R, Anbalagan R (2016) Landslide susceptibility mapping using analytical hierarchy process (AHP) in Tehri reservoir rim region, Uttarakhand. J Geol Soc India 87:271–286. https://doi.org/10.1007/s12594-016-0395-8

Kumar S, Sharma KL, Kareemulla K, Chary GR, Ramarao CA, Rao CS, Venkateswarlu B (2011) Techno-economic feasibility of conservation agriculture in rainfed regions of India. Curr Sci 101:1171–1181

Mahajan GR, Das B, Manivannan S et al (2021) Soil and water conservation measures improve soil carbon sequestration and soil quality under cashews. Int J Sediment Res 36:190–206. https://doi.org/10.1016/j.ijsrc.2020.07.009

Maji AK, Reddy GPO, Sarkar D (2010) Degraded and wastelands of India, status and spatial distribution. ICAR and NAAS Publication, New Delhi

Patil Y, Patil A, Patode HS, Gurav C (2020) Watersheds morphometric analysis of Kasari River basin in Kolhapur District, Maharashtra, India. J Appl Geol Geophys (IOSR-JAGG) 8:45–53. https://doi.org/10.9790/0990-0802014553

Poggio L, De Sousa LM, Batjes NH, Heuvelink GBM, Kempen B, Ribeiro E, Rossiter D (2021) SoilGrids 2.0: producing soil information for the globe with quantified spatial uncertainty. SOIL 7:217–240. https://doi.org/10.5194/soil-7-217-2021

Saaty TL (1980) The analytic hierarchy process, decision analysis. https://doi.org/10.3414/ME10-01-0028

Saaty TL (2004) Fundamentals of the analytic network process—multiple networks with benefits, costs, opportunities and risks. J Syst Sci Syst Eng 13(3):348–379

Saaty TL (2008) Decision making with the analytic hierarchy process. IJSSCI 1:83. https://doi.org/10.1504/IJSSCI.2008.017590

Saha S, Gayen A, Pourghasemi HR, Tiefenbacher JP (2019) Identification of soil erosion-susceptible areas using fuzzy logic and analytical hierarchy process modeling in an agricultural watershed of Burdwan district, India. Environ Earth Sci 78:649. https://doi.org/10.1007/s12665-019-8658-5

Salvati L, Zitti M (2005) Land degradation in the Mediterranean basin: linking bio-physical and economic factors into an ecological perspective. Biota 5:67–77

Sandeep P, Reddy GPO, Jegankumar R, Arun Kumar KC (2021) Modeling and assessment of land degradation vulnerability in semi-arid ecosystem of southern India using temporal satellite data, AHP and GIS. Environ Model Assess 26:143–154. https://doi.org/10.1007/s10666-020-09739-1

Strahler A (1964) Quantitative geomorphology of drainage basins and channel networks. In: Chow V (ed) Handbook of applied hydrology. McGraw Hill, New York

Tagore GS, Bairagi GD, Sharma NK, Sharma R, Bhelawe S, Verma PK (2012) Mapping of degraded lands using remote sensing and GIS. Techniques 12:9

UNCCD (1994) United Nations convention to combat desertification in those countries experiencing serious drought and/or desertification, particularly in Africa. Paris, UN

UNCCD (2002) Recommendations and conclusions of the African regional conference preparatory to the first session of the Committee for the Review of the Implementation of the United Nations Convention to Combat Desertification (UNCCD–CRIC1). Windhoek, Namibia, Secretariat of the Convention to Combat Desertification

Venkataraman K (2012) Biodiversity and its conservation. Proc Natl Acad Sci India Sect B Biol Sci 82:271–282. https://doi.org/10.1007/s40011-012-0096-z

Venkateswarlu B (2012) Carrying capacity of Indian agriculture: issues related to rainfed agriculture. Curr Sci 102:882–888

Verma RR, Manjunath BL, Singh NP et al. (2018) Soil mapping and delineation of management zones in the Western Ghats of coastal India. L Degrad Dev 29:4313–4322. https://doi.org/10.1002/ldr.3183

Wei W, Chen L, Fu B, Huang Z, Wu D, Gui L (2007) The effect of land uses and rainfall regimes on runoff and soil erosion in the semi-arid loess hilly area, China. J Hydrol 335:247–258. https://doi.org/10.1016/j.jhydrol.2006.11.016

Chapter 10
Role of Artificial Intelligence and Machine Learning for Climate Disaster Management and Earth Resource Management

Deepali Bidwai and Gijs van den Dool

Abstract Artificial Intelligence (AI) and Machine Learning (ML) are essential in tackling global challenges related to climate disaster management and earth resource management. AI-enabled systems analyse vast datasets for precise predictions of weather patterns, early warning systems for floods, hurricanes, and wildfires, and real-time decision-making during disaster response. The research aims to bridge technological advancements with ecological responsibility. This work explores the application of AI and ML technologies to predict, mitigate, and manage climate disasters while optimising the use of natural resources. ML algorithms enhance these capabilities by learning from historical data to improve accuracy and adaptability over time. AI and ML contribute to sustainable practices by monitoring and assessing the health of ecosystems with optimisation of agricultural productivity and managing water resources more efficiently. Advanced models identify patterns of deforestation, biodiversity loss, and resource depletion, offering actionable insights for policymakers and stakeholders. This study highlights the potential of AI and ML in fostering resilience against climate disasters and promoting sustainable development. It also addresses the challenges, such as data accessibility, ethical considerations, and the need for collaboration across sectors. These cutting-edge technologies can reduce environmental risks with the conservation of natural resources for a sustainable future.

Keywords Earth resource management · Climate disaster management · Artificial intelligence (AI) · Earth observations (EO) · Data-driven decision-making

D. Bidwai · G. van den Dool (✉)
Independent Researcher, Arcueil, France
e-mail: bidwaideepali@gmail.com; gijs.vandendool@gmail.com

D. Kumar, N. P. Bassill (eds.), *Artificial Intelligence and Machine Learning for Climate Disaster Management*, Disaster Resilience and Green Growth,
https://doi.org/10.1007/978-981-95-1588-2_10

10.1 Introduction

There is a requirement for efficient management of earth resources due to the rising demand for resources and the increasing frequency of climate-related disasters (Sahani et al. 2019; Talha et al. 2025). As the global demand for resources continues to surge, developing strategies for their sustainable utilisation and resilience against climate impacts become dominant (Taneja et al. 2020; Wang and Upreti 2019). Integrating artificial intelligence (AI) with Earth Resources Management (ERM) and Climate Disaster Management (CDM) suggests an optimal solution. ERM can revolutionise resource utilisation, monitoring, conservation, and preparedness for climate disasters, ensuring a more sustainable and resilient future (Dede et al. 2023; Efremova et al. 2023; Meenal et al. 2022). Artificial Intelligence (AI) with ERM and CDM offers several benefits, including improved efficiency, cost savings, enhanced sustainability, and informed decision-making (Jayaraman et al. 2009; Kocaaslan et al. 2021; Yadav 2023). AI is capable of automating various tasks, providing valuable insights, optimising resource allocation, and enhancing disaster preparedness, leading to increased operational efficiency reduced costs, and improved resilience (Fan et al. 2025; Shahvar et al. 2025). AI enables farmers to optimise crop yields, reduce water and fertiliser use, improve soil health, and anticipate climate-related risks, contributing to enhanced sustainability and disaster management in agriculture.

Traditional data collection methods are often time-consuming and resource-intensive. Acquiring precise and up-to-date data about the state of our resources and climate risks is a huge challenge. This concept presents a pathway for the efficient collection of extensive data from satellites, drones, sensors, and various sources. The capabilities of AI empower resource managers to make informed decisions and gain a comprehensive understanding of any resources beyond human capability. AI also optimises resource allocation, ensuring efficiency, minimising waste, and extending predictive capabilities of forecasting the effects of climate change, land use, human activities, and potential disasters (Alamús et al. 2018; Zhu et al. 2021).

This chapter explores the applications of an in-depth overview of AI approaches to machine learning (ML), deep learning (DL), computer vision (CV), transfer learning, and GenAI with emerging technologies in the field (Udawalpola et al. 2021; Wang and Hu 2019). It highlights the importance of diverse data sources like remote sensing, in situ measurements, field studies, and ground truthing for disaster management (Attaf et al. 2019; Ganji et al. 2019; Sunar et al. 2019). It narrates practical applications in agriculture, forestry, urban planning, climate change modelling, and biodiversity to provide insights into implementation to facilitate a more responsible, sustainable, and resilient future for mitigating climate-related disasters (Bhadra et al. 2020; Edgell 2023; Sahani et al. 2019). Artificial Intelligence (AI) has also the potential to advance sustainable resource utilisation for climate disaster management. Its other applications extend to precision agriculture, biomass inventory mapping, and land use planning for emissions forecasting, air quality monitoring, early warning systems, climate change modelling, disaster prediction and biodiversity conservation.

10.2 Overview of Artificial Intelligence (AI), Machine Learning (ML), Deep Learning (DL), Computer Vision (CV), Transfer Learning (TL), and Generative AI (GenAI)

Artificial Intelligence (AI), Machine Learning (ML), Deep Learning (DL), Computer Vision (CV), Transfer Learning (TL), and Generative AI (GenAI) are unconventional-edge tools with applications across various domains (Ding et al. 2020; Ma et al. 2024; Ren et al. 2024). These technologies propose innovative solutions to analyse, model, and predict the state of natural resources. They provide a data-driven approach to conservation, sustainable resource utilisation and climate disaster management (Butts et al. 2007; Weijs et al. 2007). These technologies play a crucial role in climate disaster management by enabling prediction and early warning systems for events such as floods, wildfires, and droughts. Figure 10.1 shows the interlinkages among AI, ML, DL and Generative AI.

Typical AI models analyse satellite data to forecast wildfire spread to offer timely evacuation and resource allocation. Machine learning/deep learning algorithms can assess flood risks by analysing rainfall patterns, topography, and historical flood data, helping communities prepare for potential flooding (Li and Sivapalan 2020; Sahani et al. 2019; Wahab et al. 2024). AI optimise water resource management by

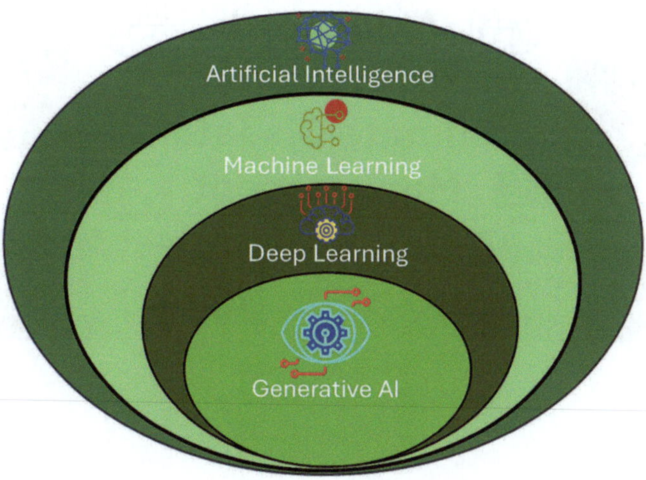

Fig. 10.1 Relationship between AI, ML, DL and Generative AI

predicting water shortages and guiding efficient irrigation practices during drought conditions. These applications not only reduce the immediate impact of disasters but also contribute to long-term resilience and sustainability.

Machine Learning (ML) emerges as a subset of Artificial Intelligence (AI) and it empowers machines to learn and forecast based on past experiences, marking a new era of predictive modelling and data-driven decision-making (Quintero et al. 2019; Zeng et al. 2024). Figure 10.2 showcases the use of AI for Earth Resource and Climate Disaster Management. Machine learning algorithms are developed through a multi-stage process, beginning with data collection and preparation (Hu et al. 2019; Zou et al. 2021). This engages selecting appropriate algorithms, training the model on labelled or unlabeled data, and fine-tuning it to achieve optimal performance. ML has a significant role in analysing extensive datasets sourced from satellites, sensors, and diverse channels to enhance the decision-making processes during climate disaster management (Sun et al. 2024; Wang et al. 2024). ML is categorised into three main categories: *Supervised Learning (SL)*, *Unsupervised Learning (UL)*, and *Reinforcement Learning (RL).* In Supervised Learning, the algorithm is provided with labelled data, allowing it to learn and make predictions, making it suitable for tasks such as regression and classification (Golparvar et al. 2021; Xu et al. 2021). Unsupervised Learning takes a self-learning approach, dealing with unlabelled data

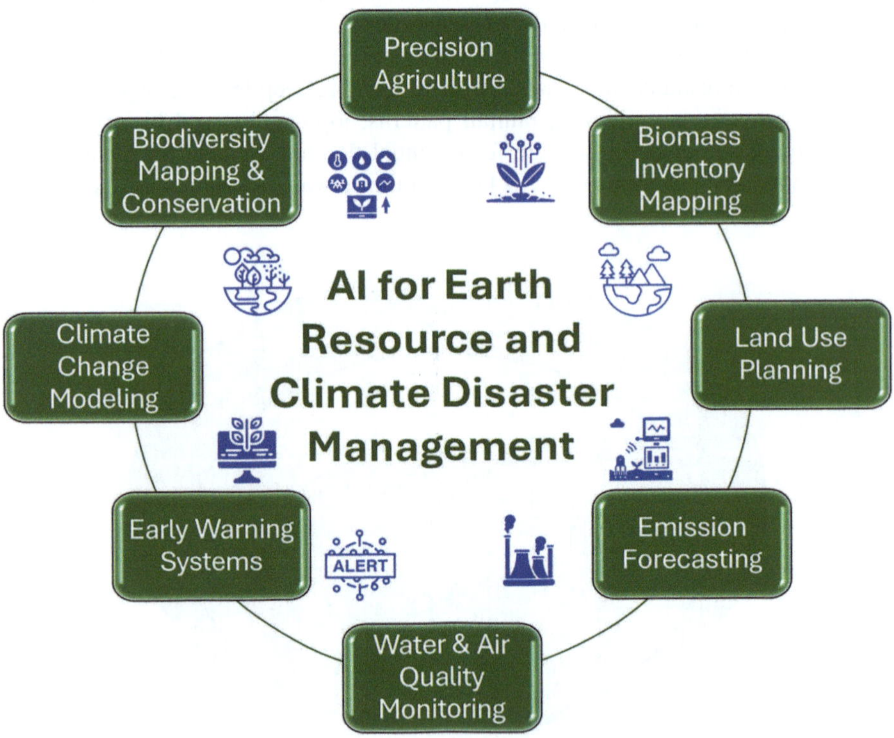

Fig. 10.2 AI for Earth Resource and Climate Disaster Management

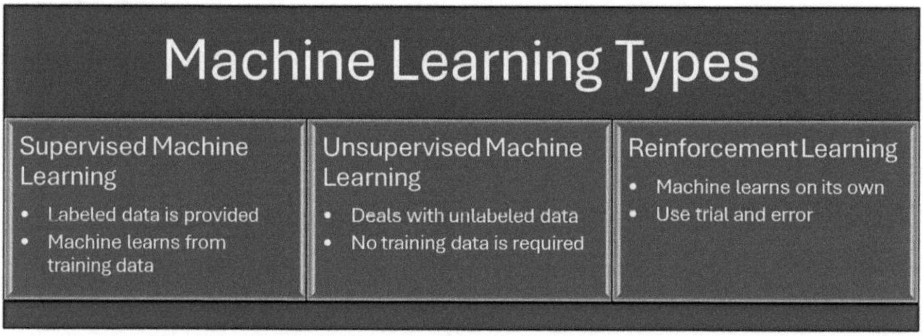

Fig. 10.3 Types of machine learning

to uncover hidden patterns, making it valuable for clustering and association tasks (Corte et al. 2021). Reinforcement Learning introduces the concept of an agent making decisions in an environment to maximise rewards, employing trial-and-error methods to achieve optimal outcomes (de Rosa and Papa 2021; Kang et al. 2024). Figure 10.3 shows the different forms of Machine Learning. Examples of supervised learning are *Linear Regression*, *Decision Trees*, *Support Vector Machines*, *Random Forests*, *k-nearest Neighbours*, *Naive Bayes*, and *Neural Networks.* Examples of *Unsupervised Learning are K-means clustering*, *Hierarchical Clustering*, *DBSCAN*, *Principal Component Analysis*, *Gaussian Mixture Models (GMM)*, *Apriori Algorithm.* Examples of Reinforcement Learning are *Q-learning*, *Deep Q Networks (DQN), and Monte Carlo Methods.*

Deep learning (DL) is also a subset of machine learning, and it represents a sophisticated approach to artificial intelligence (AI) that mimics the workings of the human brain. It revolves around neural networks, intricate structures composed of interconnected nodes, or artificial neurons. These networks consist of multiple layers, each responsible for processing specific aspects of data. The depth of these structures allows for the automatic extraction of intricate features from raw data, contributing to the model's capacity to learn complex patterns and representations (Deepa et al. 2021; Nagarajan and Wason 2019). Unlike traditional machine learning, which often requires explicit feature engineering, deep learning models can autonomously discover relevant features, making them particularly adept at handling unstructured and large-scale data. Examples of deep learning algorithms include *Convolutional Neural Networks (CNNs)*, *Recurrent Neural Networks (RNNs)*, *Long Short-Term Memory networks (LSTMs)*, *Generative Adversarial Networks (GANs)*, *Autoencoders*, and *Deep Belief Networks (DBNs).* Figure 10.4 shows the typical setup of a Deep Neural Network.

ML applications benefit precision agriculture with optimisation of crop management, monitoring soil health, and predicting yields. They also contribute to biomass inventory mapping, land use planning, early warning systems for natural disasters, climate change modelling, and biodiversity assessment (Li et al. 2019; Nedd et al. 2021). The integration of ML augments the efficiency of data analysis for adaptive strategies of conservation, sustainable use of resources, and proactive climate disaster management.

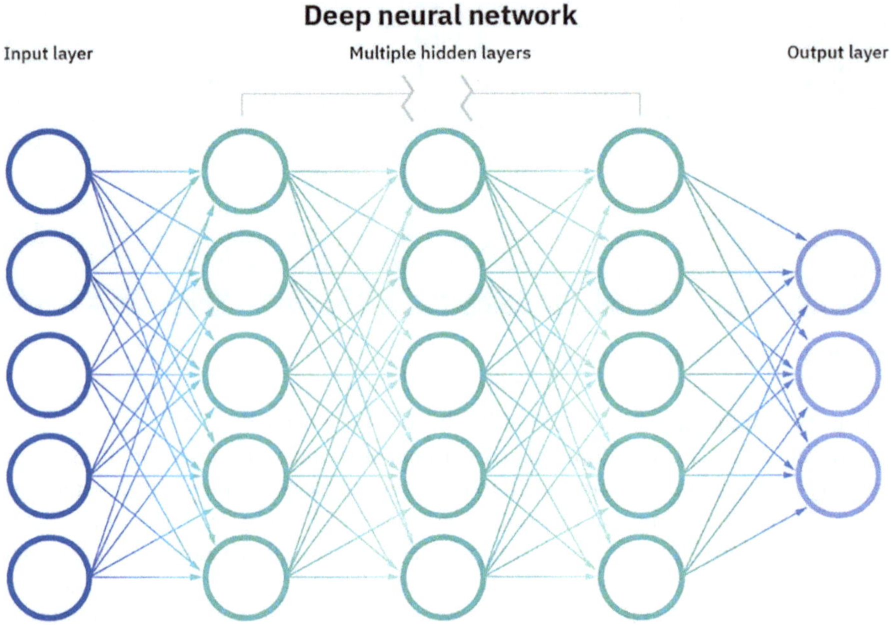

Fig. 10.4 Deep Neural Network. (Source: IBM Data and AI Team 2024)

Deep Learning is capable of addressing complex challenges of climate disaster management. Deep learning algorithms analyse extensive datasets to extract nuanced patterns in vegetation, soil composition, and land use, enabling more accurate predictions and adaptive responses to climate-related disasters. These models also contribute significantly to biodiversity assessment, climate change modelling, and early warning systems. Convolutional Neural Networks (CNNs) are deep learning architectures and process satellite imagery to accurately identify land use patterns, deforestation, and ecosystem changes for both sustainable resource use and disaster preparedness (Krishnamurthy et al. 2021; Rajkumar et al. 2024). CNNs can detect early signs of forest fires, enabling rapid response efforts, or assessing flood-prone areas to improve evacuation planning and resource allocation, thereby enhancing climate disaster management (Matsuoka and Oki 2021; Sinha et al. 2020).

Computer Vision (CV) is a field of artificial intelligence that empowers machines to interpret and comprehend visual information from the world (Du et al. 2021; Luo et al. 2023). It involves the development of algorithms and computational models designed to enable computers to extract meaningful insights, recognise patterns, and understand the visual content of images or videos. Computer Vision mimics the human ability to interpret the visual world by integrating advanced techniques from image processing, machine learning, and deep learning (Ajani et al. 2021; Luo et al. 2023). The primary goal is to equip machines with the capability to analyse and interpret visual data, allowing them to make informed decisions, recognise objects or scenes, and perform tasks that traditionally rely on human vision. Examples of CV algorithms are *CLIP (Contrastive Language-Image Pre-training), GPT-4V (Generative Pre-trained Transformer-4 Vision), ResNet (Residual Neural Network), VGG, InceptionNet, MobileNet, EfficientNet, DenseNet, YOLO (You Only Look Once) and YOLO World, Mask R-CNN, Faster R-CNN, RNN (Recurrent Neural Network), LSTM (Long Short-Term Memory),* and *GRU (Gated Recurrent Unit), AlexNet.*

Computer vision algorithms can process satellite imagery, accurately identifying and classifying different land cover types, such as forests, agricultural areas, and urban developments (Impedovo et al. 2021; Sravanthi et al. 2020; Wang et al. 2021). This capability facilitates efficient mapping and monitoring of land resources. It can be employed to monitor vulnerable areas in real time, providing early warnings and enabling rapid response efforts. Analysing satellite imagery can help detect early signs of deforestation, which could lead to increased flood risk, or identify changes in vegetation patterns that might indicate the potential for wildfires. Drones equipped with computer vision systems can survey large agricultural areas, capturing high-resolution images that assist farmers in making data-driven decisions to enhance crop yield, resource efficiency, and disaster preparedness. CV also has a role in analysing visual data to assess crop health, identify potential diseases, and optimise resource allocation.

Transfer learning is a machine learning technique to leverage knowledge gained from solving one problem and apply it to a different but related task. In traditional machine learning, models are trained from scratch for specific tasks, requiring large datasets and computational resources. This approach is especially beneficial when working with limited labelled data for a specific application. Transfer learning enhances the efficiency and effectiveness of model training for new, domain-specific challenges. Figure 10.5 presents a Transfer Learning Sequence.

Transfer learning has a crucial role in optimising the performance of machine learning models, particularly in scenarios where acquiring large labelled datasets can be challenging. A model pre-trained on general satellite imagery or environmental data can be fine-tuned for specific tasks such as land cover classification vegetation monitoring, or climate disaster management. Transfer learning contributes to the scalability and effectiveness of machine learning solutions in addressing challenges and managing climate-related risks.

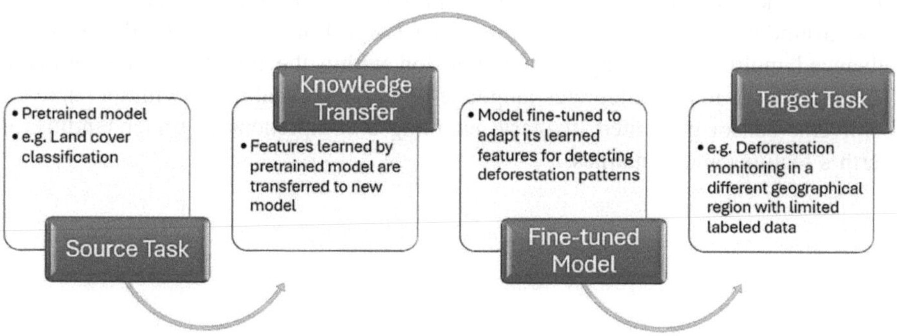

Fig. 10.5 Transfer Learning Sequence

Generative AI (GenAI), or generative artificial intelligence, includes AI to create new content such as text, images, music, audio, and videos. This technology is operated by large AI models that can multitask and perform various tasks like summarisation, Q&A, and classification with minimal training required (Google Cloud 2024). Generative AI can streamline interactions and data exploration in climate disaster management. The analysis of vast amounts of unstructured data with GenAI can predict and model climate-related disasters such as hurricanes, floods, and wildfires. It can generate detailed scenarios for disaster preparedness, helping organisations plan evacuation routes, allocate emergency resources, and optimise disaster response strategies. It can be used to stimulate the impact of various disaster management interventions, allowing for more informed decision-making in crises. GenAI can provide early warnings and tailored strategies to mitigate the effects of droughts, floods, or pest infestations, ensuring agricultural resilience in the face of climate change. GenAI requires substantial computational power to process data and generate new content. This requires high energy demand and it will contribute to carbon emissions and environmental damage. The fabrication and maintenance of generative AI systems will lead to an increase in e-waste and the depletion of natural resources.

The emerging technologies collectively contribute to the evolution of AI, enabling more powerful, interpretable, and scalable applications across diverse domains including earth resource management and climate disaster management. These technologies are likely to shape the future of intelligent systems, offering new possibilities and addressing challenges in innovative ways.

Explainable AI (XAI) is gaining traction as an essential aspect of trustworthy AI systems. As AI models become more intricate, the ability to interpret and explain their decisions becomes crucial for ensuring transparency, accountability, and user trust. Explainable AI (XAI) models are designed to provide insights into how they arrive at their decisions or predictions, making them interpretable and transparent to users.

Artificial General Intelligence (AGI) represents a significant milestone in AI development, signifying machines' capability to mimic human-like thinking and perform a wide range of tasks through techniques like transfer learning and multi-modal approaches. Large Language Models (LLMs), exemplified by GPT-3.5 utilised in ChatGPT, have garnered substantial attention due to their remarkable language understanding and generation capabilities. "EarthGPT" is an example of a new groundbreaking Universal Multi-modal Large Language Model designed for advanced multi-sensor image comprehension within the remote sensing domain. This innovative model integrates language processing capabilities with the analysis of diverse sensor-generated imagery, enabling a comprehensive understanding of Earth's features and dynamics.

10.3 Overview of Data Sources for Earth Resource Management

Data sources for earth resource management have a key role in sustainable development and conservation efforts. Key sources of data include satellite imagery and remote sensing technologies, which provide detailed data on land use, vegetation, and water bodies. Geographic Information Systems (GIS) integrate spatial data for advanced analysis. Sensor networks and IoT devices offer real-time information on soil health, weather, and resource conditions. Climate models and historical datasets help predict environmental changes and resource availability. Field surveys and community reports add localised insights. Combining these data sources enables informed decision-making, efficient resource allocation, and the development of strategies for long-term ecological balance.

There are nine different types of satellites, i.e., Communications Satellites, Remote Sensing Satellite, Navigation Satellite, Geocentric Orbit type satellites (LEO, MEO, GEO), Global positioning Satellites (GPS), Geostationary Satellites (GEOs), Drone Satellite, Ground Satellite, Polar Satellite, as well as Nano Satellites, CubeSats, and SmallSats. Satellites are designed to operate in specific orbits. Each orbit has different operational constraints, which means that certain orbits are better suited for specific operations than others. For example, the Low Earth Orbit (LEO) orbit is more popular than others, primarily because operating in this orbit gives several advantages over the higher orbits, like lower launch costs, shorter communication delays, and higher resolution imaging. When designing a satellite, two critical aspects must be considered; the horizon (how much of the Earth's surface a satellite can see) and the Field of View (FoV). Both of these factors are determined by the distance between the satellite and the Earth's surface (H). The decision to use aerial imagery rather than satellite imagery for nationwide projects involves complex trade-offs. Satellites with sub-meter ground resolution are often tasked with covering smaller areas to ensure an adequate number of cloud-free images. Managing a global or national database of all acquired images would exceed current storage and processing capacities, so tasking is optimised to balance data storage and processing efficiency.

The payloads of these satellites contain various instruments and sensors designed to collect data at different spatial, spectral, and temporal resolutions. This variety of data helps meet the diverse needs of users at local, regional, and global levels. The data from these satellites are used in several applications including agriculture, water resources, urban planning, rural development, mineral prospecting, environment, forestry, ocean resources, and disaster management. There are (broadly speaking) two classes of sensors, active and passive. These sensors serve different purposes based on their operational principles. Passive sensors are typically optical sensors that rely on an external source to illuminate the area of interest and collect the reflectance of an area. The reflected sunlight is the most common source of radiation measured by passive sensors. Active sensors do not require an external source to illuminate (or probe) the area because they are actively sending out signals (radiation) to the area and measuring the deformation of the signal when returned.

There are various options for obtaining Earth observations, each with its advantages and limitations. Satellites are well-suited for large-scale projects that require regular, systematic acquisitions with consistent imaging characteristics and broad coverage areas, making them ideal for global monitoring (e.g., tracking deforestation in the Amazon). However, there are trade-offs. High-resolution satellite data, while available, can be very expensive and require extensive processing, especially for large areas. Moreover, cloud cover can obscure observations from passive sensors, and the launch and operational costs are high. In contrast, aerial imagery campaigns typically have lower costs compared to maintaining a satellite constellation for long-term projects. The costs can vary based on data use cases and potential commercialisation. Aerial imagery offers high spatial resolution and is less affected by atmospheric conditions, but it is generally less suited for global monitoring due to its smaller coverage area. The choice between satellites and aerial imagery depends on factors such as specific data needs, budget, and desired level of detail.

10.4 Artificial Intelligence (AI) in Agriculture, Forestry, and Built Environment

Artificial Intelligence (AI) is altering agriculture, forestry, and the built environment with innovation, efficiency, and sustainability. AI enables precision farming in the agriculture sector through the use of sensors, drones, and machine learning, optimising resource allocation and enhancing crop monitoring, soil analysis, and pest control. AI-driven tools predict crop yields, improve irrigation practices, and develop climate-resilient crops, addressing food security challenges and reducing environmental impact. In forestry, AI aids in monitoring forest health, mapping deforestation patterns, and managing biodiversity. AI-enabled satellite imagery and remote sensing technologies help track illegal logging, assess carbon sequestration, and ensure sustainable forest management. These advancements contribute to global conservation efforts and mitigate the effects of climate change.

AI supports the creation of smart cities and sustainable infrastructure in the built environment, by analysing energy consumption, optimising urban planning, and enhancing building designs. AI applications, such as autonomous construction robots and predictive maintenance systems, improve project efficiency and reduce costs. AI contributes to disaster resilience by identifying structural vulnerabilities and assisting in emergency response. AI has a critical role in driving sustainable practices across agriculture, forestry, and urban development, fostering a balance between technological advancement and environmental stewardship for a resilient future.

10.4.1 AI in Agriculture for Climate Disaster Management

Agriculture has been the backbone of human civilisation and has witnessed a remarkable transformation with the introduction of Artificial Intelligence (AI) and its associated technologies. AI is revolutionising traditional farming practices across the globe, and guiding in an era of precision agriculture, data-driven decision-making, and sustainable resource management. AI in agriculture has a significant role in climate disaster management in predicting weather patterns, optimising resource use, and enhancing crop resilience. AI helps farmers monitor soil health, manage irrigation, and respond proactively to extreme weather events, ensuring sustainable food production in changing climates. AI helps in crop monitoring and management and farmers can obtain real-time insights into crop health, soil moisture levels, and pest infestations with the integration of machine learning algorithms to satellite imagery. This enables them to make informed decisions regarding irrigation schedules, fertiliser application, and pest control measures, ultimately leading to increased yields minimising the environmental impact, and disaster risk reduction. AI is making significant progress in predictive analytics and forecasting. AI models can accurately predict future trends and potential risks, allowing farmers to proactively mitigate adverse effects such as crop failures or yield losses with the analysis of historical data on weather patterns, soil conditions, and crop performance. This proactive approach minimises financial losses for farmers and enhances productivity.

AI-powered robotics and automation are transforming labour-intensive agricultural tasks like planting, harvesting, and weed control. Autonomous drones equipped with computer vision systems can survey vast expanses of farmland, identifying areas requiring attention and applying targeted treatments with precision. Also, robotic harvesters can efficiently gather crops while minimising damage and waste, leading to greater efficiency and cost savings.

In addition to on-field applications, AI is also revolutionising supply chain management and market forecasting in agriculture. The detailed analysis of market trends, consumer preferences, and logistical factors optimises distribution routes, minimises food waste, and ensures timely delivery of produce to market. This benefits farmers by maximising profits and ensures a steady supply of fresh supplies to the consumers. Figure 10.6 visualises the role of Artificial Intelligence in Agriculture.

Artificial Intelligence (AI) is revolutionising agriculture across various fronts, including crop management, soil management, water management, and livestock management. In crop management, AI-powered systems analyse satellite imagery and sensor data to monitor crop health, predict yields, and optimise irrigation and fertilisation schedules, ensuring efficient resource use and maximising yields. AI algorithms assess soil health, moisture levels, and nutrient content, guiding farmers in making informed decisions about soil conditioning, irrigation strategies, and sustainable land management practices in soil and water management. AI applications in livestock management engross monitoring animal health, behaviour, and

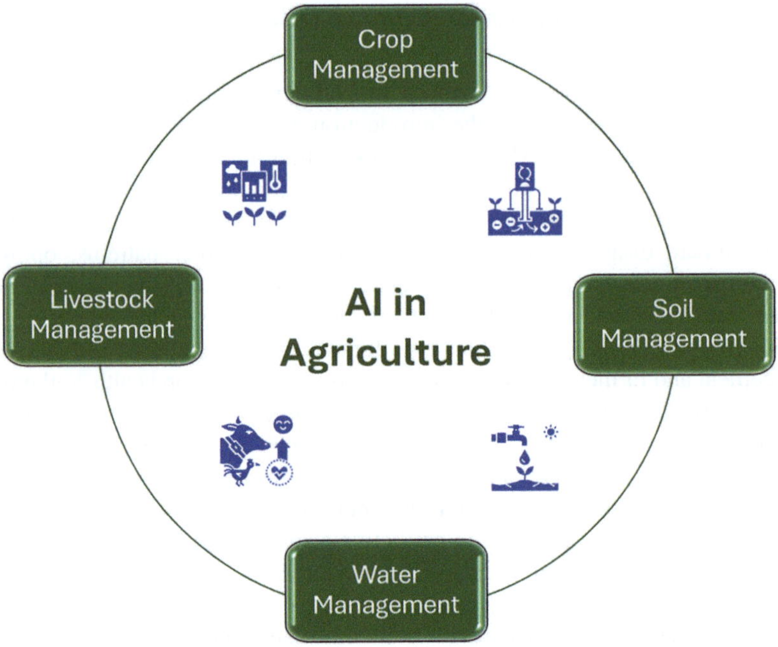

Fig. 10.6 Artificial intelligence in agriculture. (Source: Image by Authors)

productivity through wearable sensors and computer vision systems, enabling early disease detection, optimised feeding regimes, and enhanced animal welfare.

Crop monitoring involves the continuous observation and analysis of crop growth and health throughout the growing season. Advanced technologies like satellite imagery, drones, and ground-based sensors help farmers assess various aspects of crops, including vegetation indices, soil moisture levels, pest infestations, and disease outbreaks.

Figure 10.7 shows the various components of Crop Management. AI algorithms process the collected data to provide real-time insights into crop conditions, enabling farmers to make informed decisions regarding irrigation, fertilisation, pest control, and harvest timing. The farmers optimise their resources to maximise yields, and mitigate potential risks, ultimately contributing to sustainable agricultural practices and food security by monitoring crops closely.

Soil management encompasses a range of practices aimed at optimising soil health and productivity. Figure 10.8 shows different components of soil management. AI assists farmers in assessing soil health parameters such as nutrient levels, pH, and organic matter content. AI models with remote sensing technologies enable precise soil mapping and classification for tailored soil condition-specific resource management. AI also facilitates precision fertilisation, irrigation scheduling, and erosion prediction, promoting resource-efficient farming while minimising

Fig. 10.7 Components of crop management

Fig. 10.8 Elements of soil management

environmental impacts. Farmers can enhance soil fertility, water retention, and resilience to climate change, ensuring sustainable agricultural production for the future with the use of AI in soil management.

Water management encompasses strategies and practices to utilise and conserve water resources in agricultural settings efficiently. The core components of Water Management are shown in Fig. 10.9. Farmers optimise irrigation scheduling, monitor soil moisture levels, and minimise water wastage by integrating advanced technologies like sensor networks, remote sensing, and AI-driven predictive models. The farmers can enhance crop yields with precise utilisation of water as per the requirements and it will reduce the water consumption and the risk of water-related issues like drought or waterlogging. Water management techniques promote

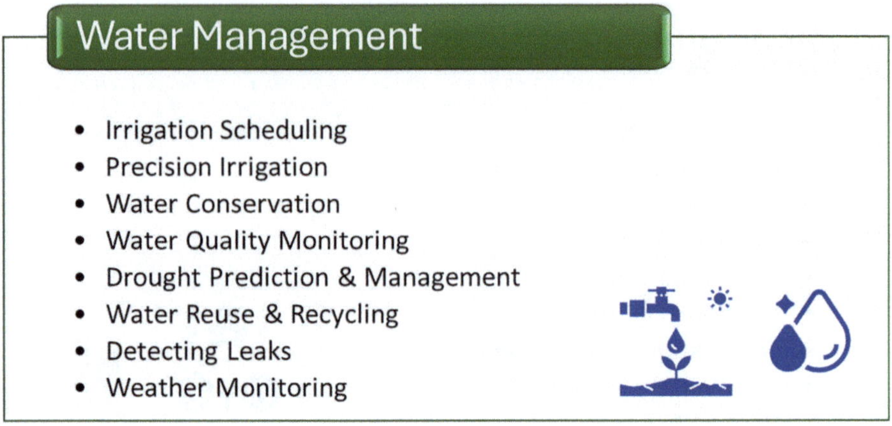

Fig. 10.9 Components of water management

Fig. 10.10 Components of livestock management

sustainability with the preservation of water quality, soil erosion, and safeguarding ecosystems. This effective water management in agriculture provides key support in ensuring the resilience and productivity of farming systems while mitigating environmental impacts.

Livestock management involves the strategic care and oversight of animals raised for agricultural purposes, encompassing aspects such as feeding, breeding, health monitoring, and overall welfare. Advanced technologies, including sensors, wearable devices, and data analytics, enable farmers to monitor livestock behaviour, health indicators, and environmental conditions in real-time. The core elements of any Livestock Management are shown in Fig. 10.10. The farmers can detect signs of illness or distress early, optimise feeding regimes, and improve breeding programs for enhanced productivity and animal welfare with the use of AI-driven algorithms and predictive models. Precision livestock farming techniques enable targeted interventions, minimising resource wastage and environmental impacts. The livestock managers can ensure the well-being of their animals while optimising production outcomes sustainably with the integration of technology and innovative management practices.

10.4.2 AI in Forestry for Climate Disaster Management

AI in forestry assists climate disaster management with monitoring of forest health, predicting wildfires, and assessing the impact of climate change on ecosystems. Machine learning algorithms analyse satellite images and environmental data to detect early signs of forest degradation, pest outbreaks, or fire risks. This enables timely intervention to prevent large-scale damage. AI models help to optimise forest management practices, promoting biodiversity and resilience. AI supports sustainable forestry practices that mitigate the effects of climate disasters and contribute to carbon sequestration efforts.

Artificial Intelligence (AI) is increasingly being used in forestry to improve forest management, data analysis, and decision-making. AI technologies like machine learning, deep learning, computer vision and satellite data analysis, are being employed to provide real-time, detailed information on forests, enabling forest owners and professionals to make more accurate and timely decisions. Figure 10.11 shows the key components of Artificial Intelligence (AI) for forestry applications. AI-based systems can analyse satellite data to detect events in the forest, such as

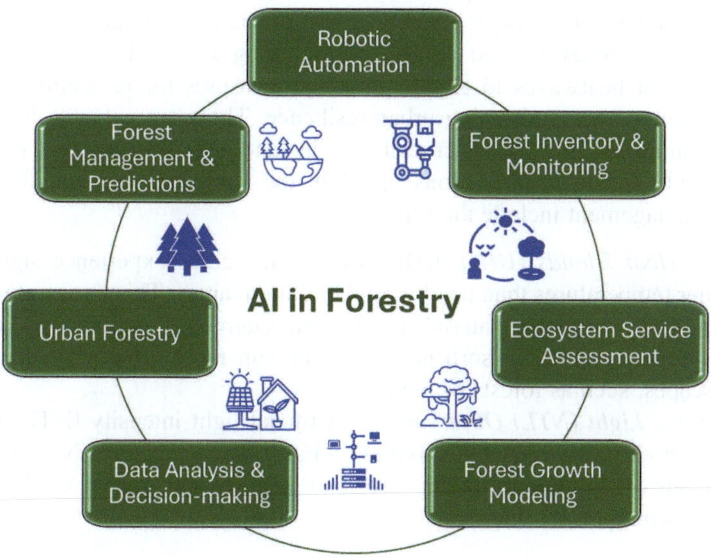

Fig. 10.11 Artificial intelligence (AI) in forestry

storm damage, drought, insect damage, and tree and canopy loss, which may not be easily visible to the naked eye. This also allows for the development of digital twins of forests, which model forest growth, health, and carbon storage, providing valuable insights for forest management and conservation.

The digital twin of a forest is a powerful tool for comprehensive forest monitoring, modelling, and decision support, contributing to proactive forest management and sustainability. A digital twin of a forest is a virtual representation of a real-world forest that is continuously updated by incoming data. It enables the visualisation of forest management activities, forest disturbance, topography, stand structure, and tree species composition. The digital twin integrates various data types and covers different use cases, operating at larger spatial scales. It is created with AI, cloud computing, earth science, and modelling, and it converts environmental, societal, and economic data into actionable insights for scientists and decision-makers. AI is used in urban forestry to assess and improve ecosystem services provided by trees in cities. The AI-based method has been developed to integrate urban forestry inventory with ecosystem service estimates, allowing for the systematic evaluation of tree conditions and the identification of efficient urban forest indicators. This approach aims to support improved urban forestry management and provide practical guidelines for promoting green cities.

10.4.3 AI in the Built Environment for Climate Disaster Management

AI in the built environment helps manage climate disasters by optimising infrastructure design, improving energy efficiency, and predicting structural vulnerabilities. The predictive modelling and real-time monitoring assess the risks like floods, earthquakes, or heatwaves to enable proactive measures for protecting buildings, reducing damage, and enhancing urban resilience. These are valuable for environmental monitoring and play a crucial role in land use planning and emergency management. The key applications of AI in the Built Environment for Climate Disaster Management include the following:

(a) *Urban Heat Islands (UHI)*: UHIs occur when cities experience significantly warmer temperatures than nearby rural or natural areas. This temperature difference results from how materials absorb and retain heat. Artificial surfaces, like concrete and asphalt, absorb heat from the sun more effectively than natural landscapes, such as forests or fields.
(b) *Nighttime Light (NTL) Observations*: Artificial light intensity (NTL) is linked with various developmental activities on Earth. The analysis of NTL data tracks urban growth, monitors energy use, and assesses the impact of disasters on infrastructure.

These applications of AI in urban environments including UHI and NTL data provide unique perspectives compared to traditional daytime observations to provide valuable information. Artificial Intelligence enhances climate disaster management within the built environment. AI is used for predictive modelling, real-time monitoring, and risk assessment for cities and infrastructure to become more resilient to climate-related hazards such as floods, heatwaves, and earthquakes. AI enables better planning, resource allocation, and adaptive design, ensuring that buildings and urban spaces can withstand extreme events. AI will improve structural safety, energy efficiency, and disaster response for creating sustainable, future-proof built environments that mitigate environmental risks effectively.

10.5 AI, ML, and DL for Climate Models

Artificial intelligence (AI) rapidly transforming various fields, offering powerful tools to tackle global challenges like Climate Change. Artificial Intelligence (AI), Machine Learning (ML), and Deep Learning (DL) have significant roles in various climate models. AI, ML, and DL are used to simulate climate scenarios, predict extreme weather events, and assess long-term environmental changes. They enable better decision-making for climate adaptation and mitigation strategies. These improve the accuracy and efficiency of climate predictions with the analysis of vast datasets to identify complex patterns for providing real-time insights. These approaches offer a more precise understanding of climate dynamics, supporting informed policymaking and sustainable resource management for climate change.

Climate models and machine learning (ML) approaches manage vast amounts of climate data from satellites and sensors to reveal hidden patterns and relationships for predicting future climate scenarios. Deep learning (DL) as a subset of ML, uses artificial neural networks to emulate complex climate systems to enable faster and more accurate simulations of cloud dynamics and ocean currents. Generative AI (GenAI) provides climate communication with natural language processing to translate complex scientific findings into accessible language with collaboration and awareness among diverse stakeholders.

Climate models are typically large-scale, global models generalising climate signals across vast areas. These models may not be suitable for regional or local applications for understanding global impacts. ML techniques are suitable for "downscaling" and these downscaling techniques bridge this gap by using local parameters to refine the large-scale signal to make it more applicable at a local level. This allows researchers to translate global variables into useful information for a specific region. Downscaling in climate modelling is often applied to both spatial and temporal dimensions due to the estimation of future climate conditions based on historical patterns. Significant computational work is required for converting General Circulation Models (GCMs) to Regional Climate Models (RCMs) with advanced Supercomputing facilities. Standardised downscaled products provide valuable insights into regional climatological changes over time. However, these regional projections may still be too coarse for city planners, so to capture the unique characteristics of urban environments, a higher level of detail is required. "Statistical Downscaling" is one technique to address this need. This approach uses statistical approaches to establish relationships between large-scale climate patterns (from global models) and local climate responses.

Deep Learning (DL) offers an alternative to statistical downscaling for predicting climate variables. Precipitation and Temperature are good candidates for DL models to predict localised precipitation and temperature change patterns. However, DL models come with their challenges. Compared to statistical downscaling methods, DL is more "data-hungry", requiring significantly larger datasets for training. This can make the deployment and transfer of these models difficult. If data is unavailable, not collected, or stored in a format incompatible with the original model, it can be challenging to adapt the model for new locations.

Deep learning techniques are increasingly being tested for downscaling purposes, achieving promising results in the present climate suite of models due to their ability to infer complex non-linear patterns from climate data. Nevertheless, the ability of these models to generalise to out-of-sample climate change conditions is still to be analysed. One of these options is to use spatio-temporal characteristics of the data. Spatial, temporal, and spatio-temporal correlations usually characterise environmental spatio-temporal data.

10.6 GenAI for Climate Models

It's important to note that Natural Language Processing (NLP) and Large Language Models (LLMs) might be used in data preparation stages. However, the core techniques for climate modelling involve Machine Learning for analysing vast datasets and Deep Learning for simulating complex systems within the climate. This focus on technical expertise ensures the models remain grounded in scientific principles. Generative AI, however, offers exciting possibilities for the future by helping translate complex scientific results into clear and understandable communication for stakeholders, policymakers, and the public.

LLMs like ChatGPT (from OpenAI) and Bart (or Gemini, from Google) excel in analysing extensive datasets, and they can be fine-tuned on documents related to climate change, policy reports, and mitigating strategies. Its natural language processing capabilities empower it to extract valuable insights from scientific literature and datasets, providing a comprehensive understanding of climate patterns and trends to researchers and policymakers. These features position LLM in a position where they aid in decision support by generating detailed reports and summaries based on complex climate data, assisting policymakers in making informed decisions about targeted and effective mitigation strategies.

10.7 AI in Biodiversity for Climate Disaster Management

Biodiversity has a rich variety of life on Earth and has a crucial role in maintaining the health and resilience of ecosystems of the planet. Mapping biodiversity is a good approach for effective resource management, conservation efforts, and disaster management. It helps monitor and assess ecosystems with the analysis of vast environmental data to enable early detection of biodiversity loss and habitat degradation for climate disaster management. Machine learning algorithms trace species movements, predict biodiversity changes, and identify potential threats such as invasive species or climate-induced shifts. AI tools support conservation efforts by optimising resource allocation and providing insights into ecosystem resilience. AI has a critical role in mitigating climate disaster impacts, preserving ecosystems, and promoting sustainable environmental practices for climate adaptation.

Biodiversity mapping stands as a powerful tool for understanding, managing, and conserving Earth's vital ecosystems. These maps provide crucial information for effective conservation strategies by pinpointing biodiversity hotspots, species distributions, and habitat fragmentation. These can improve land use planning, facilitate international cooperation, and enable climate change adaptation. Figure 10.12 shows the applications of artificial intelligence in biodiversity mapping.

Identifying Biodiversity Hotspots

Biodiversity mapping allows us to pinpoint areas with exceptionally high levels of species richness, endemism, and threatened species. These biodiversity hotspots are often prioritised for conservation efforts, as they harbour a disproportionately large number of unique and vulnerable species within a relatively small geographic area. Identifying and protecting these hotspots not only safeguards a significant portion of

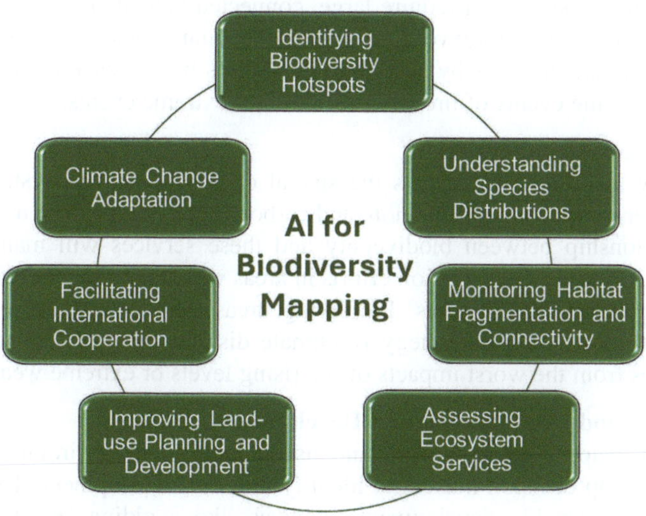

Fig. 10.12 Artificial intelligence in biodiversity mapping

the planet's biodiversity but also strengthens ecosystems against climate-related disasters. Preserving these areas can enhance flood barriers or reduce the impacts of wildfires by maintaining healthy, resilient ecosystems.

Understanding Species Distributions
Biodiversity mapping provides valuable insights into the spatial distribution of various species across different habitats and ecosystems. This information is crucial for developing effective conservation strategies, such as establishing protected areas, designing wildlife corridors, and implementing species-specific management plans. The advancements in AI and sensor technology and the availability of new, more suitable data provide the capability for researchers to integrate with the earth's observations more effectively.

Remote sensing data (aerial photography) is often combined with surveys to map the distribution of invasive species. The new, more complex, multispectral and hyperspectral images contain various data signals and relationships. Classifying and mapping the progression of the species over time is well-suited for a combination of AI techniques. The generated maps can help raise awareness and prevent the spread of invasives by enabling early detection and targeted control efforts. Additionally, understanding species distributions with climate disaster risk areas helps in the proactive relocation or protection of species that are most vulnerable to events like floods, droughts, or hurricanes, ensuring their survival amid changing climate conditions.

Monitoring Habitat Fragmentation and Connectivity
Habitat fragmentation, caused by human activities like urbanisation and deforestation, is a major threat to biodiversity. Biodiversity mapping helps identify areas where habitats are becoming fragmented, allowing for targeted interventions to maintain or restore habitat connectivity. This is essential for ensuring the long-term survival of species that require large, connected habitats for dispersal, migration, and gene flow. Moreover, maintaining habitat connectivity mitigates the impacts of climate disasters by ensuring that species have escape routes or alternative habitats in the events of fires, floods, or other extreme events.

Assessing Ecosystem Services
Biodiversity mapping also reveals the spatial distribution of ecosystem services, such as water purification, pollination, and carbon sequestration. The understanding of the relationship between biodiversity and these services will manage natural resources to prioritise conservation efforts in areas which provide critical ecosystem services to human communities. Preserving areas that offer flood regulation or coastal protection is a key strategy in climate disaster management to safeguard communities from the worst impacts of sea rising levels or extreme weather events.

Improving Land Use Planning and Development
Biodiversity maps integrated with land use planning and environmental impact assessments help decision-makers to identify areas of development. This information guides sustainable development practices like avoiding sensitive habitats, implementing mitigation measures, or exploring alternative locations. Land use

planning informed by biodiversity mapping reduces the vulnerability of both ecosystems and human communities to disasters by avoiding development in high-risk areas and ensuring natural buffers, like wetlands or forests.

Facilitating International Cooperation
International cooperation is also crucial in climate disaster management, where coordinated efforts can mitigate the impacts of disasters that affect multiple countries, such as regional droughts, floods, or hurricanes. Biodiversity mapping assists international cooperation by highlighting areas of global conservation significance which transcend national boundaries. This encourages collaborative efforts and resource-sharing among countries to protect shared natural resources and address transboundary environmental challenges.

Climate Change Adaptation
Biodiversity mapping incorporates climate data like sea level rise projections and temperature patterns to design climate-resilient protected areas and conservation strategies to account for the impacts of climate change on species distributions and habitats. The alignment of conservation efforts with climate disaster management strategies enhances the resilience of ecosystems and communities to withstand and recover from climate-related events.

The future of biodiversity mapping is possible with the integration of Artificial Intelligence to create more detailed and accurate maps with analysis of vast amounts of data from diverse sources. This includes filling data gaps and harmonising information to offer a more complete picture of ecosystems. AI models continue to develop their ability to further enhance our understanding of biodiversity. The power of biodiversity mapping with AI advancements will help to make informed decisions for sustainable development with the prioritisation of conservation efforts to ensure a future with a healthy and resilient natural world.

10.8 Future Opportunities for Climate Disaster Management

The future of climate disaster management shows numerous opportunities for innovation and improvement due to advancements in technology, data analysis, and global collaboration. AI, machine learning, and big data analytics will continue to enhance prediction models, enabling more accurate forecasts of extreme weather events, such as hurricanes, wildfires, and floods. Improved early warning systems, powered by real-time data and predictive algorithms, will allow for faster responses, minimising loss of life and property. The integration of IoT (Internet of Things) devices and remote sensing technologies will facilitate continuous monitoring of vulnerable areas, providing detailed, up-to-date insights into environmental conditions. Advances in climate resilience strategies, such as nature-based solutions and sustainable infrastructure, will also offer promising opportunities. Collaborative efforts among governments, businesses, and scientific communities, supported by AI-driven research and policymaking, will be crucial in developing scalable, adaptive solutions to combat the increasing frequency and intensity of climate-related disasters. Future expertise such as IoT, blockchain, Large Language Models (LLMs), and quantum computing hold immense potential for revolutionising earth resource and disaster management practices:

(a) *IoT (Internet of Things)*: IoT devices collect real-time data on environmental parameters like air quality, soil moisture, and water levels. A network of interconnected sensors and real-time data are assembled to monitor and manage natural resources more efficiently to improve decision-making in areas such as agriculture, water management, and conservation. This data serves as a critical input for training and validating AI models to enhance their ability to provide early warnings for environmental hazards like floods, wildfires, or droughts for more efficient climate disaster management.
(b) *Blockchain:* Blockchain technology offers transparent and secure data management, crucial for tracking the provenance of natural resources, ensuring sustainable sourcing practices, and enhancing supply chain transparency in industries like forestry and mining. Blockchains are used to ensure the integrity and transparency of relief supply chains, verifying the authenticity of aid distribution and ensuring resources reach the intended recipients.
(c) *Large Language Models (LLMs)*: LLMs, such as Google Gemini analyse vast amounts of textual data related to earth resource management including sensor data, satellite imagery, and scientific reports. These models provide insights into resource allocation strategies, environmental policies, and sustainability practices by processing and understanding complex textual information from various sources. LLMs assist in the rapid analysis of emergency communication, coordination of disaster response efforts, and the generation of actionable insights during crises.
(d) *GenAI*: GenAI is used to create simulations of future environmental conditions when coupled with LLMs to offer a powerful tool for testing the impact of different climate change mitigation strategies. This technology simulates various disaster scenarios to help policymakers and emergency responders plan and optimise their strategies in the event of natural disasters.
(e) *Quantum Computing*: Quantum computing has immense potential across various domains, promising to revolutionise energy distribution systems to facilitate the design of sustainable materials with molecular structure simulations to accurately predict the impacts of climate change, and enhance the efficiency of renewable energy systems. Quantum computing can revolutionise disaster management by providing more accurate predictions of extreme weather events, optimising emergency response strategies, to improving the resilience of critical infrastructure against climate-related disasters.

These are just a few instances of future opportunities for climate disaster management due to the continuous development of technology for more innovative applications.

10.9 Conclusion

The integration of Artificial Intelligence (AI) and Machine Learning (ML) into climate disaster management and earth resource management has proven to be a transformative force in addressing the challenges due to climate change. These technologies enable the analysis of vast amounts of environmental data, offering unprecedented insights into climate patterns, natural resource distribution, and disaster prediction. AI and ML empower decision-makers to respond more effectively to climate-induced disasters by providing real-time, actionable information that enhances preparedness, reduces risks, and mitigates damage. They facilitate the efficient use of the earth's resources, optimising agriculture, water management, and forestry practices to ensure sustainability. The ability to predict extreme weather events, assess environmental risks, and model potential scenarios helps to minimise the impact of disasters, allowing for timely interventions. Additionally, AI and ML support long-term solutions by identifying patterns that might go unnoticed by traditional methods, enabling more adaptive and resilient approaches to climate change. The collaboration of AI and ML with climate and resource management strategies will be crucial in building a more sustainable and resilient future. Continued research and development of these technologies will be key to their effective implementation across different sectors and regions worldwide.

References

Ajani TS, Imoize AL, Atayero AA (2021) An overview of machine learning within embedded and mobile devices-optimizations and applications. Sensors 21(13). https://doi.org/10.3390/s21134412

Alamús R, Pérez F, Pipia L, Corbera J (2018) Urban sustainable ecosystems assessment through airborne earth observation: lessons learned. In: W. M., J. B., & H. S (ed) International archives of the photogrammetry, remote sensing and spatial information sciences – ISPRS archives (vol 42, issue 1). International Society for Photogrammetry and Remote Sensing, pp 5–10. https://doi.org/10.5194/isprs-archives-XLII-1-5-2018

Attaf D, Djerriri K, Mansour D, Hamdadou D (2019) Mapping of burned area using presence and background learning framework on the google earth engine platform. In: C. M. S. F. Tanzi T. Altan O (ed) International archives of the photogrammetry, remote sensing and spatial information sciences – ISPRS archives (vol 42, issue 3/W8). International Society for Photogrammetry and Remote Sensing, pp 37–41. https://doi.org/10.5194/isprs-archives-XLII-3-W8-37-2019

Bhadra T, Chouhan A, Chutia D, Bhowmick A, Raju PLN (2020) Flood detection using multispectral images and SAR data. In: Machine learning, image processing, network security and data sciences. MIND 2020, Communications in computer and information science: Vol. 1240 CCIS. https://doi.org/10.1007/978-981-15-6315-7_24

Butts MB, Falk AK, Xuan Y, Cluckie ID (2007) Integrating meteorological and uncertainty information in flood forecasting: the FLOODRELIEF project, IAHS-AISH Publication, 313, pp 385–397. https://www.scopus.com/inward/record.uri?eid=2-s2.0-55749095035&partnerID=40&md5=1fcc29459c9a94ae00ed405905d12bd6

Corte I, Acevedo S, Arlego M, Lamas CA (2021) Exploring neural network training strategies to determine phase transitions in frustrated magnetic models. Comput Mater Sci 198. https://doi.org/10.1016/j.commatsci.2021.110702

de Rosa GH, Papa JP (2021) A survey on text generation using generative adversarial networks. Pattern Recogn 119. https://doi.org/10.1016/j.patcog.2021.108098

Dede J, Wewetzer D, Förster A (2023) AI in the wild: Challenges of remote deployments. In: K. M., B. fur Informatik Anna-Louisa-Karsch-Strasse 2, K. D., B. fur Informatik Anna-Louisa-Karsch-Strasse 2, W. C., B. fur Informatik Ahrstrasse 45, & W. V (ed) Lecture notes in informatics (LNI), proceedings – series of the Gesellschaft fur Informatik (GI): vol P-337. Gesellschaft fur Informatik (GI), pp 1585–1591. https://www.scopus.com/inward/record.uri?eid=2-s2.0-85181085155&partnerID=40&md5=192b4f5308144fbec8750ee955fb1adf

Deepa N, Prabadevi B, Maddikunta PK, Gadekallu TR, Baker T, Khan MA, Tariq U (2021) An AI-based intelligent system for healthcare analysis using Ridge-Adaline Stochastic Gradient Descent Classifier. https://doi.org/10.1007/s11227-020-03347-2

Ding Q, Wang L, Fu M, Huang N (2020) An integrated system for rapid assessment of ecological quality based on remote sensing data. Environ Sci Pollut Res 27(26):32779–32795. https://doi.org/10.1007/s11356-020-09424-6

Du B, Zhao Z, Hu X, Wu G, Han L, Sun L, Gao Q (2021) Landslide susceptibility prediction based on image semantic segmentation. Comput Geosci 155. https://doi.org/10.1016/j.cageo.2021.104860

Edgell DL (2023) Issues and policies that have an impact on future trends in global tourism. In: Routledge handbook of trends and issues in global tourism supply and demand. https://doi.org/10.4324/9781003260790-27

Eevala, Kalle (2024) Tekoälytyökalun hyödyntäminen kassavirran ennustuksessa

Efremova N, Foley JC, Unagaev A, Karimi R (2023) AI for sustainable agriculture and rangeland monitoring. Philos Stud Ser 152:399–422. https://doi.org/10.1007/978-3-031-21147-8_22

Fan Y-C, Yuan Y-Q, Yuan Y-C, Duan W-J, Gao Z-Q (2025) Research progress on the impact of climate change on wheat production in China. PeerJ 13(1). https://doi.org/10.7717/peerj.18569

Ganji K, Gharachelou S, Ahmadi A (2019) Urban's river flood analysing using sentinel-1 data case study: (gorganrood, aq'qala). Int Arch Photogramm Remote Sens Spatial Inf Sci 42(4/W18): 415–419. https://doi.org/10.5194/isprs-archives-XLII-4-W18-415-2019

Golparvar B, Papadopoulos P, Ezzat AA, Wang R-Q (2021) A surrogate-model-based approach for estimating the first and second-order moments of offshore wind power. Appl Energy 299. https://doi.org/10.1016/j.apenergy.2021.117286

Hu S, Yan D, An J, Guo S, Qian M (2019) Investigation and analysis of Chinese residential building occupancy with large-scale questionnaire surveys. Energ Buildings 193:289–304. https://doi.org/10.1016/j.enbuild.2019.04.007

Impedovo D, Dentamaro V, Abbattista G, Gattulli V, Pirlo G (2021) A comparative study of shallow learning and deep transfer learning techniques for accurate fingerprints vitality detection. Pattern Recogn Lett 151:11–18. https://doi.org/10.1016/j.patrec.2021.07.025

Jayaraman V, Srivastava SK, Gowrisankar D (2009) India's EO pyramid for holistic development. In: Space technologies for the benefit of human society and earth. https://doi.org/10.1007/978-1-4020-9573-3_2

Kang Y, Gao S, Roth RE (2024) Artificial intelligence studies in cartography: a review and synthesis of methods, applications, and ethics. Cartogr Geogr Inf Sci 51(4):599–630. https://doi.org/10.1080/15230406.2023.2295943

Kocaaslan S, Musaoglu N, Karamzadeh S (2021) Evaluating drought events by time-frequency analysis: a case study in Aegean region of Turkey. IEEE Access 9:125032–125041. https://doi.org/10.1109/ACCESS.2021.3110816

Krishnamurthy S, Kapeleshh KS, Dovgan E, Luštrek M, Gradišek Piletič B, Srinivasan K, Li Y-C, Gradišek A, Syed-Abdul S (2021) Machine learning prediction models for chronic kidney disease using national health insurance claim data in Taiwan. Healthcare (Switzerland) 9(5). https://doi.org/10.3390/healthcare9050546

Li B, Sivapalan M (2020) Long-term coevolution of an urban human-water system under climate change: critical role of human adaptive actions. Water Resour Res 56(11). https://doi.org/10.1029/2020WR027931

Li W, El-Askary H, Qurban MA, Li J, ManiKandan KP, Piechota T (2019) Using multi-indices approach to quantify mangrove changes over the Western Arabian Gulf along Saudi Arabia coast. Ecol Indic 102:734–745. https://doi.org/10.1016/j.ecolind.2019.03.047

Luo X, Walther P, Mansour W, Teuscher B, Zollner JM, Li H, Werner M (2023) Exploring GeoAI methods for supraglacial Lake mapping on Greenland ice sheet. In: D. M.L., R. M., E. A., K. P., & N. M.A (ed) GIS: proceedings of the ACM international symposium on advances in geographic information systems. Association for Computing Machinery. https://doi.org/10.1145/3589132.3629971

Ma L-J, Yuan J-S, Huang L (2024) Prospect of climate risks management in China under the framework of UN early warning for all initiative. Clim Change Res 20(1):48–61. https://doi.org/10.12006/j.issn.1673-1719.2023.122

Matsuoka R, Oki S (2021) Demonstration of stormwater management technology by short-term rainfall prediction and real-time runoff analysis system using small X-band radar. J Disaster Res 16(3):403–409. https://doi.org/10.20965/jdr.2021.p0403

Meenal R, Binu D, Ramya KC, Michael PA, Vinoth Kumar K, Rajasekaran E, Sangeetha B (2022) Weather forecasting for renewable energy system: a review. Arch Comput Methods Eng 29(5): 2875–2891. https://doi.org/10.1007/s11831-021-09695-3

Nagarajan A, Wason JV (2019) Machine learning approach to predict lung cancer using CT scan images. Int J Adv Trends Comput Sci Eng 8(6):2974–2976. https://doi.org/10.30534/ijatcse/2019/48862019

Nedd GA, Oyedotun TDT, Simard M (2021) Evaluation of spatio-temporal dynamics of Guyana's mangroves using SAR and GEE. Earth Syst Environ 7. https://doi.org/10.1007/s41748-021-00277-8

Quintero N, Viedma O, Urbieta IR, Moreno JM (2019) Assessing landscape fire hazard by multitemporal automatic classification of landsat time series using the Google Earth Engine in West-Central Spain. Forests 10(6). https://doi.org/10.3390/f10060518

Rajkumar N, Viji C, Latha PM, Vennila VB, Shanmugam SK, Pillai NB (2024) The power of AI, IoT, and advanced quantum based optical systems in smart cities. Opt Quant Electron 56(3). https://doi.org/10.1007/s11082-023-06065-0

Ren Q, Liu D, Liu Y, Liu Y (2024) Spatio-temporal dynamics and socio-ecological determinants of ecosystem service interplays in Shandong Province's coastal region (2000–2020): implications for environmental protection and sustainable ecosystem management. Environ Res 243. https://doi.org/10.1016/j.envres.2023.117824

Sahani J, Kumar P, Debele S, Spyrou C, Loupis M, Aragão L, Porcù F, Shah MAR, Di Sabatino S (2019) Hydro-meteorological risk assessment methods and management by nature-based solutions. Sci Total Environ 696. https://doi.org/10.1016/j.scitotenv.2019.133936

Shahvar MP, Scuderi D, Valenti D, Collura A, Miccichè S, Farina V, Marsella G (2025) Climate change multi-risk assessment for mango cultivation in Sicily, Italy, by using Bayesian network. Acta Horticulturae 1(1415). https://doi.org/10.17660/ActaHortic.2025.1415.15

Sinha DD, Singh AN, Setiyono TD, Singh US (2020) Targeting dissemination of submergence tolerant rice in Assam, India: a geomatics approach. 40th Asian conference on remote sensing, ACRS 2019: progress of remote sensing technology for smart future

Sravanthi GL, Devi MV, Sandeep KS, Naresh A, Gopi AP (2020) An efficient classifier using machine learning technique for individual action identification. https://doi.org/10.14569/IJACSA.2020.0110664

Sun Y, Huang S-K, Zhao X (2024) Predicting hurricane evacuation decisions with interpretable machine learning methods. Int J Disaster Risk Sci 15(1):134–148. https://doi.org/10.1007/s13753-024-00541-1

Sunar AF, Yagmur N, Dervisoglu A (2019) Flood analysis with remote sensing data – a case study: Maritsa River, Edirne. In: C. M. S. F. Tanzi T. Altan O (ed) International archives of the photogrammetry, remote sensing and spatial information sciences - ISPRS archives (vol 42, issue 3/W8). International Society for Photogrammetry and Remote Sensing, pp 497–502. https://doi.org/10.5194/isprs-archives-XLII-3-W8-497-2019

Talha M, Nejadhashemi AP, Moller K (2025) Soft computing paradigm for climate change adaptation and mitigation in Iran, Pakistan, and Turkey: a systematic review. Heliyon 11(2). https://doi.org/10.1016/j.heliyon.2025.e41974

Taneja K, Ahmad S, Ahmad K, Attri SD (2020) Spatio-temporal analysis of urban air quality: a comprehensive approach toward building a Smart City. In: Smart cities—opportunities, Lecture notes in civil engineering, vol 58. https://doi.org/10.1007/978-981-15-2545-2_27

Udawalpola M, Hasan A, Liljedahl AK, Soliman A, Witharana C (2021) Operational-scale geoai for pan-arctic permafrost feature detection from high-resolution satellite imagery. In: S. K. & A. M (ed) International archives of the photogrammetry, remote sensing and spatial information sciences – ISPRS archives (vol 44, issues M-3). International Society for Photogrammetry and Remote Sensing, pp 175–180. https://doi.org/10.5194/isprs-archives-XLIV-M-3-2021-175-2021

Wahab A, Muhammad M, Ullah S, Abdi G, Shah GM, Zaman W, Ayaz A (2024) Agriculture and environmental management through nanotechnology: eco-friendly nanomaterial synthesis for soil-plant systems, food safety, and sustainability. Sci Total Environ 926. https://doi.org/10.1016/j.scitotenv.2024.171862

Wang J, Hu Y (2019) Are we there yet? Evaluating state-of-the-art neural network based geoparsers using EUPEG as a benchmarking platform. In: M. B., M. L., & M.-F. P (ed) Proceedings of the 3rd ACM SIGSPATIAL international workshop on geospatial humanities, GeoHumanities 2019. Association for Computing Machinery, Inc. https://doi.org/10.1145/3356991.3365470

Wang Z-H, Upreti R (2019) A scenario analysis of thermal environmental changes induced by urban growth in Colorado River Basin, USA. Landsc Urban Plan 181:125–138. https://doi.org/10.1016/j.landurbplan.2018.10.002

Wang Z, Nie F, Zhang C, Wang R, Li X (2021) Joint nonlinear feature selection and continuous values regression network. Pattern Recogn Lett 150:197–206. https://doi.org/10.1016/j.patrec.2021.06.035

Wang B, Hua L, Al-Mohaimeed AM, Zhao N (2024) Ocean carbon emission prediction and management measures based on artificial intelligence remote sensing estimation in the context of carbon neutrality. Environ Res 251. https://doi.org/10.1016/j.envres.2024.118591

Weijs S, Van Leeuwen E, Van Overloop P-J, Van De Giesen N (2007) Effect of uncertainties on the real-time operation of a lowland water system in The Netherlands, IAHS-AISH Publication, 313, pp 463–470. https://www.scopus.com/inward/record.uri?eid=2-s2.0-55749109886&partnerID=40&md5=0b1668dcae891eff66ff3af524220885

Xu C, Hystad P, Chen R, Van Den Hoek J, Hutchinson RA, Hankey S, Kennedy R (2021) Application of training data affects success in broad-scale local climate zone mapping. Int J Appl Earth Obs Geoinf 103. https://doi.org/10.1016/j.jag.2021.102482

Yadav A (2023) Leveraging artificial intelligence for sustainable development and environmental resilience. In: Exploring ethical dimensions of environmental sustainability and use of AI. https://doi.org/10.4018/9798369308929.ch008

Zeng Y-F, Chang M-J, Lin G-F (2024) A novel AI-based model for real-time flooding image recognition using super-resolution generative adversarial network. J Hydrol 638. https://doi.org/10.1016/j.jhydrol.2024.131475

Zhu H, Samtani S, Brown RA, Chen H (2021) A deep learning approach for recognizing activity of daily living (adl) for senior care: exploiting interaction dependency and temporal patterns. MIS Q 45(2). https://doi.org/10.25300/MISQ/2021/15574

Zou L, Li G, Xu S (2021) A novel method for optimizing air temperature estimation and quantifying canopy layer heat Island intensity in Eastern and Central China. Adv Space Res 68(8): 3291–3301. https://doi.org/10.1016/j.asr.2021.06.023

Chapter 11
From Data to Planning: Innovations Through LiDAR Technology and Data Science in Sustainable Urban Resource Management

Jayati Vijaywargiya and Anandakumar M. Ramiya

Abstract To realize the United Nations Sustainable Development Goals (SDGs), especially those focused on promoting sustainable urban development and improving community well-being, the integration of digital technologies has become essential for modern cities. These technological innovations play a vital role in the planning, implementation, and administration of urban spaces, ensuring alignment with tenets of sustainability and resilience. Recent advancements in sensor technology and data processing methodologies have enabled precise modelling of the physical urban environment, which is essential for sustainable urban development initiatives. In urban areas with dense vegetation and vertical urbanization, it is essential to capture and analyse the urban fabric in three dimensions. Advanced remote sensing technologies like Airborne Laser Scanning (ALS) can generate detailed three-dimensional representations of landscapes through point cloud datasets. However, current urban representation often overlooks vegetation, focusing on buildings, limiting overall comprehensiveness. This chapter seeks to establish a framework and methodology for a comprehensive near-realistic representation of urban areas to enhance governance and support sustainable development. The study focuses on developing methods for efficiently representing complex urban objects, extracting geometric properties from diverse elements, and implementing a flexible schema to ensure consistency and compatibility for application across multiple domains. As a case study, the proposed framework is implemented on ALS data over Thiruvananthapuram, Kerala, India, a city characterized by a blend of diverse vegetation and urban development. The chapter also demonstrates how the proposed framework can contribute to achieving the three pillars of Sustainable Development Goal 11: mapping and monitoring, community involvement, and data-driven decision-making.

J. Vijaywargiya · A. M. Ramiya (✉)
Department of Earth and Space Sciences, Indian Institute of Space Science and Technology, Thiruvananthapuram, Kerala, India
e-mail: ramiya@iist.ac.in

D. Kumar, N. P. Bassill (eds.), *Artificial Intelligence and Machine Learning for Climate Disaster Management*, Disaster Resilience and Green Growth,
https://doi.org/10.1007/978-981-95-1588-2_11

Keywords SDG-11 · Urban representation · Sustainable development · point clouds · Empirical decision-making

11.1 Introduction

11.1.1 Background

Urbanization represents a significant global trend that is altering socio-economic and environmental frameworks, with over 56% of the global population currently living in urban settings (Randolph and Storper 2023). India, recognized as one of the nations experiencing rapid urban growth, is witnessing a substantial transformation, with projections indicating that its urban population will increase from 34% in 2020 to nearly 50% by 2050 (Patel and Patel 2024). The United Nations introduced Sustainable Development Goal 11 (SDG 11) as part of the 2030 Agenda for Sustainable Development, which seeks to tackle the challenges posed by urbanization (Kalfas et al. 2023). The successful attainment of SDG 11 hinges on three fundamental pillars: mapping and monitoring, community engagement, and empirical decision-making, all of which rely on the inclusivity and comprehensiveness of urban representation.

The concept of digital urban representation encompasses the visualization, analysis, and management of urban spaces through the utilization of sophisticated digital technologies. This evolution in urban representation incorporates tools such as Geographic Information Systems (GIS), three-dimensional urban modelling, and digital twins (Vijaywargiya and Ramiya 2024). The origins of GIS can be traced back to the 1960s, emerging from the necessity to map land use, infrastructure, and environmental conditions while analysing spatial patterns (Song and Changshan 2021). In the early twenty-first century, three-dimensional urban modelling gained traction due to advancements in computer graphics and simulation technologies, transitioning from merely visualizing architectural designs to offering realistic and immersive representations of urban infrastructure. Meanwhile, the notion of digital twins, which initially found its roots in the manufacturing industry, began to be adapted for urban contexts in the late 2010s (Mylonas et al. 2021). Digital twins serve as virtual representations of physical entities, systems, or processes, enabling real-time simulation, prediction, and performance optimization. A thorough digital representation of urban environments facilitates micro-morphology analysis, thereby enabling data-driven decision-making that aligns with the goals of Sustainable Development Goal 11.

For accurate representation of complex urban landscapes, 3D remote sensing is crucial, particularly in areas where traditional 2D methods fall short. For example, in densely vegetated regions, houses or smaller structures can be completely or partially hidden beneath tree canopies, as shown in Fig. 11.1a. When using conventional satellite imagery, these structures might not be accurately represented or even detected, leading to incomplete or misleading data in urban planning and management. Furthermore, as known, "all greens are not the same"—trees, shrubs, and green spaces have distinct characteristics which can be assessed efficiently using 3D, as shown in Fig. 11.1b.

Fig. 11.1 (**a**) Houses obscured by vegetation (**b**) Levels of vegetation

3D remote sensing, such as LiDAR (Light Detection and Ranging), addresses these issues by capturing detailed spatial data in three dimensions. For instance, in an urban area with mixed-use green spaces and residential buildings, LiDAR can distinguish between a tall tree, a small shrub, and a building by measuring the height, shape, and geometrical intricacies of each object. This allows for the creation of highly accurate models that reflect the true layout of the urban landscape, including structures that are obscured by vegetation. This level of detail is critical for urban planners who need to assess the impact of vegetation on infrastructure, manage land use effectively, and ensure that all structures are accounted for in city development plans. Airborne LiDAR (Light Detection and Ranging) scanning operates by emitting laser pulses from instruments on aerial platforms like aircraft towards the Earth's surface and measuring the time it takes for the pulses to return to the sensor (Salehi-Dorcheabedi et al. 2024). This results in highly detailed and accurate three-dimensional point clouds over the spatial landscape. This method allows for the rapid and efficient collection of extensive areas, providing a level of precision that traditional surveying techniques struggle to match. The ability to capture such detailed elevation data makes LiDAR an invaluable tool for creating precise derivatives like building footprint, digital elevation models (DEMs), and digital surface models (DSMs), essential for effective urban planning and management (Oniga et al. 2023). Airborne LiDAR overcomes the challenges of obscured houses and vegetation representation by penetrating through gaps within the vegetation canopy to map the surface beneath and providing clear, detailed representations of both natural and constructed features. Despite its many advantages, airborne LiDAR scanning presents challenges related to the voluminous and unstructured nature of point cloud data. The sheer volume of data generated can be overwhelming, necessitating substantial storage and processing resources (Aljumaily et al. 2023). Additionally, the unstructured format of point cloud data makes it complex to derive actionable insights, interpret, and analyse effectively (Fig. 11.2).

Current urban reconstruction techniques, especially for large-scale urban environments, aim for high detail and automation. Consequently, standard models such as CityGML have been developed to represent urban infrastructure at different levels of detail (LoDs). However, most of these models are primarily concerned with building representation. The comprehensiveness of urban modelling is compromised when essential elements like vegetation and other structures like poles and power lines are not represented. These components are essential to effectively depict the urban environment.

Fig. 11.2 Airborne LiDAR scanning of urban landscape.

This can be attributed to two major challenges encountered in urban modelling. Firstly, the sheer volume of data presents a significant obstacle. Urban environments consist of a multitude of diverse elements, ranging from buildings to vegetation. Capturing and processing this vast amount of data can overwhelm conventional computational resources and software systems, leading to difficulties in effectively managing and utilizing the data from all the respective classes for modelling purposes.

Secondly, representing objects that are not discretely identifiable poses another challenge. Unlike buildings, which can be easily recognized and delineated, certain urban elements such as trees, high vegetation, or shrubs may lack distinct boundaries and exhibit complex morphologies. This ambiguity makes it challenging to accurately model and represent these objects within the digital environment. Consequently, existing modelling techniques may struggle to capture the characteristics of such elements, resulting in incomplete or inaccurate representations of the urban landscape.

Moreover, the challenges in comprehensive urban representation manifest at three critical stages. Separating and isolating individual entities from these cluttered scenes is a difficult task due to overlapping structures, occlusions, and variations in size, shape, and orientation. Once entities are segmented and isolated, representing them in a digital environment requires capturing their intricate geometries and semantic attributes. The derived entities must have meaningful attributes and information assigned to them for applications to comprehend and utilize them. Attributes enrich the semantic understanding of urban representation and enable their effective use in urban planning, management, and analysis. However, maintaining consistency in metadata for diverse entities is challenging. Real-world urban structures exhibit diverse entities, necessitating the designing of a robust schema to depict the urban scene. Further, the inherent noise, incompleteness, and sparsity of data sources can also hinder the accurate representation of urban entities, compromising the fidelity of digital models. Overcoming these challenges is essential to ensure the reliability and usefulness of digital urban representation for diverse applications.

This work aims to present a comprehensive panoptic urban illustration utilizing airborne LiDAR scanning technology, to achieve a near-real representation of urban landscapes. This system is envisioned to encapsulate all objects within the scene as individual entities or contextual segments, thereby facilitating all-inclusive analysis for urban planning, governance, and sustainable development initiatives. In pursuit of this concept, the chapter addresses the following questions to establish a research foundation:

(a) How can we construct near-realistic urban representations from annotated urban point clouds to effectively depict the physical environment and its various components?
(b) What methodologies can be employed to efficiently represent objects within the urban environment, even those that may not be easily identifiable or separable as distinct entities in the representation?
(c) How can we develop a robust approach to extract geometric properties from the diverse elements present in the urban scene?
(d) Can the implementation of a flexible schema ensure the consistency and compatibility of a city information system to meet the growing demand for applications across multiple domains?

By addressing these research questions, the chapter aims to contribute to the advancement of urban information systems, enabling more accurate, comprehensive, and insightful analysis of urban environments for various stakeholders and decision-makers. The case study was conducted on the TALD dataset due to several compelling reasons. The dataset represents a complex urban landscape, providing a challenging environment that tests the robustness of analytical methods. To illustrate the practical application of this approach, a micro-morphology analysis was performed on the vegetation class, demonstrating how this framework can contribute to sustainable urban planning by extracting detailed spatial and structural characteristics from complex urban environments (Fig. 11.3).

11.2 Point Cloud to Actionable Insights

Converting a point cloud into actionable insights and application-ready data involves a two-phase process. The first phase begins with the acquisition of raw point cloud data, obtained through technologies like LiDAR or photogrammetry. This data, while rich in 3D spatial information, is often unstructured and noisy, requiring significant preprocessing. Techniques such as filtering, down-sampling, and noise reduction are applied to clean the data. Once refined, the data is annotated, a process

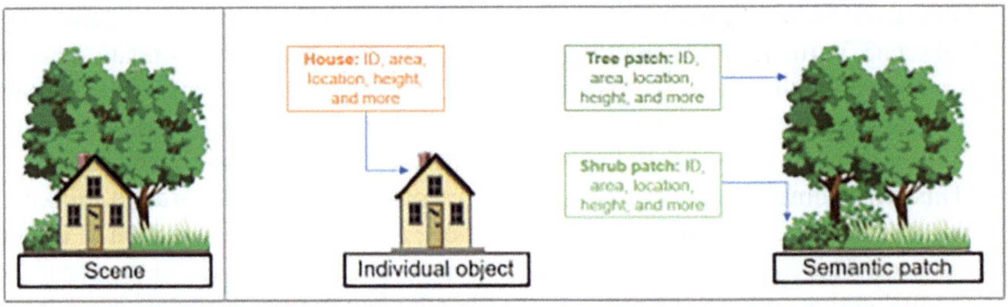

Fig. 11.3 Comprehensive urban representation showing a scene as individual object and semantic patches with attributes

Fig. 11.4 Aerial LiDAR point clouds coloured (**a**) by elevation (**b**) by RGB colour (**c**) by classification

that involves labelling points with semantic information. Point cloud annotation/labelling is a crucial process in transforming raw LiDAR data into useful information for various applications. There are three primary methods for annotating point clouds. The first method is filtering, which involves applying algorithms to remove noise and irrelevant points, thus isolating the features of interest (Chmelar et al. 2020). The second method is software-aided manual labelling, where specialized software tools assist human operators in identifying and labelling different elements within the point cloud, such as buildings, vegetation, and infrastructure. This approach, while labour-intensive, ensures high accuracy and precision. The third method leverages pre-trained machine learning or deep learning models, which can automatically classify, and label point cloud data based on patterns learned from vast datasets. These models significantly accelerate the annotation/labelling process and improve efficiency, though they require extensive training data and computational resources. Combining these methods can enhance the overall quality and speed of point cloud annotation/labelling, facilitating more accurate and detailed analyses for various applications (Fig. 11.4).

The classification of ALS (Airborne LiDAR Scanning) point cloud data has undergone significant evolution over the years with the advent of machine learning (ML) and deep learning (DL) techniques. These advancements have greatly enhanced the accuracy and efficiency of categorizing different elements in urban environments, such as buildings, vegetation, and roads. The methods for point cloud classification can be broadly categorized into point-based and object-based techniques (Grilli et al. 2021). Point-based methods classify each point individually based on its features, whereas object-based methods group points into meaningful objects before classification.

The advent of deep learning in the 2010s brought about a paradigm shift in point cloud classification. Convolutional Neural Networks (CNNs), initially popular in image processing, were adapted for point cloud data by converting it into voxel grids or using point-based networks (Fareed et al. 2023). CNNs are capable of learning hierarchical feature representations, making them highly effective in capturing spatial hierarchies and patterns within the data. The development of Graph Neural Networks (GNNs) and PointNet in 2017 marked a significant advancement in directly working with raw point cloud data without requiring voxelization, revolutionizing classification by maintaining geometric details and ensuring high accuracy (Li et al. 2024; Haznedar et al. 2023). The introduction of Transformer models, such as PCT, further advanced the field by leveraging attention mechanisms to capture long-range dependencies, making them highly effective for classifying complex datasets (Guo et al. 2021).

Post-labelling LiDAR-derived products like Digital Elevation Models (DEMs), building footprints, and vegetation statistics are crucial for empirically aiding in sustainability efforts. LiDAR technology has also been instrumental in heritage conservation by providing detailed 3D documentation of historic landscapes for preservation and restoration. LiDAR-derived building footprints have been utilized for assessing urban heat islands for climate-related urban studies. These footprints provide detailed information on building dimensions and spatial arrangement, enabling researchers to analyse surface temperatures and thermal variations across urban areas which are pivotal for developing strategies to enhance climate resilience and improve urban microclimate conditions (Chen et al. 2022). One study used LiDAR data to analyse the spatial structure of urban canyons, contributing to urban climate studies (Mayalu et al. 2020). Furthermore, LiDAR-derived terrain models are indispensable for flood risk assessment within urban regions by identifying flood-prone areas, modelling flood inundation scenarios, and guiding effective disaster management and mitigation strategies (Saifi and Anandakumar 2024).

The existing research on urban reconstruction has focused on four aspects: building reconstruction, tree delineation and vegetation analysis, DEM extraction from ground points, and asset mapping, for example, powerline mapping (Shen et al. 2024). Different methods have been attempted for varied purposes. A few of the research into utilizing point clouds convert them into 3D object models such as CityGML and CityJSON (Yarroudh et al. 2023). However, these models often lose the detailed geometrical characteristics of the original point clouds and are limited in scope to specific object classes, such as buildings. Digital twin city models have been developed integrating building information modelling and energy visualization systems, facilitating improved energy analysis and decision-making.

Despite significant advancements in the utilization of LiDAR and 3D semantics in urban studies, notable research gaps persist, particularly concerning sustainable urban development. Most existing models predominantly represent the urban landscape through buildings and built structures, often neglecting the comprehensive representation necessary for a holistic understanding of the urban landscape. This narrow focus overlooks crucial elements such as vegetation, which play significant roles in urban ecosystems and contribute to environmental sustainability. While these applications are valuable, they do not fully exploit the potential of LiDAR data for micromorphology analysis, which can yield relevant insights into fine-scale urban dynamics crucial for sustainable development.

A more inclusive approach that incorporates all possible representative elements of urban landscapes would provide a richer and more accurate portrayal of urban landscapes, facilitating better-informed decision-making in sustainable urban planning and management practices. By leveraging LiDAR data for micromorphological analysis, decision-makers can gain a deeper understanding of the interactions between various urban elements to support the development of more efficient urban policies and interventions for achieving sustainable urban development goals. Addressing these research gaps by adopting a more comprehensive urban representation and utilizing LiDAR data for detailed micromorphological analysis along with 3D city models for visualizations and simulations is crucial for advancing the applicability of this dataset.

11.3 Urban Representation for Aid in SDG-11

The framework and methodology aim to extract a comprehensive urban representation from classified airborne LiDAR point clouds over large urban scenes to aid in achieving the Sustainable Development Goal. The process involves the following steps: defining entity representation, unsupervised instance segmentation, extracting geometry and attributes, integrating GIS information and spatial morphological analysis for sustainable development, as shown in Fig. 11.5. This comprehensive urban representation can be leveraged for urban informatics, governance, and planning.

11.3.1 Defining Representation

The urban landscape is inherently complex, comprising a diverse array of objects and entities that interact in intricate ways. Some objects, like buildings, poles, and roads, are individually identifiable due to their distinct and recognizable shapes and structures. These objects can be easily segmented based on their clear boundaries and unique characteristics. On the other hand, certain elements of the urban landscape, such as vegetation, shrub patches, and fences, do not possess distinct individual identities and are often considered macro-entities for analysis. These elements are more challenging to segment due to their continuous and often overlapping

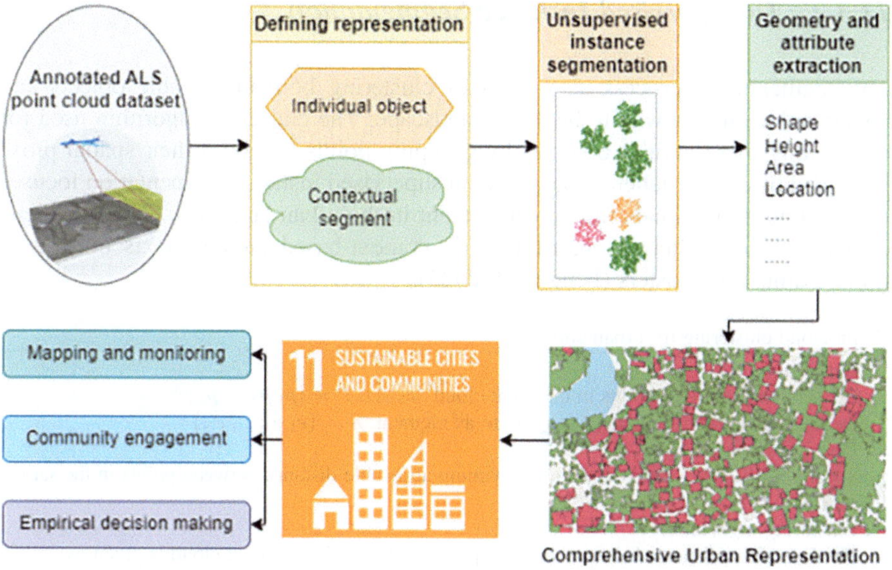

Fig. 11.5 Comprehensive urban representation: methodology and application

nature. Vegetation, for example, can vary greatly in density and distribution, making it difficult to delineate clear boundaries between different patches.

Given this complexity, creating a comprehensive representation that includes all types of objects within an urban scene is a challenging task. The representation must account for both clearly defined, individual objects and more amorphous, unidentifiable elements. To achieve this, the urban landscape is segmented using a context-aware element representation approach.

In this approach, the urban scene is divided into two main categories: identifiable objects and contextual segments. Identifiable objects are those entities that can be distinctly recognized and segmented, such as buildings, poles, and other infrastructure. Contextual segments, on the other hand, are collections of unidentifiable objects that are grouped based on their contextual similarity.

Mathematically, let P be the set of points in the point cloud dataset: $P = \{p_1, p_2, \ldots, p_n\}$ Each point p_i is defined by its spatial coordinates and other attributes. The segmentation function S maps each point to an urban element E_j: $S: P \rightarrow Ej$.

The urban elements, E, are categorized into object entities O and contextual segments CS: $O = \{o_1, o_2, \ldots, o_k\}$, $CS = \{cs_1, cs_2, \ldots, cs_l\}$.

The complete set of elements is then: $E = O \cup CS$.

This approach ensures a comprehensive, all-inclusive representation of the urban landscape, capturing the complexity and diversity by contextualizing each element in its representation. This provides a robust framework for analysing and understanding the intricate dynamics of urban spaces.

11.3.2 Unsupervised Instance Segmentation

The segmentation process also involves clustering both identifiable objects and contextual segments within the urban landscape. The clustering algorithm used for segmenting these entities operates by grouping points based on their spatial proximity and their prior label of class. The unsupervised instance segmentation focuses on isolating point clouds from each element in the urban landscape. The Euclidean distance-based clustering algorithm with context-based parameters is utilized to group points into clusters (Feng et al. 2023).

Point cloud clustering for urban entities
1. *Initialization*: Let P be the set of all points in the point cloud dataset: $P = \{p_1, p_2, \ldots, p_n\}$ Define E as the set of points belonging to an element: $E = \{e_1, e_2, \ldots, e_m\}$ 2. *Distance threshold*: Specify a distance threshold dth for maximum allowable distance between points in the same cluster. 3. *Clustering algorithm*: The Euclidean distance between any two points pi and pj in the point cloud is given by: $d(p_i, p_j) = (x_i - x_j)^2 + (y_i - y_j)^2 + (z_i - z_j)^2$ For each point p_i: Initialize a new cluster C_k if p_i is not already assigned to an existing cluster Add p_i to Ck For each point p_j, belonging to the same class as p_i, that is not already assigned to a cluster: Calculate the distance $d(p_i, p_j)$ If $d(p_i, p_j) \leq d$th, add p_j to Ck 4. *Iterative process*: Repeat the clustering process for each point in E until all points are assigned to clusters. 5. *Resulting clusters*: The algorithm outputs a set of clusters $\{C_1, C_2, \ldots, C_l\}$, where each cluster C_k represents an isolated urban element

11.3.3 Extracting Geometry and Attributes

After identifying individual clusters, the alpha shape algorithm is used to define the building boundaries (Widyaningrum et al. 2020).

$$\alpha(B) = \left\{ \left(p_{ei}, p_{ej}, p_{ek}\right) \in \text{Delaunay}(E) \mid \text{circumradius} \leq \alpha \right\}$$

The alpha parameter is adjusted to create detailed shapes from the set of points. Geometric features such as area A, centroid location C, maximum elevation $Z_{\max}$, and perimeter P are extracted. The height H of each element is adjusted using DEM obtained from ground points: $H = Z_{\max} - \text{DEM}$.

This process yields a comprehensive set of quantitative descriptors, offering a detailed characterization of the spatial attributes of each element. The comprehensive urban representation comprises accurate urban element footprints with geometric features and accurate height information, forming a foundational dataset for subsequent analyses and information extraction.

11.3.4 RDBMS-Based Storage and Spatio-Semantic Queries for SDGs

In the database of urban representation, each element and its associated attributes are stored as tuples in a structured element table. These tables are categorized based on the class of the urban entities they represent. When combined, these tables create a holistic and integrated spatial database for comprehensive urban representation, offering a detailed and interconnected view of the various urban entities and their attributes across the entire cityscape.

The proposed framework showcases its versatility through various queries designed to assess the spatial implications of urban development. It can evaluate the impact of potential new construction or road expansion on existing structures as well as vegetation. The output of these queries includes spatial morphological parameters such as aerial distribution, vertical spread, urban density, and volumetric magnitude to provide insights into the impact of urban development. This analysis helps urban planners and policymakers make informed decisions in the context of the Sustainable Development Goals (SDGs), particularly SDG 11. This representation monitors environmental impact, plans for disaster resilience, and supports inclusive urban planning. It can further identify areas lacking essential green infrastructure, and aid community engagement in the development on green spaces and biodiversity. This comprehensive, data-driven approach supports the achievement of sustainable and resilient urban environments as outlined in the SDGs.

11.4 Case Study: Implementation and Results

11.4.1 Dataset

The dataset used for implementation is the Thiruvananthapuram Airborne LiDAR Dataset (TALD), which covers a substantial 9-km^2 area within the Thiruvananthapuram district of Kerala, India (Vijaywargiya and Ramiya, 2025). This dataset is annotated into four distinct semantic classes: ground terrain, shrubs, trees, and houses. It consists of 123,097,587 LiDAR points, offering a detailed and extensive representation of urban, peri-urban, and suburban environments. The ground terrain class includes 28,675,395 points, the shrub class contains 15,467,085 points, the trees class comprises 59,773,029 points, and the houses class consists of 19,182,078 points. The dataset is distributed in 9 tiles of 1 km

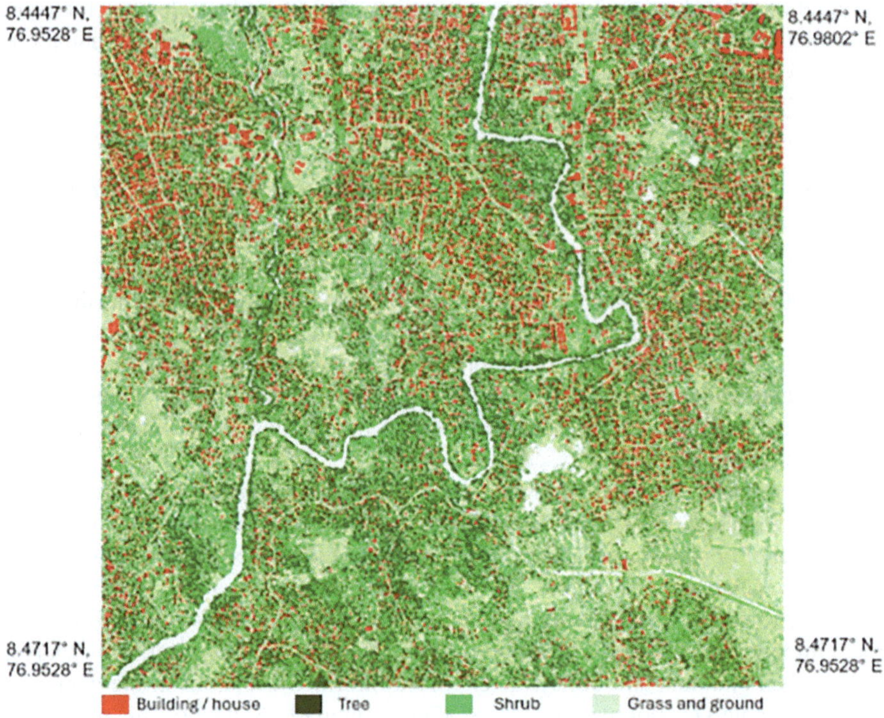

Fig. 11.6 Classified TALD point cloud dataset with four classes: buildings or houses (red), trees (dark green), shrubs (green), and grass or ground (light green).

each, as shown in Fig. 11.6. The TALD dataset is valuable for its broad coverage, making it ideal for the development of advanced urban information systems.

11.4.2 Implementation and Results

The comprehensive urban representation over the region of the TALD dataset was developed, as shown in Table 11.1. Individual houses/buildings were clustered as shown in Fig. 11.7, and their attributes are presented in Fig. 11.8. The shrubs and trees were also represented as contextual segments, as shown in Fig. 11.9.

11.4.3 Comprehensive Urban Representation for SDG

To demonstrate the applicability of the proposed urban representation, example queries are designed that address the three pillars of Sustainable Development

Table 11.1 Table shows the Urban element distribution present in the region.

#Buildings/houses	#Tree segments	#Shrub segments
18,742	59,288	25,456

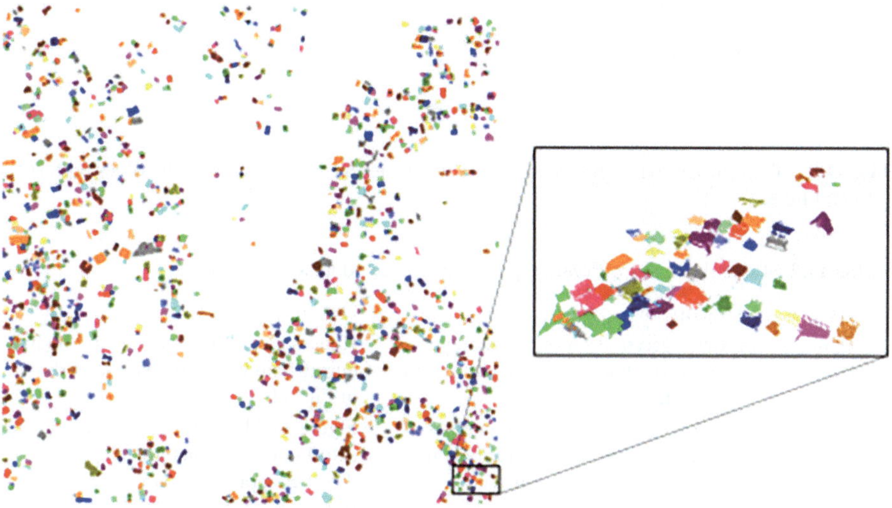

Fig. 11.7 Building clusters representing individual building present in the scene. Each colour represents different buildings.

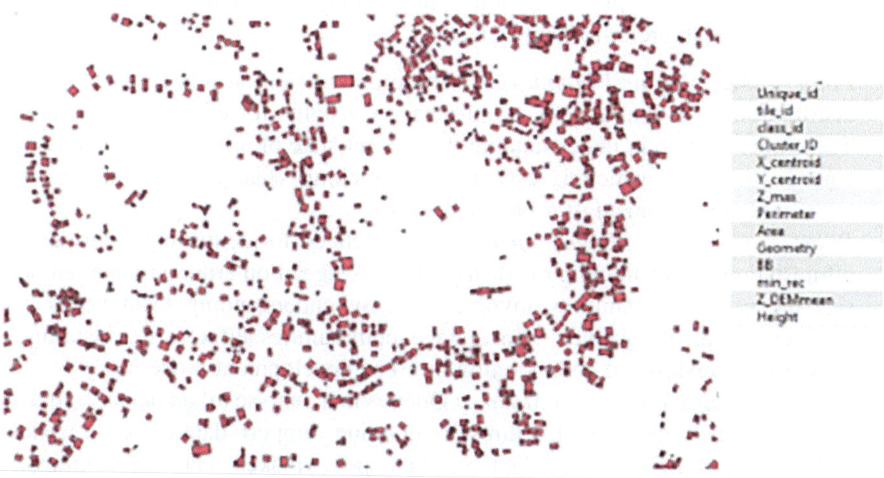

Fig. 11.8 Figure illustrates building footprint and its associated attributes.

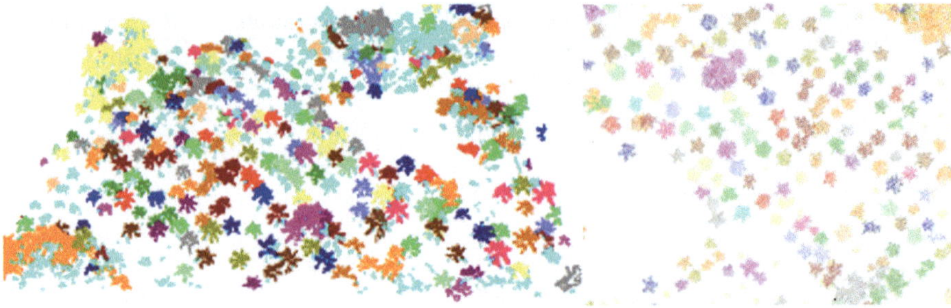

Fig. 11.9 Tree clusters representing individual trees present in the scene. Each colour represents different trees.

Table 11.2 Table illustrates different types of queries to address the SDG

Query	Description	Output
Mapping and monitoring	Extract geolocation of one-storey houses which are in low-lying regions from Attukal ward	Houses are most vulnerable to flooding during heavy rainfall events, as shown in Fig. 11.10
Community engagement	Ward-wise morphological analysis of urban green (trees) for priority and gap-based afforestation drive	Figure 11.11 shows the morphological distribution of urban green spaces in the overlapping region of data and ward
Empirical decision-making	Ward wise of urban density to monitor sustainable growth and prevent overdevelopment	Figure 11.12 shows the statistics of the current density and sprawl of urban areas which overlap with the region of the dataset

Goal 11 (SDG-11): mapping and monitoring, community engagement, and empirical decision-making refer to Table 11.2.

- *Pillar 1: Mapping and Monitoring:* These queries aim at assessing the spatial distribution, and size, and focus on detailed insights for effective governance. These queries help in understanding the current state of urban infrastructure and natural resources, which is essential for tracking changes and ensuring the sustainable management of urban landscapes.
- *Pillar 2: Community Engagement:* Queries related to identifying priorities for afforestation and examining the distribution of vegetation around houses emphasize the role of community involvement. By understanding local needs and preferences, urban planners can engage communities effectively, fostering a sense of ownership and collaboration in urban development processes.
- *Pillar 3: Empirical Decision-Making:* Queries that analyse urban density, sprawl, and the vulnerability of structures to flooding support data-driven decision-making. By utilizing comprehensive data, policymakers can make informed decisions that enhance urban resilience, guide sustainable growth, and address potential risks effectively.

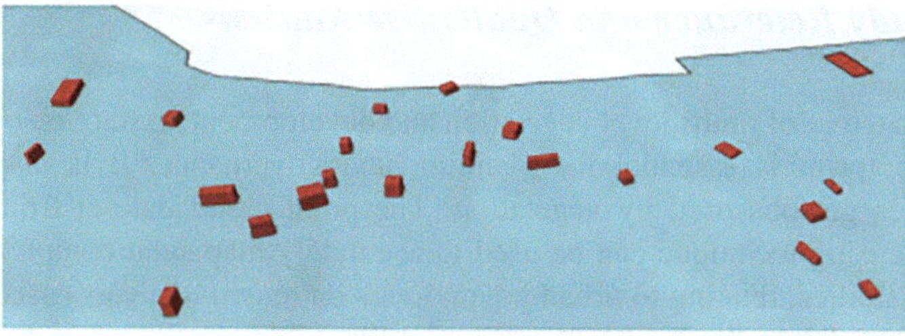

X_centroid	Y_centroid
715665.393866	935233.210752
715735.881224	935603.878890
715583.672098	935518.736789
715564.670532	935543.396483
715501.592158	935509.453763

Fig. 11.10 Application of comprehensive urban representation in mapping and monitoring. Figure shows houses most vulnerable to flooding.

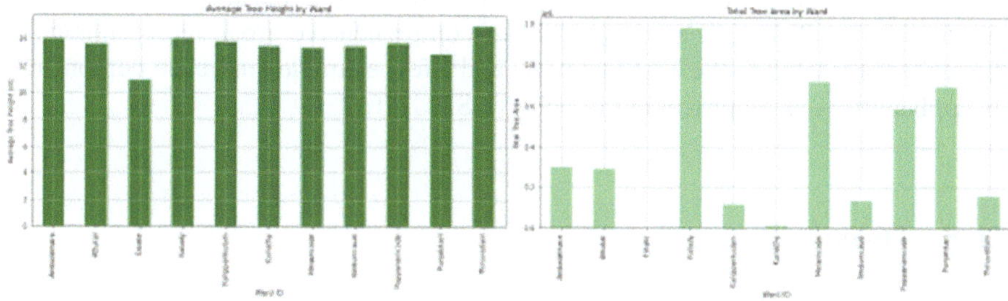

Fig. 11.11 Application of comprehensive urban representation in community engagement. Figure represents the morphological distribution of trees in different wards in the study area.

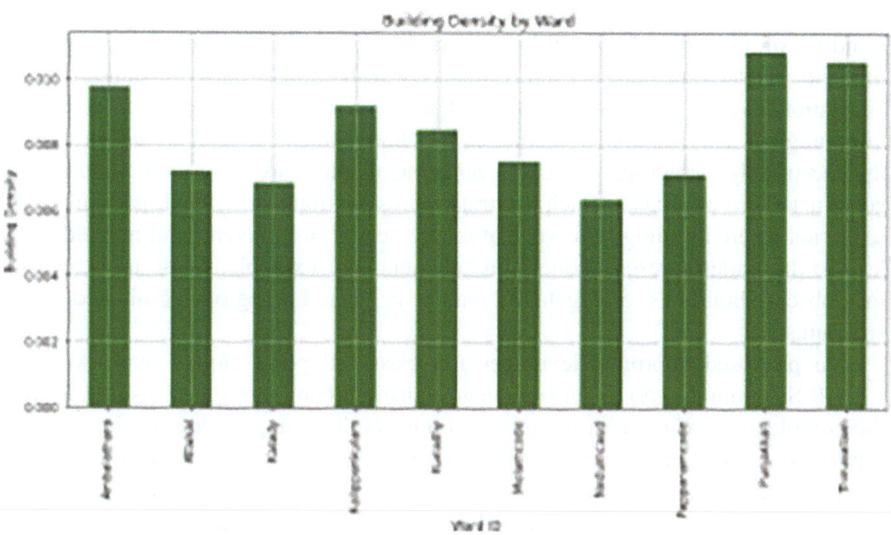

Fig. 11.12 Application of comprehensive urban representation in empirical decision-making. Figure represents the building densities in different ward.

11.4.4 Case Study Relevance and Qualitative Analysis

In urban regions with a mix of multi-level vegetation and development, extraction of accurate building footprint is essential for planning and governance. It is also challenging as houses get obscured by vegetation. The point cloud dataset from ALS, a 3D remote sensing technique can be used to accurately map such complex regions. Most of the studies, relating to urban green, focus on macro-analysis based on various indices that define the health of vegetation based on greenness in the region. The macro-analysis gives a broader glimpse, but for achieving SDG-11, accurate mapping and representation of urban green at micro-scale is also necessary. In comparison, the LiDAR-derived footprints of buildings/houses inculcate all big and small structures, which are not adequately presented in existing urban representation/footprint. This case study prepared a comprehensive urban representation of the region presented in the TALD dataset and a few example queries to demonstrate its applicability, as empirical aid to achieve SDG-11.

11.5 Discussion

The integration of remote sensing and data science plays a pivotal role in the advancement of Sustainable Development Goal (SDG) 11, particularly through the detailed monitoring and analysis of urban environments. This integration enhances urban planning, governance, and decision-making processes. Utilizing 3D remote sensing techniques allows for the capture of elevation and geometric features of various urban components, thereby providing a more precise and holistic understanding of the urban landscape. Airborne LiDAR (ALS) technology generates intricate point cloud datasets that effectively differentiate between buildings and vegetation, even in densely populated urban areas where these elements are closely intermingled. This capability facilitates the development of realistic urban models that accurately depict the spatial arrangement and density of both natural and constructed environments. Such accurate representations are crucial, particularly in scenarios where buildings and vegetation are nearby, as conventional methods may lead to the underrepresentation of one element. For example, dense tree canopies may obscure buildings, or vegetation may be neglected in regions dominated by tall structures.

The proposed approach leverages the classified point clouds from Airborne LiDAR Scanning (ALS) to generate near-realistic urban representations that address the complexities of modern urban environments. This approach is designed to be robust, adaptable, and comprehensive, ensuring that it can meet diverse needs and key attributes like scalability, flexibility, and compatibility. These models are crucial for supporting sustainable urban development, particularly in areas where the interplay between natural and built environments is complex and dynamic. Scalability is a key strength of the proposed approach, allowing it to effectively address urban landscapes of different sizes and complexities. Whether applied to small neighbourhoods or vast metropolitan areas, the methodology ensures that the resulting urban representations are consistent and reliable across various scales. This adaptability was demonstrated in a case study covering a 9 km^2. area in Thiruvananthapuram, Kerala, India, with the potential for the approach to be extended to larger regions and applied in diverse urban contexts worldwide. The flexibility of the proposed framework is evident in its ability to adapt to the unique characteristics of various urban environments. Urban areas mix a diverse set of entities, each presenting its challenges. The concept of encompassing individual objects and contextual segments ensures that the framework can accurately represent these diverse elements, making it a highly versatile tool for comprehensive urban representation. Thus, if the annotated point clouds include additional classes, such as poles or powerlines, these can be seamlessly integrated into the urban representation. Whether in densely built areas or regions with significant natural cover, the approach can be tailored to the specific needs of the region, ensuring that all relevant features are comprehensively captured and represented.

A key strength of the proposed framework is its compatibility with existing systems, particularly integrating ALS-derived data with current GIS layers enhances its utility. Additionally, the framework is built to accommodate updates and expansions, ensuring that the models remain consistent with current systems while being adaptable to new developments. Thus, the framework is both sustainable and practical, offering a robust and integrative urban representation that significantly enhances the ability to make data-driven decisions. The applicability of the proposed framework and method extends across multiple domains, including infrastructure planning, environmental monitoring, and disaster management. The detailed urban representation can inform zoning regulations, guide sustainable land use planning, and support the implementation of policies aimed at achieving the goals of SDG 11. Moreover, the models can be used to engage communities by providing visual representations of the current statistics, fostering a deeper understanding of the impacts of urbanization. By seamlessly combining ALS-derived data with existing GIS layers, it supports the three pillars of Sustainable Development Goal 11 (SDG 11): mapping and monitoring, community engagement, and empirical decision-making. Its broad applicability makes it a powerful asset for advancing sustainable urban development in diverse and complex environments, ensuring that urban planning and governance are informed, inclusive, and effective.

Although there have been significant advancements in remote sensing and data science, various limitations impede their complete integration into urban planning. A primary obstacle is the availability of data. Technologies such as LiDAR and 3D remote sensing yield intricate information; however, their application is frequently restricted to particular areas or projects. Furthermore, incorporating these technologies into current urban planning frameworks necessitates a transformation in governance practices, highlighting the importance of adopting more data-driven and evidence-based decision-making approaches. While remote sensing and data science present considerable opportunities for promoting Sustainable Development Goal

11, it is essential to tackle issues related to data availability and governance to fully exploit these technologies in fostering sustainable urban environments.

11.6 Conclusion

In the context of the Fourth Industrial Revolution, the proposed framework for urban representation through Airborne LiDAR Scanning (ALS) represents a significant advancement in urban planning and management. By harnessing the power of 3D remote sensing and data science, this approach effectively bridges the gap between physical landscapes and their digital counterparts, supporting Sustainable Development Goal 11 (SDG 11), which focuses on creating inclusive, safe, resilient, and sustainable cities.

The framework demonstrates its feasibility in generating near-realistic urban representation by addressing the complexities of modern urban environments. Its key attributes—scalability, flexibility, and compatibility—ensure that it can be applied to various urban contexts, from small neighbourhoods to extensive metropolitan areas. The ability to scale and adapt the framework, as demonstrated in the case study of Thiruvananthapuram, Kerala, highlights its robustness and versatility. The framework's capacity to integrate additional classes of data, such as poles or powerlines, further enhances its applicability and ensures comprehensive urban representation. Moreover, the framework's compatibility with existing GIS systems amplifies its utility, enabling seamless integration with current geographic data and supporting a wide range of applications, including infrastructure planning, environmental monitoring, and disaster management. By aligning with the three pillars of SDG 11—mapping and monitoring, community engagement, and empirical decision-making—the framework contributes to informed, inclusive, and effective urban governance. However, challenges related to data availability and shifts in planning and governance practices remain. The current emphasis on developing digital twins for smart cities in India presents an opportunity to scale comprehensive urban representations, yet the integration of these technologies into existing frameworks requires a commitment to data-driven decision-making and evidence-based planning.

References

Aljumaily H, Laefer DF, Cuadra D, Velasco M (2023) Point cloud voxel classification of aerial urban LiDAR using voxel attributes and random forest approach. Int J Appl Earth Obs Geoinf 118:103208

Chen C, Bagan H, Yoshida T, Borjigin H, Gao J (2022) Quantitative analysis of the building-level relationship between building form and land surface temperature using airborne LiDAR and thermal infrared data. Urban Clim 45:101248

Chmelar P, Rejfek L, Nguyen TN, Ha D-H (2020) Advanced methods for point cloud processing and simplification. Appl Sci 10(10):3340

Fareed N, Flores JP, Das AK (2023) Analysis of UAS-LiDAR ground points classification in agricultural fields using traditional algorithms and PointCNN. Remote Sens 15(2):483

Feng T, Wang W, Wang X, Yang Y, Zheng Q (2023) Clustering based point cloud representation learning for 3D analysis. In: Proceedings of the IEEE/CVF international conference on computer vision, pp 8283–8294

Grilli E, Poux F, Remondino F (2021) Unsupervised object-based clustering in support of supervised point-based 3D point cloud classification. Int Arch Photogramm Remote Sens Spatial Inf Sci 43:471–478

Guo M-H, Cai J-X, Liu Z-N, Tai-Jiang M, Martin RR, Shi-Min H (2021) PCT: point cloud transformer. Comput Vis Media 7:187–199

Haznedar B, Bayraktar R, Ozturk AE, Arayici Y (2023) Implementing PointNet for point cloud segmentation in the heritage context. Heritage Sci 11(1):2

Jayati, Vijaywargiya Anandakumar M, Ramiya (2025) Thiruvananthapuram Aerial LiDAR Dataset (TALD): A Benchmark for Complex Urban Point Cloud IEEE Access 1342350-42363. https://doi.org/10.1109/ACCESS.2025.3546628

Kalfas D, Kalogiannidis S, Chatzitheodoridis F, Toska E (2023) Urbanization and land use planning for achieving the sustainable development goals (SDGs): a case study of Greece. Urban Sci 7(2): 43

Li D, Lu C, Chen Z, Guan J, Zhao J, Jixiang D (2024) Graph neural networks in point clouds: a survey. Remote Sens 16(14):2518

Mayalu A, Kochersberger K, Jenkins B, Malassenet F (2020) Lidar data reduction for unmanned systems navigation in urban canyon. Remote Sens 12(11):1724

Mylonas G, Kalogeras A, Kalogeras G, Anagnostopoulos C, Alexakos C, Muñoz L (2021) Digital twins from smart manufacturing to smart cities: a survey. IEEE Access 9:143222–143249

Oniga V-E, Loghin A-M, Macovei M, Lazar A-A, Boroianu B, Sestras P (2023) Enhancing LiDAR-UAS derived digital terrain models with hierarchic robust and volume-based filtering approaches for precision topographic mapping. Remote Sens 16(1):78

Patel F, Patel RB (2024) Urbanization, challenges and planning with special reference to India: implications and GOI initiatives. Educ Adm Theory Pract 30(5):3994–4003

Randolph GF, Storper M (2023) Is urbanisation in the Global South fundamentally different? Comparative global urban analysis for the 21st century. Urban Stud 60(1):3–25

Saifi S, Anandakumar RM (2024) Web-based visualization and rendering of aerial LiDAR point cloud for urban flood simulation. Int J Disaster Resilience Built Env 16:260

Salehi-Dorcheabedi M, Asgari J, Amiri-Simkooei A, Nasrabadi SBF (2024) Improving LiDAR height precision in urban environment: low-cost GNSS ranging prototype for post-mission airborne laser scanning enhancement. Remote Sens Appl Soc Environ 35:101251

Shen Y, Huang J, Wang J, Jiang J, Li J, Ferreira V (2024) A review and future directions of techniques for extracting powerlines and pylons from LiDAR point clouds. Int J Appl Earth Obs Geoinf 132:104056

Song W, Changshan W (2021) Introduction to advancements of GIS in the new IT era. Ann GIS 27(1):1–4

Vijaywargiya J, Ramiya AM (2024) Information extraction system for urban planning and governance using LiDAR based 3D repository. J Spat Sci 69(1):203–223

Widyaningrum E, Peters RY, Lindenbergh RC (2020) Building outline extraction from ALS point clouds using medial axis transform descriptors. Pattern Recogn 106:107447

Yarroudh A, Nys G-A, Hajji R (2023) 3D modeling of road infrastructures according to CityGML 3.0 and its CityJSON encoding. Int Arch Photogramm Remote Sens Spatial Inf Sci 48:63–70

Chapter 12
Beyond the Office Walls: Exploring Nature Connectedness for Climate Stress and Disaster Management

Simran Mittal and Richa Manocha

Abstract Workplace anxiety has increased significantly during the past decades, and it has a significant detrimental effect on the well-being of individuals. One can differentiate between the impacts of workplace tension on employees (such as depressive disorders and anxiousness) and the firm (such as absences and efficiency). It has been discovered that mindfulness practices are helpful stress-reduction strategies that may also be applied to handling challenging work-related situations. Besides, engaging in mindfulness exercises in the outdoors promotes wellness and reduces stress. This work tries to study the relationship between nature connectedness and stress levels amongst employees in the workplace to outline effective ways to integrate nature into the work environment. The study discusses the use of AI and VR tools in adopting the natural setup in the workplace. This explores the nexus between job-related stress management and nature. It discovers nature's capability to mitigate stress. It becomes significant due to the requirement to improve worker productivity and general health. It explores the concept of nature for improving the employees' potential at work. This work reveals the critical requirement to create workplaces with natural landscapes with an emphasis on outdoor settings. It investigates the effects of workplace green space on workers' well-being to highlight the subsequent gains for companies seeking a better quality of professional life. This concludes by fixing multiple viewpoints and offering a crisp distillation of methods professional organizations can employ to foster work environments that encourage enhanced productivity and collaboration.

Keywords AI and VR · Mental Health · Resilient · employee efficiency · Stress Level

---

S. Mittal · R. Manocha (✉)
Amity School of Business, Amity University Uttar Pradesh (AUUP), Noida, Uttar Pradesh, India
e-mail: simran.mittal1@s.amity.edu; rmanocha@amity.edu

D. Kumar, N. P. Bassill (eds.), *Artificial Intelligence and Machine Learning for Climate Disaster Management*, Disaster Resilience and Green Growth,
https://doi.org/10.1007/978-981-95-1588-2_12

12.1 Introduction

Stress, a common occurrence in contemporary life, can take numerous shapes and has a big impact on both working conditions and personal well-being. Within the framework of job opportunities, work-related stress refers to the expectations and difficulties that people face in their roles as professionals. Workplace stress, whether it be from hectic schedules, pressure to complete tasks on time, or interpersonal problems, can harm one's physical and emotional well-being. Companies must comprehend the complex nature of anxiety in the job environment if they are to establish work cultures that foster productivity and delight.

In today's ever-changing workplace, companies that want to create an efficient and resilient work environment are placing a greater emphasis on the welfare of their employees. Organizations are focusing on comprehending and handling stress, a prevalent issue that can affect the general quality of work life, to maximize production. Vischer JC explores the complex connection between the environment, stress at work, and the quality of work–life balance. It also investigates the advantages of incorporating natural aspects into work settings.

The idea of introducing nature into the workplace is supported by several studies looking into the relationship between environment and wellness. This suggests that disparities in the availability of environmental factors rather than differences between surroundings and unhealthy behaviour are the actual cause of the wellness gap between urban and rural areas. The environment may provide rejuvenation alongside mental and physical wellness, which is highly helpful at work. People who are around ecosystems, for instance, show higher perseverance in jobs requiring rational thought, and a 40-second glimpse of a roof covered in vegetation can help them regain focus. Additionally, studies have shown that those who are exposed to the outdoors recover physiologically and emotionally from anxiety more quickly.

Being in nature enhances a state of mind, focused attention, and dedication. Having a view of the outdoors and having vegetation around the office causes arterial pressure to drop swiftly after attention-demanding tasks, which increases employee efficiency. Others have seen a connection between having access to office vegetation and decreased anxiety as well as behaviour at work.

12.2 Correlation Between Nature Connectedness and Stress Levels Among Employees

Employees face high levels of stress and deteriorating mental health for a variety of reasons in the industry; one of the most significant factors that impacts mental health is a chaotic and stressful work environment that also is increasingly devoid of green space. Several studies have found that psychological stress, anxiety, and depression dropped and connection to the environment increased for individuals over the WFH (Work from Home) experience. The psychological well-being of the fashion industry employees suffered less during the time of WFH as they were close to nature, one of the major reasons was that they took breaks and interacted with the environment. There is much speculation regarding the steps that companies are considering taking to help mitigate the negative effects of digital pressures on employees' mental health in the fashion industry. Data that shows being involved in outdoor pursuits is pivotal to the health and well-being of human beings.

It is quite common that "technostress" among workers paves the way to mental instabilities and diseases. Perhaps the most common thread amid deteriorating mental health—the primary factor influencing the same—is the absence of nature. Changes in behaviour, architectural office design, and managerial practices (e.g., increased nature exposure) can promote mental health and well-being in the working industry, at least. A person is less frazzled after a trip to the park or greenery at a commutable distance. Green space can help reduce the stress you feel personally and physically. People who visited green spaces had significantly less stress than those who visited indoor workout centres or city outdoor spaces.

There may be a drop in stress levels as the mind and body get more at ease by being in peaceful surroundings. More serene and happier emotions have been linked to the feeling of more greenery and more organic surroundings. The setting is perfect for meditation and/or working out, which also lowers your stress. There is a mixture of sounds and sight signals that can also aid in lowering anxiety and helping you to feel more at ease and serene. Employee stress and its relationship with natural environments and surroundings has become crucial as people have become a subject of increased priority in recent years. These studies show that natural environments may increase individuals' sense of connection with their surroundings and provide restorative benefits as well. According to studies done by Kuo and Sullivan's (2001) and Hartig et al.'s (2010) natural environments leads to reduced anxiety. It has been shown that low levels of cortisol and low levels of perceived stress are associated with more time spent in natural environments, such as a forest or greenhouse. By placing green areas in or near businesses, workers have a chance to interact with environmental settings, which has an anxiety-reducing effect. Several other stress-alleviation procedures that involve the environment have demonstrated success. An effective strategy for stress reduction is participation in activities that involve being in the environment such as walking in green spaces or watching an environmental setting.

Responsive resilience is the capability to recognize and surmount obstacles. There is a positive connection between this concept and a sense of connection to the environment. Studies by Capaldi et al. and Nisbet et al. indicate that people with a high level of connectedness to the environment show evidence of resolute coping strategies. In other words, nature aids in enhancing employees` ability to tackle strains at work.

12.3 Effective Ways to Integrate Nature into the Work Environment

Employee emotional well-being is significantly influenced by the presence of interior vegetation and flowers. Ulrich (1983) conducted research indicating that small interactions with indoor plants have a large impact on anxiety rates. Improved ventilation, higher concentration rehabilitation, and an overall greater sense of emotional well-being all appear regarding indoor plants (Lohr et al. 1996). Natural sunlight is another major contributor to workplace anxiety. Referred to the importance of daylight in promoting lower levels of anxiety and happier moods in the studies. Enhanced general health leads to increased overall employee health and happier staff at work based on lighting in the work environment.

Biophilic design aims to design spaces that promote contact with the natural world. In a study conducted in 2014, Ryan et al. and Browning et al. found that exposure to actual and simulated natural settings (e.g., views of green spaces and the use of natural materials) has been shown to improve worker productivity and lower stress levels. A recent innovation is the use of modular offices that allow workers to interact with their surroundings even while at their desks. As Lee et al. and van den Berg et al. explained, rooftop plantings and green corridors are examples of natural settings that can give workers a way to decompress and recover during periods that would otherwise be stressfully short. Although biophilic architecture has seen success, it is not without challenges. Beatley and Yang et al. have identified several of them; personal preference, the logistical difficulties of managing live plants, and the need for careful planning among others. Overcoming these obstacles will be key to maximizing the stress-reducing potential of biophilic architecture.

The concept of regenerative environments making its mark in today's office designs comes from the fact that these workspaces are now being redesigned with attributes that offer tranquillity and help people find relief from work-based stress. Research from Ulrich (1983) shows that introducing environmental light and water to a workstation can make people feel less stressed and more satisfied with their work, for starters; Occupational relationship with nature is directly linked with employee engagement, researchers Otto and Pensini and Rigolon et al. introduce among their research programs "How crucial it is to incorporate nature into your company's culture". Also being considered is whether anxiety can be reduced in day-to-day work situations by appealing to multiple senses. Findings from Heschong and Tennessen and Cimprich have shown that by incorporating elements such as waterfall displays or the sounds of nature, a space can produce a calm that takes in a range of sensations. Understanding what the employees favour is key to making a space work. Today's workspaces have adopted customized models that enable people to incorporate natural elements at their workspace; discovering that it seems to create a sense of autonomy that crosses over to make a positive impact at levelling off stress.

12.4 Relationship Between Nature Connectedness and Workplace Productivity

Our natural human affinity with the natural world leads to mental agility on the job. Researchers Joye, and Mayer and Frantz, have found that folks who feel closely connected to nature tend to have greater mental agility, which includes improved reasoning, sharper imagination, and better concentration. Those are productivity makers for sure. Several studies including Berto, and Largo-Wight et al., found that time in a natural setting tends to reduce the degree to which individuals feel stress. Likely, nature-connected workers would experience less stress. That takes you to a better place mentally for focusing your brain and, therefore, at doing your job, overall.

Howell et al. investigated the impact of connectivity to nature on a wide variety of wellness indicators in their study. The team of explorers utilized an ongoing approach to study the mental health and sense of connectivity to the natural world of individuals over a period. The researchers found an upward trajectory in the degree of mental wellness of the individuals — more mental wellness was consistently observed among those individuals who reported that they felt more connected to the natural world.

Standard measures were used to evaluate an individual's overall well-being, which includes domains like life satisfaction, positive affect, and a sense of purpose. Howell et al. found that emotional and intellectual connectedness to the environment around them played a major role in determining individuals' overall well-being, based on their study's results. This indicates that efforts to foster urban connections may well have trickle-down benefits to a person's mental health. A more thorough investigation of the connection between nature connectedness and mental health was conducted by Capaldi et al., who also explored potential mechanisms through which this connection might occur. The study used a longitudinal methodology to assess participants' relationships with the environment car over time, using a blend of environmental momentary assessments and self-report questionnaires. The results of Capaldi et al. confirmed those of Howell et al., with mental health showing a positive association with nature connectedness. The additional analysis of potential mechanisms underlying this association was instructive. Perceptions of closeness, competence, and autonomy were found to mediate the positive effect of nature connectedness on well-being.

The outcomes has far-reaching effects on, psychological assessment, psychological treatment, and worker wellness programs. Office layouts may reduce tension, while city planners could layout cities in a manner that enhances psychological well-being and recovery spaces for enhancing well-being Adding nature-based therapies can help in this direction. Companies and regulators who would like to enhance worker well-being should think about ways to facilitate psychological well-being to enhance association with nature.

12.5 The Potential of Technology in Fostering Nature Connectedness and Stress Reduction

Virtual reality (VR) technology allows users to become truly immersed in three-dimensional computer representations. The displays have individuals believing that they've been transported to forests, coastlines, mountains, and others without leaving their office. Prepared results from studies reveal the biological and emotional impacts of virtual reality-induced environment encounters' anxiety reduction, increased happiness, and overall well-being. VR apps allow wildlife encounters that are entirely realistic and can be everything from idyllic beaches to cloistered woodlands. This digital naturalness, as research has put it, allows people to get lost in an environment and react as they do in the real world. If they are fully capable of enjoying intimate virtual reality (VR) experiences, people can get lost in this sense of being present and also experience reality.

Research has assessed the biological and emotional efficacy of VR treatments, using measures like cardiovascular fluctuation, hormone levels, and subjective stress ratings. The overview refers to data indicating that plunging into natural settings using virtual reality (VR) could significantly reduce anxiety and furnish workers with plenty of opportunities to zone out during the workday.

Another cutting-edge approach to introducing natural encounters into interior workspaces is through the employment of augmented reality (AR). AR provides an alternate yet similar means of synthesizing the digital with the real in interior workspaces by superimposing electronic data over the real surroundings as opposed to virtual reality, wherein augmented reality programs produce a hybrid experience that incorporates actual jobs with simulated environmental aspects. By way of a process of visual searching that involves looking through surroundings for fascia — or visual tracking — employees can communicate with virtual components laid over the real in their actual environment. With their mobile gadget or AR (augmented reality) lenses, workers can look at and communicate with virtual ecosystems that are represented over their workstations or office settings and/or digital creatures. The standards applied to this type of interface analysis include convenience for users, functionality, and level of engagement in the enhanced world. There are a variety of work contexts benefitting from virtual reality technology and augmented reality, per the evaluation. Such advancements can be executed across numerous settings relatively easily, for instance. They can promote context-induced anxiety reduction as well as nature connectedness.

12.6 Conclusion

The purpose of this study was to examine the benefits of lush vegetation for employees' health at scientific centre companies. According to the research, having physical exposure to these natural areas from interior places as well as using them both promote worker satisfaction. The results displayed here also imply that perspectives from workers' desks can influence happiness more strongly than a hands-on encounter with a office natural environment. By promoting staff wellness and efficiency, optimizing landscapes with soothing horticulture and the larger environmental landscape through structural architecture may therefore help to encourage possibilities for healing and may be advantageous to both people and organizations.

This suggests that when developing such locations, environment and structure architecture should be considered from the beginning. Most importantly, the specific greenery elements that workers can observe from their places of employment seem to impact satisfaction consequences; greater degrees of self-identified contentment among technology centre workers were linked to viewing more vegetation, grass, shrubs, or blossoming flora in the glass view. This study strengthens the commercial argument for investing in superior quality natural environments on corporate campuses; determining the financial effects of the advantages to workers' good health on company output may be a crucial subsequent stage.

The location wherein individuals devote most of their days is their place of employment. Therefore, an organization ought to be able to produce satisfaction alongside functioning as a centre of productivity. We outlined the benefits of mindfulness and being in nature about professional achievements in this paper. Additionally, we emphasized how mindfulness environmental approaches might be therapeutically used in work settings. Thus, the study highlights the relationship between nature connectedness and the stress levels of the employees working in the organization. The major objective of the study includes outlining effective ways which can be implemented and help in integrating nature into the work environment considering elements such as green spaces, natural lighting or incorporating nature-inspired design elements. The study aims to investigate the relationship between nature connectedness and workplace productivity which examines the potential of technology such as virtual nature experiences or nature-themed applications in fostering nature connectedness and stress reduction for employees working in indoor or urban environments.

References

Browning WD, Ryan CO, Clancy JO (2014) 14 patterns of Biophilic design. Terrapin Bright Green, New York

Hartig T, Van Den Berg AE, Hagerhall CM, Tomalak M, Bauer N, Hansmann R, ... & Waaseth G (2010) Health benefits of nature experience: Psychological, social and cultural processes. In Forests, trees and human health (pp. 127–168). Dordrecht: Springer Netherlands.

Kuo, F. E., & Sullivan, W. C. (2001) Environment and crime in the inner city: Does vegetation reduce crime?. Environment and behavior 33(3):343–367

Lohr VI, Pearson-Mims CH, Goodwin GK (1996) Interior plants may improve worker productivity and reduce stress in a windowless environment. J Environ Hortic 14:97–100

Ríos-Rodríguez ML, Testa Moreno M, Moreno-Jiménez P (2023) Nature in the office: a systematic review of nature elements and their effects on worker stress response. Healthcare 11(21):2838. https://doi.org/10.3390/healthcare11212838

Ulrich RS (1983) Aesthetic and affective response to natural environment. In: Altman I, Wohlwill JF (eds) Behavior and the natural environment. Springer, Boston, pp 85–125

Van der Klink JJ, Blonk RW, Schene AH, van Dijk FJ (2001) The benefits of interventions for work-related stress. Am J Public Health 91:270–276

Vischer JC (2007) The effects of the physical environment on job performance: towards a theoretical model of workspace stress. Stress Health 23:175–184

Chapter 13
Challenges and Future Trends in Climate Disaster Management

Deepak Kumar and Nick P. Bassill

Abstract The increasing frequency and intensity of climate disasters necessitate more effective management strategies, and artificial intelligence (AI) and machine learning (ML) offer promising solutions. There is a need to explore the challenges and future trends in applying AI and ML for climate disaster management. Key challenges include data quality and availability, where incomplete or inconsistent data can hinder model accuracy. Computational requirements cause hurdles to the AI and ML approaches, as they require substantial processing power and energy consumption. Model interpretability and transparency are critical issues for AI-driven systems as most of the approaches function like "black boxes," which makes it difficult to validate predictions and build trust. Ethical and social considerations like safeguarding equity and fairness, also play a significant role in the responsible deployment of these technologies. Despite these challenges, several trends highlight a huge potential for AI and ML in climate disaster management. Enhanced data integration and processing capabilities will enable more comprehensive analysis for better-informed decision-making. The development of higher-accuracy climate models and real-time monitoring systems will improve predictive capabilities for more effective early warning systems and impact assessments. The establishment of regulatory and ethical frameworks will guide the responsible

D. Kumar (✉)
Atmospheric Sciences Group, Department of Geosciences, College of Arts & Sciences, Texas Tech University, Lubbock, Texas, USA

Center of Excellence in Weather & Climate Analytics, Atmospheric Sciences Research Center (ASRC), University at Albany, State University of New York (SUNY), Albany, New York, USA

N. P. Bassill
Center of Excellence in Weather & Climate Analytics, Atmospheric Sciences Research Center (ASRC), University at Albany, State University of New York (SUNY), Albany, New York, USA

State Weather Risk Communication Center (SWRCC), Atmospheric Sciences Research Center (ASRC), University at Albany, State University of New York (SUNY), Albany, New York, USA

D. Kumar, N. P. Bassill (eds.), *Artificial Intelligence and Machine Learning for Climate Disaster Management*, Disaster Resilience and Green Growth,
https://doi.org/10.1007/978-981-95-1588-2_13

development and application of AI and ML in this field. The integration of AI and ML in climate disaster management has significant opportunities to enhance predictive models and support mitigation and adaptation efforts for a resilient and sustainable future.

Keywords AI And ML · Remote Sensing · Climate Management · Sustainable Future · Disaster Management

13.1 Introduction

Climate change has escalated the frequency and intensity of natural disasters worldwide and it creates unprecedented challenges in disaster management (Shabahang et al. 2019; Talha et al. 2025). The consequences of a rapidly changing climate demand urgent and innovative responses ranging from devastating floods to extreme wildfires and powerful hurricanes (Parente et al. 2019; Wang et al. 2024). As communities grapple with the effects of these disasters, the field of climate disaster management faces significant hurdles in preparedness, mitigation, and recovery efforts (Attaf et al. 2019; Sunar et al. 2019). In addition to the physical and logistical challenges, social, economic, and political factors further complicate effective disaster response. The future of climate disaster management centers on our ability to adapt to evolving environmental conditions and embrace new technologies and strategies to promote resilience. This includes improving early warning systems, strengthening infrastructure, and ensuring equitable recovery efforts for vulnerable populations (Ezangina and Gromyshova 2020; Obrist et al. 2021). This explores the key challenges faced in managing climate-induced disasters and examines emerging trends and technologies that will shape the future of disaster response. The analysis explores the current gaps in policy, resources, and collaboration to provide insights into how the global community can prepare, respond, and recover from the growing impact of climate-related disasters.

13.1.1 Background and Significance

Climate-induced disasters have become an increasingly critical concern in the twenty-first century, driven by the accelerating effects of global climate change (Gupta et al. 2024; Selvam and Al-Humairi 2025). Rising temperatures, changing weather patterns, and more frequent extreme events such as hurricanes, droughts, wildfires, and floods are reshaping the global landscape and posing significant risks to human lives, ecosystems, and economies (Cao and Shi 2021; Richards and Belcher 2020). These disasters are not only devastating in terms of immediate damage but also have long-term consequences, further exacerbating vulnerabilities, especially in underdeveloped and developing nations. The traditional approaches to disaster management are effective in some instances and are increasingly inadequate to address the scale and complexity of climate-related events (Capolupo et al. 2021; Icaza et al. 2016). Rising sea levels, population growth, and urbanization create new risks, while many regions face resource constraints that hinder effective disaster preparedness and response (Sahani et al. 2019; Talha et al. 2025). The understanding of challenges faced due to climate change in disaster management are crucial for framing future strategies for mitigation, adaptation, and recovery. This study seeks to identify the emerging trends that could define future management practices, such as the integration of artificial intelligence in early warning systems, the role of community-based disaster risk management, and the potential for green infrastructure in enhancing resilience (Fadlullah and Kato 2021; Thangavel et al. 2023). The significance of this work lies in its potential to inform both policymakers and practitioners about the evolving landscape of climate disaster management, providing insights that can lead to more effective and sustainable solutions. The current barriers and the innovative trends shaping disaster management in a changing climate contribute to the development of strategies that not only reduce the impact of disasters but also build the resilience of communities worldwide (C. R. Singh and Manoharan 2024). It will help shape policies and practices that are more adaptive, inclusive, and forward-thinking in the face of the climate crisis.

13.1.2 Challenges in Climate Disaster Management

Climate disaster management is a complex and multi-faceted issue, and it raises significant challenges due to the rapidly changing climate and evolving risks (Ayub et al. 2024; Cilli et al. 2022; Dtissibe et al. 2024). These challenges are not only technical and logistical but also social, political, and economic, and they require a multi-disciplinary approach to address effectively. The key challenges in climate disaster management include the following:

(a) **Increasing Frequency and Intensity of Disasters:** As the climate warms, extreme weather events such as storms, floods, droughts, and wildfires are becoming more frequent and severe. The unpredictability and escalation of these events make it difficult for governments and organizations to plan, prepare, and respond effectively. The sheer scale and intensity of these disasters can overwhelm existing infrastructure, emergency services, and resources.

(b) **Inadequate Infrastructure and Resources:** Many regions, particularly in developing countries, lack the necessary infrastructure to withstand climate-related disasters. Poorly constructed buildings, limited access to emergency services, and underdeveloped early warning systems exacerbate the impact of disasters. Financial and human resources for disaster response and recovery are often insufficient, further hindering effective management.

(c) **Poor Coordination and Governance:** Climate disaster management involves multiple stakeholders, including local, national, and international agencies, non-governmental organizations (NGOs), and the private sector. The lack of coordinated efforts and clear communication channels between these entities often leads to inefficient disaster response, delays in recovery, and duplication of efforts. In many cases, governance structures are fragmented, and there is a lack of clear policies, roles, and responsibilities.

(d) **Vulnerability of Marginalized Populations:** Vulnerable populations, like low-income communities, women, children, and indigenous groups, are disproportionately affected by climate disasters. These groups often have limited access to resources, are in high-risk areas, and face social and economic inequalities that hinder their ability to recover from disasters. Ensuring these communities are included in disaster preparedness, response, and recovery plans is a major challenge.

(e) **Insufficient Data and Early Warning Systems:** Accurate data on climate risks and early warning systems are essential for effective disaster management. However, many regions still lack robust data collection, monitoring, and forecasting capabilities. Early warning systems are often underfunded or poorly implemented, and the dissemination of alerts to vulnerable communities can be slow or ineffective, leaving them unprepared for the disaster ahead.

(f) **Climate Change Uncertainty:** Climate change introduces a level of uncertainty in predicting disaster events and their impacts. While climate models can provide projections, they often cannot predict the exact timing, location, and magnitude of disasters. This uncertainty makes it difficult to prepare adequately and allocate resources efficiently, as authorities are unsure about which risks will manifest and when.

(g) **Long-Term Recovery and Resilience Building:** Rebuilding infrastructure, restoring livelihoods, and addressing the long-term health and social consequences of disasters require significant resources, planning, and coordination. While emergency responses to disasters may be immediate, long-term recovery and rebuilding efforts often take years, and sometimes even decades. In many cases, the focus on short-term relief can overshadow the need for sustained efforts to build resilience for future disasters.

(h) **Political and Economic Challenges:** Political instability, corruption, and competing priorities often hinder effective climate disaster management. In some cases, the lack of political will to implement climate adaptation strategies or allocate resources to disaster preparedness leads to ineffective disaster management policies. Furthermore, economic constraints may force governments to prioritize short-term development goals over long-term disaster risk reduction.

(i) **Climate-Induced Migration:** As climate disasters become more frequent and severe, large numbers of people are being forced to migrate from their homes, creating a new set of challenges. Climate-induced migration puts pressure on both the displaced populations and the areas receiving them, as both struggle to provide adequate shelter, resources, and services. Managing this displacement requires coordination between countries and the implementation of policies to ensure the protection of displaced individuals.

(j) **Public Awareness and Engagement:** Public awareness and engagement are essential for effective climate disaster management. Many people remain unaware of the risks posed by climate change and climate-induced disasters. Public education campaigns, community outreach programs, and the inclusion of local knowledge are needed to ensure that communities are prepared and responsive to disasters. However, achieving this level of awareness and involvement is often challenging, particularly in areas with low literacy rates or limited access to information.

These challenges require a comprehensive and coordinated approach and involve the collaboration of governments, NGOs, the private sector, and communities. It also necessitates innovative solutions like the integration of advanced technologies in disaster management, improved climate data and forecasting, and the development of policies that prioritize both immediate response and long-term resilience (Ekka et al. 2024; Wai et al. 2022). A holistic approach mitigates the impact of climate-induced disasters and better protects vulnerable communities in an increasingly unpredictable climate.

13.2 Climate Disaster Parameters for Predictive Analytics

Predictive analytics is becoming an essential tool in climate disaster management to allow authorities and organizations to forecast and prepare for potential disasters more effectively (Ma et al. 2024; Wagner et al. 2014). Several key climate disaster parameters are used to build predictive models for accurate forecasting and early warning systems and these include the following:

(a) **Temperature and Heat Extremes:** Temperature anomalies, such as heat waves and extreme cold events, are critical parameters for predictive analytics in climate disaster management. Elevated temperatures lead to the exacerbation of wildfires, droughts, and heat-related illnesses. Monitoring temperature fluctuations and identifying patterns of extreme heat can help predict the likelihood of heat waves and the associated risks for vulnerable populations and ecosystems.

(b) **Precipitation Patterns and Rainfall Forecasting:** Changes in precipitation patterns, including the frequency, intensity, and duration of rainfall, are vital indicators for forecasting floods, landslides, and other water-related disasters. Predictive models analyze historical rainfall data and future projections to assess the likelihood of heavy rainfall or prolonged droughts. This helps in preparing for flash floods, riverine floods, and drought conditions, allowing for the timely allocation of resources and flood defence measures.

(c) **Wind Speed and Storm Intensity:** For tropical storms, hurricanes, and cyclones, wind speed is a crucial parameter in determining the severity of the storm. Predictive analytics models use satellite data and atmospheric pressure readings to monitor storm development and predict changes in intensity and track. This allows for early warning systems that can issue alerts and facilitate evacuation plans in at-risk areas before the storm makes landfall.

(d) **Sea Level Rise and Coastal Flooding:** Coastal communities are increasingly threatened by rising sea levels and storm surges, which contribute to coastal flooding, erosion, and saltwater intrusion. Predictive models that incorporate data on sea level rise trends, tidal patterns, and extreme weather events can assess the potential for coastal flooding. These predictions are essential for informing adaptation strategies and protecting coastal infrastructure and populations.
(e) **Soil Moisture and Drought Conditions:** Soil moisture levels play a critical role in forecasting droughts, especially in agriculture-dependent regions. Predictive models use satellite data and ground-based monitoring to measure soil moisture content and detect early signs of drought conditions. The analysis of these parameters, and predictive analytics help to identify areas at risk of water shortages, crop failure, and food insecurity, facilitating early intervention.
(f) **Snowfall and Glacial Melt:** Snowpack levels and the rate of glacial melt are important indicators for forecasting flooding, especially in regions that rely on snowmelt-fed rivers. A sudden increase in glacial melt or a rapid decline in snowpack can trigger downstream flooding, particularly in mountainous regions. Predictive models assess these conditions, helping to anticipate seasonal runoff and mitigate flood risks associated with melting snow and glaciers.
(g) **Atmospheric Pressure and Humidity:** Atmospheric pressure and humidity levels are key variables in weather prediction, particularly for the development and intensity of storms and extreme weather. Low atmospheric pressure is often associated with the formation of tropical storms and cyclones, while high humidity levels can increase the likelihood of heavy rainfall and thunderstorms.
(h) **Ocean Surface Temperatures (SST) and El Niño/La Niña Events:** Changes in ocean surface temperatures are strongly linked to the occurrence and severity of climate events such as El Niño and La Niña. These events influence global weather patterns, including the frequency of hurricanes, droughts, and flooding. Predictive analytics models use ocean temperature data to monitor the development of El Niño and La Niña conditions, allowing for more accurate long-term forecasting of climate disasters.
(i) **Vegetation Health and Land Use Changes:** Vegetation health, monitored through satellite imagery and remote sensing, is an important indicator of drought conditions and fire risk. Changes in land use, such as urbanization, deforestation, and agricultural expansion, can also contribute to an increased vulnerability to climate disasters. Predictive models track vegetation stress and land use changes to assess risks such as wildfires, desertification, and landslides.
(j) **Air Quality and Pollution Levels:** Changes in air quality and pollutant concentrations can indicate the onset of certain climate-related health risks, such as respiratory diseases during heatwaves or wildfire events. Monitoring particulate matter (PM) levels, carbon monoxide, and other pollutants is essential for public health and safety in disaster-prone areas. Predictive models use this data to assess potential health risks and advise on evacuation and safety measures.
(k) **Seismic Activity and Volcanic Eruptions (Indirect Climate Disasters):** While not directly linked to climate change, seismic activity and volcanic eruptions can

have cascading effects on climate disasters, such as tsunamis. Predictive analytics be used to monitor seismic data and volcanic activity to provide early warnings of related disasters, which may be aggravated by changing climate conditions.

(l) **Ecosystem Stress and Biodiversity Loss:** The health of ecosystems is an important parameter in predicting the resilience of communities to climate disasters. Monitoring biodiversity loss, forest degradation, and marine ecosystem stress helps assess long-term climate risks and potential cascading effects. Predictive models that incorporate ecological data help determine the impact of climate disasters on natural resources, food security, and livelihoods.

Integrating these climate-related disaster parameters into predictive analytics models, authorities will help more informed decisions on disaster preparedness, mitigation, and response (Alotaibi and Nassif 2024; David et al. 2024; Liu et al. 2021). Predictive analytics help forecast climate-related events and assist in managing risk and allocating resources more efficiently, ultimately saving lives and minimizing the long-term impact of disasters. However, the accuracy of predictions depends on the quality and timeliness of data, as well as the ability to incorporate complex interactions between these various parameters.

13.3 Impact of Climate Disaster Management

Effective climate disaster management plays a key role in reducing the impact of climate-related events on communities, economies, and ecosystems. It saves lives, protects infrastructure, enhances resilience, and promotes sustainable recovery in the face of increasingly frequent and severe disasters (Alotaibi and Nassif 2024; Jitkajornwanich et al. 2024). The impact of climate disaster management can be viewed from several critical angles:

(a) **Reduction in Loss of Life and Injuries**

One of the most significant impacts of effective climate disaster management is the reduction in the loss of life and injuries during extreme events. Early warning systems, timely evacuations, and well-prepared emergency response plans save countless lives. In the prediction of disasters such as hurricanes, floods, and heatwaves, authorities can take proactive measures to protect at-risk populations, especially vulnerable groups such as the elderly, children, and low-income communities. Furthermore, the implementation of resilient healthcare systems ensures that medical support is available to those in need during and after a disaster.

(b) **Minimization of Economic Losses:**

Climate disasters often cause extensive economic damage, including the destruction of infrastructure, homes, crops, and businesses. The financial toll is minimized by strengthening infrastructure to withstand extreme weather events, ensuring effective insurance mechanisms, and creating contingency funds for rapid response and recovery. Preparedness plans focus on protecting critical infrastructure such as transportation, energy, and water systems to reduce the long-term economic disruptions that follow a disaster.

(c) **Protection of Ecosystems and Biodiversity:**

Climate-induced disasters can have severe effects on natural ecosystems and biodiversity, such as coastal ecosystems damaged by storm surges, forests destroyed by wildfires, and agricultural land eroded by floods. Proactive disaster management strategies help ecosystems to safeguard measures like preserving wetlands for flood control, establishing firebreaks in forests, and implementing sustainable land management practices. Moreover, disaster management encourages the restoration of ecosystems following a disaster, helping to reduce the long-term environmental impacts.

(d) **Enhanced Resilience of Communities:**

Disaster management is essential for building resilience in vulnerable communities. The integration of climate adaptation strategies into local planning and fostering community-based disaster risk management helps to respond and recover from climate-induced disasters.

(e) **Improved Coordination and Governance:**

Effective climate disaster management nurtures stronger collaboration and coordination between local governments, national agencies, international organizations, and non-governmental organizations (NGOs). This multi-stakeholder approach leads to more efficient resource allocation, faster response times, and a more comprehensive understanding of risks and needs. Better governance through transparent policies, clear responsibilities, and data sharing ensures that disaster response is aligned with the needs of affected populations and the capacities of various stakeholders.

(f) **Faster Recovery and Reconstruction:**

A well-planned disaster management strategy accelerates recovery by providing clear guidelines and support mechanisms for reconstruction efforts. The incorporation of risk reduction measures into recovery plans makes them more resilient to future disasters. In the long term, these efforts contribute to sustainable development by reducing vulnerability and enhancing the capacity of regions to bounce back quickly after a disaster, thus minimizing the economic and social costs of recovery.

(g) **Promotion of Climate Change Mitigation and Adaptation:**

Climate disaster management also drives broader efforts to combat climate change. It encourages the adoption of mitigation strategies like reducing greenhouse gas emissions and transitioning to renewable energy sources, to help prevent the worst effects of climate change. It promotes adaptation efforts by identifying and implementing strategies that reduce vulnerability to future climate risks, such as enhancing water management, improving crop resilience, and protecting biodiversity.

(h) **Enhanced Public Awareness and Empowerment:**

Climate disaster management often involves public education and awareness campaigns that help people understand the risks posed by climate change and the importance of disaster preparedness. This empowerment enables individuals and communities to take proactive steps to protect themselves, such as building disaster-resistant homes, storing emergency supplies, and participating in local preparedness drills. Raising awareness about the causes and effects of climate change also contributes to greater public support for climate action policies.

(i) **Strengthened International Cooperation:**

Climate disaster management has a significant impact on fostering international cooperation and solidarity. Climate-induced disasters often affect multiple countries, especially in regions prone to hurricanes, floods, and droughts. Collaborative efforts in disaster management, such as sharing early warning systems, providing humanitarian aid, and supporting joint research, are essential for building a global response network. Moreover, international cooperation in financing climate adaptation and disaster risk reduction programs helps support developing nations that may lack the resources to manage the impacts of climate change on their own.

(j) **Long-Term Sustainability and Climate Resilience:**

The ultimate impact of climate disaster management is the promotion of long-term sustainability and climate resilience. The prioritization of disaster risk reduction, investing in green infrastructure (such as flood barriers, stormwater management systems, and reforestation), and adopting sustainable land-use practices, communities reduce their exposure to future climate hazards. This approach fosters a transition toward more sustainable development practices, balancing immediate disaster recovery with long-term environmental and social goals.

The impact of climate disaster management is far-reaching, and it influences the immediate response to climate-related events for long-term resilience of communities and ecosystems (Jitkajornwanich et al. 2024; Liao et al. 2024). Effective disaster management is fundamental to achieving a sustainable future amid the growing uncertainties posed by climate change. It plays a critical role in reducing loss of life, minimizing economic damages, safeguarding biodiversity, and strengthening governance frameworks. These interconnected objectives collectively enhance the resilience of societies and ecosystems, enabling them to better withstand and recover from environmental hazards (Ma et al. 2024; Wang et al. 2024). It ensures that vulnerable populations are supported and that recovery efforts contribute to building back better, enhancing global efforts to address the complex challenges posed by climate change.

13.4 Advancements in Climate Disaster Management

In recent years, advancements in technology, data analytics, and global collaboration have significantly improved the field of climate disaster management (Ma et al. 2024; Selvam and Al-Humairi 2025). These innovations have led to more effective and proactive strategies, enhancing preparedness, response, and recovery efforts. As the impacts of climate change continue to intensify, the adoption of new tools and approaches is crucial for managing the risks associated with climate-induced disasters. The key advancements in climate disaster management include the following:

(a) **Early Warning Systems (EWS) and Predictive Analytics**

Early warning systems have evolved from simple weather alerts to highly sophisticated tools powered by real-time data and predictive analytics. Using satellite imagery, weather forecasting models, and machine learning algorithms, modern EWS predict the onset and intensity of natural disasters such as hurricanes, floods, heatwaves, and wildfires with greater accuracy (Ceccato et al. 2018; Doblas et al. 2020; Fan et al. 2025). These systems provide authorities with advanced warnings, allowing them to take proactive measures like evacuations, resource mobilization, and public awareness campaigns before a disaster strikes.

Example: The development of integrated flood warning systems that combine hydrological models with real-time rainfall data to predict flood events up to several days in advance.

(b) **Geospatial Technology and Remote Sensing**

Geospatial technologies, comprising Geographic Information Systems (GIS), satellite imagery, and drones, have revolutionized disaster management by providing real-time situational awareness during and after a disaster (Atek et al. 2022; Balestra et al. 2021; H. Wang et al. 2020). These tools allow for detailed mapping of affected areas, enabling more precise damage assessments, resource allocation, and decision-making (Gupta et al. 2024). They have a crucial role in monitoring environmental changes, such as deforestation, sea-level rise, and land degradation, which are vital for long-term disaster risk reduction.

Example: Remote sensing technology has been used to track the progression of wildfires and monitor deforestation, providing critical information to firefighting teams and policymakers.

(c) **Climate-Resilient Infrastructure**

Advancements in engineering and construction techniques have led to the development of climate-resilient infrastructure that can withstand extreme weather events (Bieber et al. 2018; Singh 2017; Wong and Jim 2017). These innovations focus on making buildings, roads, bridges, and other critical infrastructure more robust and adaptable to climate change impacts. It includes flood-resistant structures, green roofs, and seawalls designed to protect coastal communities from rising sea levels and storm surges.

Example: Floating houses and flood-resistant urban infrastructure have been built in flood-prone areas like the Netherlands, offering a model for climate adaptation.

(d) **Artificial Intelligence (AI) and Machine Learning**

AI and machine learning are increasingly used to improve disaster response and recovery efforts by processing large datasets quickly and identifying patterns or anomalies that might otherwise go unnoticed (Chiloane et al. 2021;

A. Wang et al. 2021). These technologies analyse weather data, social media feeds, and historical disaster records to predict disaster events, optimize emergency responses, and assist with damage assessments.

Example: AI models predict fire outbreaks and help fire departments allocate resources efficiently based on real-time data and environmental conditions.

(e) **Community-Based Disaster Risk Management (CBDRM)**

Empowering local communities to take an active role in disaster risk reduction and management is a growing trend (Kim et al. 2021; Singh and Manoharan 2024). Community-based disaster risk management (CBDRM) recognizes that local knowledge and participation are critical in building resilience and addressing the specific needs of vulnerable populations (Buscail et al. 2012; Hossain and Easson 2016). The engagement of communities in preparedness, response, and recovery efforts for climate disaster management becomes more inclusive and effective.

Example: Community-led evacuation planning in areas prone to tsunamis or landslides, where residents help identify safe evacuation routes and shelters.

(f) **Green Infrastructure and Nature-Based Solutions**

Green infrastructure and nature-based solutions are gaining traction as sustainable, cost-effective approaches to disaster risk reduction. These solutions utilize natural systems, such as wetlands, forests, and mangroves, to reduce the impacts of climate disasters, including floods, storm surges, and erosion (Khirfan and El-Shayeb 2020; O'Donnell et al. 2020). The investment in ecosystem restoration and the protection of natural habitats, and communities enhances their resilience to climate change.

Example: Coastal mangrove restoration in countries like Bangladesh and the Philippines to reduce the impact of storm surges and protect coastal communities from flooding.

(g) **Data-Driven Decision-Making and Big Data:**

The use of big data in climate disaster management allows more informed decision-making with the analysis of vast amounts of information from various sources, such as weather data, demographic data, infrastructure maps, and social media (de Oliveira et al. 2022; Kumari and Rana 2023; Pramanik 2024). Decision support systems (DSS) integrate big data to help authorities prioritize interventions, allocate resources efficiently, and assess vulnerabilities in real time.

Example: Data-driven platforms that track real-time population movements, resource availability, and disaster impact, providing decision-makers with up-to-date information for disaster management.

(h) **Climate Finance and Risk Financing Tools**

Climate finance, including innovative risk financing tools, has become essential for ensuring resources for disaster response and recovery, especially in low-income and vulnerable regions (Thammadi et al. 2022; Tubridy 2021). Instruments like catastrophe bonds, insurance, and the Green Climate Fund help countries manage the financial risks associated with climate disasters, enabling quicker recovery and reducing the burden on governments and communities.

Example: The use of disaster insurance, where countries like the Caribbean nations purchase insurance that pays out immediately following a disaster, ensuring fast financial support for recovery.

(i) **Resilient Agricultural Practices and Food Security**

Climate change poses significant risks to global food security in regions highly dependent on agriculture. Advancements in climate-resilient agricultural practices like drought-resistant crops, precision farming, and climate-smart irrigation systems, help to mitigate the impacts of climate disasters on food production (Dinesha et al. 2024; Zimmerman et al. 2018). These practices support long-term sustainability and reduce vulnerability to climate-related disruptions.

Example: The development of drought-tolerant crop varieties by organizations like the ensure food security in arid regions.

(j) **Public–Private Partnerships (PPP)**

Public–private partnerships are playing an increasingly important role in climate disaster management with the collaboration between governments, businesses, and non-governmental organizations (NGOs). These partnerships bring together resources, expertise, and technology to enhance disaster preparedness, response, and recovery (Chirwa and Amodu 2021; Swanson and Sakhrani 2020). Businesses can contribute with their technological innovations, logistics capabilities, and financial resources.

Example: Companies like Microsoft and IBM collaborate with governments to develop climate disaster management platforms using AI, cloud computing, and big data analytics.

Advancements in climate disaster management are reshaping how the world approaches disaster preparedness, response, and recovery. These innovations not only improve the efficiency and effectiveness of disaster management systems but also help build more resilient communities and ecosystems in the face of climate change (Crnčević and Orlović Lovren 2018; Sreekeshava et al. 2024). As new technologies and strategies continue to evolve, there is an increasing opportunity to prevent, mitigate, and recover from climate-related disasters more sustainably and equitably. However, these advancements require continued investment in research, collaboration, and capacity-building to ensure that they benefit the most vulnerable populations and regions at risk.

13.5 Climate Disaster Monitoring, Management, and Response

Climate disaster monitoring, management, and response form a critical triad of activities that help communities, governments, and organizations anticipate, address, and recover from climate-induced disasters (Kow et al. 2024; Mliyeh et al. 2024).

These processes rely on the use of advanced technologies, effective strategies, and well-coordinated efforts to mitigate the impacts of extreme weather events such as hurricanes, floods, droughts, heat waves, and wildfires (Boumans et al. 2014; Ma et al. 2024; Mosavi et al. 2018). Each component plays a distinct role in reducing vulnerability, protecting lives and infrastructure, and fostering resilience in the face of changing climate patterns.

13.5.1 Climate Disaster Monitoring

Monitoring is the first step in the disaster management cycle, allowing for the early detection and tracking of potential hazards. Through continuous observation and data collection, monitoring systems provide the foundation for timely and accurate predictions that inform decision-making.

13.5.1.1 Key Techniques in Disaster Monitoring

(a) **Remote Sensing and Satellite Data:** Satellites offer real-time data on weather patterns, sea surface temperatures, land cover, and environmental changes that are critical for monitoring the onset of climate hazards. Remote sensing technologies provide detailed images and data on the extent of damage caused by disasters after they occur.
Example: NASA's Earth Observing System provides satellite images that track storms, wildfires, and flooding, helping to inform disaster response teams.
(b) **Weather Forecasting and Modeling:** Advanced weather models use historical data, atmospheric analysis, and machine learning to predict the likelihood and severity of climate-related disasters, such as hurricanes, heat waves, or floods. These models help generate early warnings for at-risk areas.
Example: The Global Forecasting System (GFS) produces daily weather forecasts that inform early warning systems worldwide.
(c) **Environmental and Climate Monitoring Stations:** Ground-based sensors and monitoring stations are deployed to measure air quality, temperature fluctuations, rainfall, and humidity levels. These systems provide localized data that can detect changes in climate patterns and help identify emerging disaster risks.
Example: Flood gauges placed along rivers monitor water levels in real-time to warn of imminent flood threats.
(d) **Social media and Crowdsourcing:** Social media platforms and mobile applications have become valuable tools for disaster monitoring, especially during real-time events. Crowdsourced data from citizens helps emergency responders track localized hazards and the needs of affected populations.
Example: Platforms like Ushahidi collect crowd-sourced data during natural disasters to map real-time crises and coordinate response efforts.

13.5.2 Climate Disaster Management

Disaster management focuses on reducing risks and enhancing resilience through preparation, prevention, and mitigation (Yadav 2023; Zhou and Zhou 2021). Climate disaster management ensures that countries, regions, and communities are equipped with the necessary tools and strategies to anticipate, mitigate, and manage climate-related events.

13.5.2.1 Key Techniques in Disaster Management

(a) **Disaster Risk Reduction (DRR):** DRR encompasses measures taken to reduce the impact of disasters through policies, strategies, and programs that minimize vulnerability and exposure to climate risks. It includes risk assessments, building regulations, and environmental conservation practices.
Example: Risk zoning and land-use planning are used to prevent the construction of homes in flood-prone areas, thereby reducing exposure to disaster risks.

(b) **Preparedness and Training:** Preparedness planning is vital for ensuring communities to respond to climate disasters. This includes creating disaster response plans, conducting regular drills, and training local responders and citizens on effective disaster mitigation practices.
Example: Emergency evacuation drills, public education campaigns, and community-based disaster management programs ensure that local populations are informed and prepared to act when needed.

(c) **Infrastructure Resilience:** The development of resilient infrastructure helps communities withstand the effects of extreme climate events. This includes building climate-resistant roads, bridges, homes, and flood barriers, as well as retrofitting existing infrastructure to withstand increasing climate risks.
Example: Coastal cities like New York are investing in seawalls and flood barriers to protect against rising sea levels and storm surges.

(d) **Nature-Based Solutions:** Managing and restoring ecosystems such as wetlands, mangroves, and forests mitigate the effects of climate disasters by acting as natural buffers against floods, storms, and droughts.
Example: Coastal restoration projects to protect mangrove forests that reduce the impacts of tsunamis, storm surges, and coastal erosion.

(e) **Community-Based Disaster Risk Management (CBDRM):** Empowering local communities to assess, plan, and implement disaster risk management strategies is essential for long-term resilience. Involving communities ensures that disaster management plans address local needs and leverage local knowledge.
Example: Localized risk assessments that engage communities in identifying vulnerabilities and planning adaptive measures, such as flood mitigation infrastructure and early warning systems.

13.5.3 Climate Disaster Response

Response focuses on delivering immediate assistance to those affected by climate disasters, including rescue, relief, and rehabilitation efforts (Valle-Cruz et al. 2019; Yadav 2023). An efficient and well-coordinated disaster response can mitigate the loss of life, reduce economic damage, and expedite recovery processes.

13.5.3.1 Key Techniques in Disaster Response

(a) **Emergency Response Teams:** Specialized teams, including search-and-rescue, medical, and humanitarian teams, are deployed to affected areas immediately following a disaster. These teams provide critical support in areas such as first aid, emergency shelter, water, and food distribution.
Example: The deployment of the United Nations Disaster Assessment and Coordination (UNDAC) teams to assess immediate needs and coordinate emergency aid efforts following a disaster.

(b) **Rapid Damage and Needs Assessment:** A quick and accurate damage assessment is essential for informing response strategies. Teams use technology like drones, satellite imagery, and GIS to rapidly assess the extent of damage to infrastructure, homes, and vital services such as water and electricity.
Example: Post-hurricane assessments using drones to evaluate damage to power lines, buildings, and roads, enabling quicker decision-making for resource allocation.

(c) **Relief Distribution:** Providing immediate relief to affected populations, including food, water, medicine, and shelter, is a key aspect of disaster response. Efficient logistics and coordination between humanitarian organizations, local governments, and the private sector ensure that aid reaches those in need as quickly as possible.
Example: The use of pre-positioned emergency supplies in disaster-prone regions, ensuring quick distribution after an event.

(d) **Evacuation Plans and Shelters:** Evacuations are critical to saving lives in the face of imminent climate disasters such as floods or hurricanes. Evacuation plans should be well-practiced and inclusive, ensuring that vulnerable populations (e.g., children, the elderly, and the disabled) have access to safe evacuation routes and shelters.
Example: Establishing multi-use emergency shelters that provide temporary housing, medical care, and food to evacuees during a disaster.

(e) **Psychosocial Support:** Addressing the mental health needs of individuals affected by climate disasters is essential for long-term recovery. Providing psychosocial support helps disaster survivors cope with trauma, loss, and displacement.
Example: Setting up counselling centers and hotlines to provide emotional support and mental health services to those affected by the disaster.

Climate disaster monitoring, management, and response are interconnected elements of a comprehensive disaster risk management framework (Dyca et al. 2020; Li et al. 2024; Thammadi et al. 2022). Monitoring provides essential early warnings that enable informed management and preparation, while management ensures that communities, governments, and organizations are equipped to face and mitigate climate risks. The response focuses on providing immediate relief and support to those affected, helping them recover and rebuild their lives. Collaborating across sectors, and fostering community involvement, these techniques form the foundation for creating climate-resilient societies to navigate the challenges of an increasingly volatile climate.

13.6 Future Trends in Climate Disaster Analysis and Adaptation

Future trends in climate disaster analysis and adaptation are expected to be modelled with the advancements in technology, evolving environmental policies, and the growing recognition of the need for coordinated global action. The key trends that are likely to emerge are the following:

(a) **Advanced Climate Modelling and Predictive Analytics**

AI and Machine Learning: Artificial intelligence (AI) and machine learning will continue to improve climate modelling, allowing for more accurate and localized predictions of weather patterns, extreme events, and their potential impacts. These models will better inform disaster preparedness and risk mitigation strategies.

Big Data Integration: Combining large datasets, such as satellite imagery, social media posts, weather patterns, and historical climate data, will provide deeper insights into potential climate hazards and help predict their effects on specific regions.

(b) **Resilient Infrastructure and Design**

Climate-Resilient Architecture: New construction technologies will focus on designing buildings, roads, and other infrastructures that can withstand more frequent and intense climate events, such as hurricanes, floods, and wildfires.

Green and Blue Infrastructure: The development of natural solutions, such as urban forests, green roofs, and wetlands, will be integrated into city planning to act as buffers against flooding and heat waves while promoting biodiversity and ecosystem health.

(c) **Early Warning Systems and Real-Time Monitoring**

IoT and Sensors: Internet of Things (IoT) sensors will be deployed in real-time monitoring systems to detect climate hazards, such as rising river levels, soil moisture, and air quality, providing faster early warnings and more precise responses.

Mobile Apps and Digital Platforms: These technologies will allow citizens and governments to access real-time disaster data, evacuation routes, and recovery resources, improving disaster management and community preparedness.

(d) **Data-Driven Policy and Decision-Making**

Climate Risk Assessment Tools: Governments and businesses will increasingly rely on advanced climate risk assessment tools to guide their policies, insurance decisions, and investments in resilience.

Climate Adaptation Strategies: There will be a focus on creating adaptation plans that are flexible, sustainable, and tailored to local contexts. These will be based on data-driven insights and will involve community participation in decision-making processes.

(e) **Insurance and Financial Instruments**

Parametric Insurance Models: In response to the rising frequency of climate disasters, insurance companies will increasingly offer parametric insurance, which pays out based on the occurrence of predefined events (e.g., a specific hurricane or flood level), rather than relying on traditional claims processes.

Climate Bonds and Green Financing: Financial markets will continue to invest in climate resilience, with green bonds and climate bonds becoming more prevalent, helping to fund adaptation projects.

(f) **Ecosystem-Based Solutions**

Nature-Based Adaptation: There will be an increased focus on ecosystem-based adaptation (EbA) measures, such as mangrove restoration, reforestation, and wetland conservation, which provide natural barriers against storms and flooding while enhancing biodiversity.

Carbon Sequestration: Nature-based solutions will also contribute to carbon sequestration, helping to mitigate climate change while adapting to its effects.

(g) **Community-Led Adaptation**

Local Knowledge Integration: Climate adaptation strategies will incorporate traditional ecological knowledge and indigenous practices, which have long relied on local observations of climate patterns and sustainable land management techniques.

Social Justice and Equity: Adaptation measures will focus on the most vulnerable populations, ensuring that adaptation strategies are inclusive and equitable, addressing the needs of marginalized communities who are often disproportionately affected by climate disasters.

(h) **Climate-Resilient Agriculture**

Precision Agriculture: The use of sensors, drones, and data analytics will enable farmers to monitor weather patterns and soil conditions more accurately, helping them adapt to changing climate conditions by optimizing irrigation, crop selection, and pest management.

Climate-Smart Agriculture: There will be a greater emphasis on climate-resilient crop varieties, sustainable farming practices, and water-efficient techniques to ensure food security in the face of climate change.

(i) **Global Collaboration and Knowledge Sharing**

International Partnerships: As climate disasters transcend borders, there will be greater collaboration among countries, particularly in sharing disaster response strategies, knowledge, and technologies. This will be crucial in ensuring a coordinated global approach to adaptation.

(j) **Climate Diplomacy:** Climate change and disaster adaptation will be key agendas for diplomatic discussions, with governments working together to address transnational risks and ensure mutual support in times of crisis.

(k) **Climate Justice and Human Rights**

Forced Migration and Displacement: As climate disasters increase, more communities will be displaced. Future strategies will need to address climate-induced migration and displacement, ensuring the protection of human rights and supporting the integration of displaced populations into new environments.

Humanitarian Aid and Resilience Building: Efforts will shift from simply providing aid during disasters to empowering communities with the tools and resources they need to build resilience over the long term, ensuring sustainable recovery.

These trends will transform how we approach climate disaster preparedness, response, and adaptation, with a greater focus on sustainability, equity, and technology-driven solutions (Chavan and Mulay 2023; Moshozi and Bazrafshan 2018; Shaikh et al. 2022). It will minimize the impacts of climate change while supporting communities in their ability to adapt and thrive in an increasingly unpredictable world.

13.7 Summary

Climate disaster management faces several challenges including increasing frequency and intensity of extreme weather events, and limited resources. The impacts of climate disasters are extreme, and these often lead to loss of life, widespread destruction, displacement, and significant economic damage. Effective climate disaster management and response require fast and efficient action, including preparedness plans, timely evacuations, and collaboration across local, national, and international levels. Advancements in disaster management are being made with the use of advanced technologies like AI, satellite imaging, and early warning systems to predict and monitor climate events. There is a growing focus on integrating climate change adaptation strategies into disaster management in the future. This includes enhancing resilience through sustainable infrastructure, community-based approaches, and developing better predictive models to anticipate climate impacts. The future of climate disaster management will depend on improving data collection, fostering international cooperation, and emphasizing long-term adaptation to mitigate the effects of climate change. The challenges in climate disaster management focus on advancements in our ability to predict, prepare, and respond to climate-related disasters. The adoption of comprehensive, inclusive, and innovative approaches will help to build more resilient and sustainable societies in the face of climate change.

References

Alotaibi E, Nassif N (2024) Artificial intelligence in environmental monitoring: in-depth analysis. Discover Artif Intell 4(1). https://doi.org/10.1007/s44163-024-00198-1

Atek S, Pesaresi C, Eugeni M, De Vito C, Cardinale V, Mecella M, Rescio A, Petronzio L, Vincenzi A, Pistillo P, Bianchini F, Giusto G, Pasquali G, Gaudenzi P (2022) Satellite-based earth cognitive decision support system for COVID-19 pandemic management. In: Proceedings of the international Astronautical congress, IAC. https://www.scopus.com/inward/record.uri?eid=2-s2.0-85166918652&partnerID=40&md5=3ee68d890fd98dbb1c327d5bfdd08ca3

Attaf D, Djerriri K, Mansour D, Hamdadou D (2019) Mapping of burned area using presence and background learning framework on the google earth engine platform. In: Tanzi CMSF, Altan TO (eds) International archives of the photogrammetry, remote sensing and spatial information sciences – ISPRS archives, vol 42, Issue 3/W8. International Society for Photogrammetry and Remote Sensing, pp 37–41. https://doi.org/10.5194/isprs-archives-XLII-3-W8-37-2019

Ayub S, Uday P, Vaibhav S, Udaykumar P (2024) Advancing AI models for drought forecasting, challenges and future directions. Disaster Advances 17(12):58–71. https://doi.org/10.25303/1712da058071

Balestra M, Chiappini S, Malinverni ES, Galli A, Marcheggiani E (2021) A machine learning approach for mapping Forest categories: an application of Google earth engine for the case study of Monte Sant'Angelo, Central Italy. In: Lecture notes in computer science (including subseries lecture notes in artificial intelligence and lecture notes in bioinformatics), 12955 LNCS, pp 155–168. https://doi.org/10.1007/978-3-030-87007-2_12

Bieber N, Ker JH, Wang X, Triantafyllidis C, van Dam KH, Koppelaar RHEM, Shah N (2018) Sustainable planning of the energy-water-food nexus using decision making tools. Energy Policy 113:584–607. https://doi.org/10.1016/j.enpol.2017.11.037

Boumans RJM, Phillips DL, Victery W, Fontaine TD (2014) Developing a model for effects of climate change on human health and health-environment interactions: heat stress in Austin, Texas. Urban Clim 8:78–99. https://doi.org/10.1016/j.uclim.2014.03.001

Buscail C, Upegui E, Viel J-F (2012) Mapping heatwave health risk at the community level for public health action. Int J Health Geogr 11. https://doi.org/10.1186/1476-072X-11-38

Cao Z, Shi X (2021) A systematic literature review of entrepreneurial ecosystems in advanced and emerging economies. Small Bus Econ 57(1):75–110. https://doi.org/10.1007/s11187-020-00326-y

Capolupo A, Monterisi C, Barletta C, Tarantino E (2021) Google earth engine for land surface albedo estimation: comparison among different algorithms. In: M. A. Neale CMU (ed) Proceedings of SPIE - the international society for optical engineering, vol 11856. SPIE. https://doi.org/10.1117/12.2597666

Ceccato P, Ramirez B, Manyangadze T, Gwakisa P, Thomson MC (2018) Data and tools to integrate climate and environmental information into public health. Infect Dis Poverty 7(1). https://doi.org/10.1186/s40249-018-0501-9

Chavan S, Mulay P (2023) Assessing the impact of GeoAI in the world of spatial data and energy revolution. In: Risk detection and cyber security for the success of contemporary computing. IGI Global. https://doi.org/10.4018/978-1-6684-9317-5.ch008

Chiloane C, Dube T, Shoko C (2021) Impacts of groundwater and climate variability on terrestrial groundwater dependent ecosystems: a review of geospatial assessment approaches and challenges and possible future research directions. Geocarto Int 37. https://doi.org/10.1080/10106049.2021.1948108

Chirwa D, Amodu N (2021) Economic, social and cultural rights, sustainable development goals, and duties of corporations: rejecting the false dichotomies. Bus Human Rights J 6(1):21–41. https://doi.org/10.1017/bhj.2020.34

Cilli R, Elia M, D'Este M, Giannico V, Amoroso N, Lombardi A, Pantaleo E, Monaco A, Sanesi G, Tangaro S, Bellotti R, Lafortezza R (2022) Explainable artificial intelligence (XAI) detects wildfire occurrence in the Mediterranean countries of southern Europe. Sci Rep 12(1). https://doi.org/10.1038/s41598-022-20347-9

Crnčević T, Orlović Lovren V (2018) Displacement and climate change: improving planning policy and increasing community resilience. Int J Climate Change Strat Manag 10(1):105–120. https://doi.org/10.1108/IJCCSM-05-2017-0103

David AO, Ndambuki JM, Muloiwa M, Kupolati WK, Snyman J (2024) A review of the application of artificial intelligence in climate change-induced flooding—susceptibility and management techniques. Civil Eng 5(4):1185–1198. https://doi.org/10.3390/civileng5040058

de Oliveira GC, Bertone E, Stewart RA (2022) Optimisation modelling tools and solving techniques for integrated precinct-scale energy–water system planning. Appl Energy 318. https://doi.org/10.1016/j.apenergy.2022.119190

Dinesha S, Hosur SR, Toushif PK, Bodiga D, Dechamma NLD, Ashwath MN, Pradhan D (2024) Sustaino-resilient agroforestry for climate resilience, food security and land degradation neutrality. In: Land and environmental management through forestry. Wiley. https://doi.org/10.1002/9781119910527.ch9

Doblas J, Shimabukuro Y, Sant'anna S, Carneiro A, Aragão L, Almeida C (2020) Optimizing near real-time detection of deforestation on tropical rainforests using sentinel-1 data. Remote Sens 12(23):1–31. https://doi.org/10.3390/rs12233922

Dtissibe FY, Ari AAA, Abboubakar H, Njoya AN, Mohamadou A, Thiare O (2024) A comparative study of machine learning and deep learning methods for flood forecasting in the far-north region, Cameroon. Sci Afric 23. https://doi.org/10.1016/j.sciaf.2023.e02053

Dyca B, Muldoon-Smith K, Greenhalgh P (2020) Common value: transferring development rights to make room for water. Environ Sci Policy 114:312–320. https://doi.org/10.1016/j.envsci.2020.08.017

Ekka P, Patra S, Upreti M, Kumar G, Kumar A, Saikia P (2024) Land degradation and its impacts on biodiversity and ecosystem services. In: Land and environmental management through forestry. Wiley. https://doi.org/10.1002/9781119910527.ch4

Ezangina IA, Gromyshova OS (2020) Directions for improving the monitoring system of state programs of socio-economic development of Russia. Fin Theory Pract 24(5):112–127. https://doi.org/10.26794/2587-5671-2020-24-5-112-127

Fadlullah ZM, Kato N (2021) On smart IoT remote sensing over integrated terrestrial-aerial-space networks: an asynchronous federated learning approach. IEEE Netw 35(5):129–135. https://doi.org/10.1109/MNET.101.2100125

Fan Y-C, Yuan Y-Q, Yuan Y-C, Duan W-J, Gao Z-Q (2025) Research progress on the impact of climate change on wheat production in China. Peer J 13(1). https://doi.org/10.7717/peerj.18569

Gupta RK, Jain A, Wang J, Pateriya RK (2024) Reshaping environmental science through machine learning and IoT. IGI Global. https://doi.org/10.4018/979-8-3693-2351-9

Hossain AKMA, Easson G (2016) Potential impacts of the growth of a mega city in Southeast Asia: a case study on the city of Dhaka, Bangladesh. In: Handbook of climate change mitigation and adaptation, vol 2, 2nd edn. Springer. https://doi.org/10.1007/978-3-319-14409-2_68

Icaza LE, Van Den Dobbelsteen A, Van Der Hoeven F (2016) Integrating urban heat assessment in urban plans. Sustainability (Switzerland) 8(4). https://doi.org/10.3390/su8040320

Jitkajornwanich K, Vijaranakul N, Jaiyen S, Srestasathiern P, Lawawirojwong S (2024) Enhancing risk communication and environmental crisis management through satellite imagery and AI for air quality index estimation. MethodsX 12. https://doi.org/10.1016/j.mex.2024.102611

Khirfan L, El-Shayeb H (2020) Urban climate resilience through socio-ecological planning: a case study in Charlottetown, Prince Edward Island. J Urban 13(2):187–212. https://doi.org/10.1080/17549175.2019.1650801

Kim J, Kamari M, Lee S, Ham Y (2021) Large-scale visual data-driven probabilistic risk assessment of utility poles regarding the vulnerability of power distribution infrastructure systems. J Constr Eng Manag 147(10). https://doi.org/10.1061/(ASCE)CO.1943-7862.0002153

Kow P-Y, Liou J-Y, Yang M-T, Lee M-H, Chang L-C, Chang F-J (2024) Advancing climate-resilient flood mitigation: utilizing transformer-LSTM for water level forecasting at pumping stations. Sci Total Environ 927. https://doi.org/10.1016/j.scitotenv.2024.172246

Kumari J, Rana DP (2023) GeoAI-based Covid-19 Prediction Model. In: Lecture notes in networks and systems, vol 653 LNNS, pp 669–684. https://doi.org/10.1007/978-981-99-0981-0_51

Li X, Yue J, Wang S, Luo Y, Su C, Zhou J, Xu D, Lu H (2024) Development of geographic information system architecture feature analysis and evolution trend research. Sustainability (Switzerland) 16(1). https://doi.org/10.3390/su16010137

Liao Y, Liao W, Wu D, Zhang C, Xu J, Majumdar S (2024) Inspired applications of artificial intelligence in disaster-resilient regional construction for climate risk. Urban Sustainability, Part F3945:305–315. https://doi.org/10.1007/978-981-97-9658-8_16

Liu D, Liao T, Sun H, Ren F (2021) Research progress and development direction of Chinese remote sensing software: taking PIE as an example; [中国遥感软件研制进展与发展方向-以像素专家PIE为例]. J Image Graphics 26(5):1169–1178. https://doi.org/10.11834/jig.200125

Ma L-J, Yuan J-S, Huang L (2024) Prospect of climate risks management in China under the framework of UN early warning for all initiative. Climate Change Res 20(1):48–61. https://doi.org/10.12006/j.issn.1673-1719.2023.122

Mliyeh MM, Ait Brahim Y, Koutsovili E-I, Tzoraki O, Zian A, Aqnouy M, Benaabidate L (2024) Multi-index approach to assess and monitor meteorological and agricultural drought in the Mediterranean region: case of the upper Oum Er Rabia watershed, Morocco. Water (Switzerland) 16(21). https://doi.org/10.3390/w16213104

Mosavi A, Ozturk P, Chau K-W (2018) Flood prediction using machine learning models: literature review. Water (Switzerland) 10(11). https://doi.org/10.3390/w10111536

Moshozi GZ, Bazrafshan O (2018) Impact of climatic signals on the wet and dry season precipitation (case study: Persian gulf and Oman Sea watersheds). J Earth Space Phy 44(2):333–349. https://doi.org/10.22059/jesphys.2018.231949.1006893

O'Donnell E, Maskrey S, Everett G, Lamond J (2020) Developing the implicit association test to uncover hidden preferences for sustainable drainage systems. Philos Trans R Soc A Math Phys Eng Sci 378(2168). https://doi.org/10.1098/rsta.2019.0207

Obrist MD, Kannan R, Schmidt TJ, Kober T (2021) Decarbonization pathways of the Swiss cement industry towards net zero emissions. J Clean Prod 288. https://doi.org/10.1016/j.jclepro.2020.125413

Parente J, Amraoui M, Menezes I, Pereira MG (2019) Drought in Portugal: current regime, comparison of indices and impacts on extreme wildfires. Sci Total Environ 685:150–173. https://doi.org/10.1016/j.scitotenv.2019.05.298

Pramanik S (2024) AI's function in sustainable development's renewable energy planning. In: Next generation materials for sustainable engineering. IGI Global. https://doi.org/10.4018/979-8-3693-1306-0.ch016

Richards DR, Belcher RN (2020) Global changes in urban vegetation cover. Remote Sens 12(1). https://doi.org/10.3390/RS12010023

Sahani J, Kumar P, Debele S, Spyrou C, Loupis M, Aragão L, Porcù F, Shah MAR, Di Sabatino S (2019) Hydro-meteorological risk assessment methods and management by nature-based solutions. Sci Total Environ 696. https://doi.org/10.1016/j.scitotenv.2019.133936

Selvam AP, Al-Humairi SNS (2025) Environmental impact evaluation using smart real-time weather monitoring systems: a systematic review. *Innovative infrastructure*. Solutions 10(1). https://doi.org/10.1007/s41062-024-01817-7

Shabahang S, Gjerde M, Vale B, Balador Z (2019) The problem of lack of green space and rise in surface temperature in the City of Mashhad. In: Smart innovation, systems and technologies, vol 131. Springer. https://doi.org/10.1007/978-3-030-04293-6_26

Shaikh TA, Mir WA, Rasool T, Sofi S (2022) Machine learning for smart agriculture and precision farming: towards making the fields talk. Arch Comput Methods Eng 29(7):4557–4597. https://doi.org/10.1007/s11831-022-09761-4

Singh A (2017) Disaster law: emerging thresholds. https://doi.org/10.4324/9781315102931

Singh CR, Manoharan G (2024) Strengthening resilience: AI and machine learning in emergency decision-making for natural disasters. In: Internet of things and AI for natural disaster management and prediction. IGI Global. https://doi.org/10.4018/9798369342848.ch012

Sreekeshava KS, Kolathayar S, Menon VC, N., & Bhargavi.C. (2024) Civil engineering for multi-hazard risk reduction-an introduction. Lecture Notes Civil Eng 457:1–10. https://doi.org/10.1007/978-981-99-9610-0_1

Sunar AF, Yagmur N, Dervisoglu A (2019) Flood analysis with remote sensing data – A case study: Maritsa River, Edirne. In: Tanzi CMSF, Altan TO (eds) International archives of the photogrammetry, remote sensing and spatial information sciences – ISPRS archives, vol 42, Issue 3/W8. International Society for Photogrammetry and Remote Sensing, pp 497–502. https://doi.org/10.5194/isprs-archives-XLII-3-W8-497-2019

Swanson R, Sakhrani V (2020) Appropriating the value of flexibility in ppp megaproject design. J Manag Eng 36(5). https://doi.org/10.1061/(ASCE)ME.1943-5479.0000770

Talha M, Nejadhashemi AP, Moller K (2025) Soft computing paradigm for climate change adaptation and mitigation in Iran, Pakistan, and Turkey: A systematic review. Heliyon 11(2). https://doi.org/10.1016/j.heliyon.2025.e41974

Thammadi S, Nagabhatla N, Pisini S, Koza S, Mahmood A (2022) Regional trends in social-ecological-technological (SET) approaches to sustainable urban planning: focus on Asia. In: Blue-green infrastructure across asian countries: improving urban resilience and sustainability. https://doi.org/10.1007/978-981-16-7128-9_2

Thangavel K, Spiller D, Sabatini R, Amici S, Sasidharan ST, Fayek H, Marzocca P (2023) Autonomous satellite wildfire detection using hyperspectral imagery and neural networks: A case study on Australian wildfire. Remote Sens 15(3). https://doi.org/10.3390/rs15030720

Tubridy D (2021) Co-financing green resilient infrastructures in Copenhagen: integrated or superficial design? Landsc Res 46(2):261–272. https://doi.org/10.1080/01426397.2020.1850664

Valle-Cruz D, Sandoval-Almazan R, Ruvalcaba-Gomez EA, Ignacio Criado J (2019) A review of artificial intelligence in government and its potential from a public policy perspective. ACM Int Conf Proc Ser:91–99. https://doi.org/10.1145/3325112.3325242

Wagner M, Chhetri N, Sturm M (2014) Adaptive capacity in light of hurricane Sandy: the need for policy engagement. Appl Geogr 50:15–23. https://doi.org/10.1016/j.apgeog.2014.01.009

Wai CY, Muttil N, Tariq MAUR, Paresi P, Nnachi RC, Ng AWM (2022) Investigating the relationship between human activity and the urban heat Island effect in Melbourne and four other international cities impacted by COVID-19. Sustainability (Switzerland) 14(1). https://doi.org/10.3390/su14010378

Wang H, Hu Y, Tang L, Zhuo Q (2020) Distribution of urban blue and green space in Beijing and its influence factors. Sustainability (Switzerland) 12(6). https://doi.org/10.3390/su12062252

Wang A, Xu L, Li Y, Xing J, Chen X, Liu K, Liang Y, Zhou Z (2021) Random-forest based adjusting method for wind forecast of WRF model. Comput Geosci 155. https://doi.org/10.1016/j.cageo.2021.104842

Wang Y, Wang H, Cui P, Chen D, Tang J, Ou T, Hao J, Wang J, Zhang G, Lei Y, Lei Y, Wu C (2024) Disaster effects of climate change and the associated scientific challenges. Kexue Tongbao/Chinese Sci Bullet 69(2):286–300. https://doi.org/10.1360/TB-2023-0325

Wong GKL, Jim CY (2017) Urban-microclimate effect on vector mosquito abundance of tropical green roofs. Build Environ 112:63–76. https://doi.org/10.1016/j.buildenv.2016.11.028

Yadav A (2023) Leveraging artificial intelligence for sustainable development and environmental resilience. In: Exploring ethical dimensions of environmental sustainability and use of AI. https://doi.org/10.4018/9798369308929.ch008

Zhou G, Zhou L (2021) Ecological meteorology: origin, concept, and prospects. Kexue Tongbao/Chinese Sci Bullet 66(2):210–218. https://doi.org/10.1360/TB-2020-1274

Zimmerman R, Zhu Q, Dimitri C (2018) A network framework for dynamic models of urban food, energy and water systems (FEWS). Environ Prog Sustain Ener 37(1):122–131. https://doi.org/10.1002/ep.12699

[illegible]

MIX
Papier aus verantwortungsvollen Quellen
Paper from responsible sources
FSC® C105338

If you have any concerns about our products,
you can contact us on
ProductSafety@springernature.com

In case Publisher is established outside the EU,
the EU authorized representative is:
Springer Nature Customer Service Center GmbH
Europaplatz 3, 69115 Heidelberg, Germany

Printed by Libri Plureos GmbH
in Hamburg, Germany